Vol. 58 No. 2

Mars-avril · 2003 · March-April

AF769824

Oil & Gas Science and Technology

Revue de l'Institut Français du Pétrole

NUMÉRO SPÉCIAL • SPECIAL ISSUE

Contributions à la géologie pétrolière

Insights into Petroleum Geochemistry

Sous la direction de/Edited by
B. Colletta

ISSN 1294-4475 OGSTFA 58 (2) 167-324 (2003)

Editions TECHNIP 27 rue Ginoux, 75737 PARIS Cedex 15, FRANCE

quelques articles à paraître dans notre prochain numéro

some papers to be published in our next issue

D. DECROOCQ
Energy Conservation and CO_2 Emissions in the Processing and Use of Oil and Gas

P. BÉREST and B. BROUARD
Safety of Salt Caverns Used for Underground Storage

P. MAGNOUX, G. JOLY, S. BRIMAUD and K. RAGIL
Formation de coke à partir de propène sur silicalite

Rédaction / *Editorial offices*

Institut français du pétrole, 1 et 4, avenue de Bois-Préau,
92852 Rueil-Malmaison Cedex - France
Tél. 01 47 52 60 00 – Tél. International +33 1 47 52 60 00
Fax 01 47 52 70 00 – Fax international +33 1 47 52 70 00
e-mail : patrick.boisserpe@ifp.fr
http://www.ifp.fr

Tarif abonnement / *Subscription rates*

1 an, 6 numéros / *1 year, 6 issues*
Abonnement uniquement à une année complète de janvier à décembre
Subscription to a complete year from January to December only

Abonnement : année 2003 ***Subscription:*** *year 2003*	420 €
Vente au numéro / *Single issues* – 1946-2001 – 2002-2003	60 € 70 €

Administration / *Publisher*

Éditions Technip, 27 rue Ginoux,
75737 Paris Cedex 15, France
Tél. 01 45 78 33 80 – Tél. international +33 1 45 78 33 80
Fax 01 45 75 37 11 – Fax international +33 1 45 75 37 11
e-mail : info@editionstechnip.com
http://www.editionstechnip.com

Les opinions émises par les auteurs dans leurs articles n'engagent que leur propre responsabilité.
Tous droits de reproduction, de traduction et d'adaptation réservés pour tous pays.

Opinions expressed in articles are the sole responsibility of their authors.
All reproduction, translation and adaptation rights reserved for all countries.

© 2003 Institut français du pétrole, Rueil-Malmaison et/*and* Éditions Technip, Paris

Oil & Gas Science and Technology – Rev. IFP, Vol. 58 (2003), No. 2, pp. 167-170
Copyright © 2003, Éditions Technip

Éditorial

Bernard Tissot au Fort de Flatters au Sahara lors d'une campagne de prospection géologique en 1959.

La géochimie organique appliquée à l'exploration pétrolière est, depuis près de cinquante ans, une des principales activités de la division géologie-géochimie de l'*Institut français du pétrole.* Ce numéro spécial de la revue *Oil & Gas Science and Technology*, consacré à cette discipline, a été réalisé en hommage à Bernard Tissot pour saluer son entrée à l'Académie des Sciences et lui témoigner, à travers ce recueil d'articles, notre reconnaissance et notre estime. Bernard Tissot, chef de file des équipes de géochimie l'*Institut français du pétrole,* a fait partie des pionniers de la géochimie organique en exploration pétrolière et il a participé, de manière significative, à écrire les pages de l'histoire de cette "jeune" discipline. C'est son approche quantitative qui a ouvert la voie à la modélisation de bassin, laquelle constitue aujourd'hui un outil incontournable de l'exploration moderne.

Dans l'introduction, Jean Dercourt, secrétaire perpétuel de l'Académie des Sciences et président du Conseil scientifique de l'*IFP* brosse les activités actuelles de Bernard Tissot à la direction de la *Commission nationale d'évaluation des recherches pour la gestion des déchets radioactifs (CNE).* Lucien Montadert, ancien directeur de l'objectif Exploration-Gisement à l'IFP et qui fut le promoteur de la modélisation de bassin dans la division Géologie-Géochimie, explique comment le jeune géologue "alpin" Bernard Tissot s'est intéressé aux phénomènes physicochimiques de génération des hydrocarbures et a créé la géochimie pétrolière qui a fait sa renommée et celle de l'*IFP*.

Ce numéro spécial débute par une version en anglais d'un des articles de Bernard Tissot publié pour la première fois en 1969 dans la *Revue de l'Institut Français du Pétrole,* et qui est à l'origine de la géochimie pétrolière et en particulier de tous les travaux de modélisation de bassin.

Viennent ensuite une série d'articles faisant le point sur la géochimie pétrolière depuis ses origines jusqu'à nos jours.

Bernard Durand, ancien directeur de la division Géologie-Géochimie, retrace avec beaucoup de précision et de rigueur l'émergence de cette nouvelle discipline au début du XXe siècle, puis évoque les travaux décisifs des équipes *IFP* dont il fut lui-même l'un des acteurs importants. Cet historique constitue la référence pour tout explorateur ou producteur cherchant à comprendre l'évolution des idées et des concepts ayant abouti à cette nouvelle discipline.

Alain Huc montre ensuite, dans un article synthétique, l'apport majeur des méthodes géochimiques à l'échelle du bassin sédimentaire et à l'échelle du réservoir. Il souligne également le rôle majeur que peut jouer la géochimie organique pour l'exploration et la production de l'offshore profond. Il analyse les axes de recherche à développer pour faire de la géochimie un outil de prédiction quantitative.

Philippe Ungerer, qui a aussi largement contribué au développement de la modélisation de bassin à l'*IFP*, présente différents aspects de la géochimie à travers les travaux de géochimie pétrolière qui ont permis de prendre en compte le craquage secondaire, la cinétique compositionnelle et la migration. Il aborde également le problème des gisements haute pression et les problèmes d'équilibre de phase entre huile et gaz par le biais de la thermodynamique. Enfin, il ouvre d'importantes perspectives vers la simulation moléculaire comme outil de prédiction quantitative notamment pour le traitement des gaz riches en H_2S.

Mireille Vandenbroucke révèle toute la partie expérimentale qui a été mise au point pour caractériser le kérogène par l'analyse chimique et en particulier par la pyrolyse. Elle montre également comment le couplage de la modélisation moléculaire et de la statistique a permis d'approcher la structure très complexe des kérogènes et, ainsi, de visualiser les différents types de kérogènes des roches sédimentaires.

Alain Prinzhofer et Anne Battani viennent compléter la panoplie des outils de la géochimie organique par une approche nouvelle, se fondant sur l'analyse combinée des isotopes du gaz naturel et des isotopes des gaz rares. Cette nouvelle voie de recherche offre des résultats prometteurs pour mieux contraindre l'origine et l'âge des hydrocarbures.

Dans le dernier article, Frédéric Schneider présente l'un des derniers prototypes en modélisation de bassin développé à l'*IFP*. Il s'agit d'un outil 2D simulant l'écoulement de fluides (3 phases) dans des structures géométriquement très complexes, telles que les zones de chevauchements ou les structures diapiriques. En particulier, le rôle des failles est pris en compte au cours de l'évolution du bassin. Ce prototype est un premier jalon vers le modèle de bassin du futur qui travaillera dans les quatre dimensions (trois dimensions de l'espace + le temps) et qui pourra être utilisé tant en exploration pétrolière que pour l'évaluation et le suivi des stockages souterrains.

Bernard Colletta
Directeur de la division Géologie-Géochimie

Editorial

Bernard Tissot in "Fort de Flatters" during a geological survey in Sahara in 1959.

Organic geochemistry applied to petroleum exploration has been for almost fifty years one of the main activities of the Geology-Geochemistry Department at the *Institut français du pétrole*. This special issue, dedicated to this subject was prepared to pay homage to Bernard Tissot, to celebrate his admission to the *Académie des Sciences*, and testify through this collection of papers our gratitude and respect. Bernard Tissot, leader of the geochemistry team of the *Institut français du pétrole,* is one of the pioneers of organic geochemistry applied to petroleum exploration and he participated significantly in writing the main pages of the story of this "young" discipline. His quantitative approach opened the way to basin modeling which today constitutes the reference tool for modern exploration.

In the introduction, Jean Dercourt, *Secrétaire perpétuel de l'Académie des sciences* and *President* of the *IFP* Scientific Council describes the present activities of Bernard Tissot as President of the *Commission nationale d'évaluation des recherches pour la gestion des déchets radioactifs (CNE).* Lucien Montadert, former Head of the Exploration-Production Centre in *IFP*, and who strongly promoted the development of basin modeling, explains how the young "Alpine" geologist Bernard Tissot took an active interest in the physics and chemistry of hydrocarbon generation and created the petroleum geochemistry discipline which contributed to his fame and to the fame of *IFP*.

This special issue begins by an english version of a paper by Bernard Tissot that was first published in 1969 in the *Revue de l'Institut Français du Pétrole* and established the bases of the petroleum geochemistry and of the basin modelling studies.

Then, a series of papers gives an overview of the petroleum geochemistry from its origins to the present day.

Bernard Durand, a former Director of the *IFP* Geology-Geochemistry Department, presents with many details and rigor the onset of this new discipline at the beginning of the 20th century; then, he highlights the decisive works by the *IFP* teams in which he was an important actor. This review constitutes a reference for all explorers or developers who are interested in understanding the evolution of ideas and concepts that gave birth to this new discipline.

Alain Huc shows afterward in a synthetic paper the major contribution of the geochemistry methodology at basin and reservoir scale. He also underlines the major role that organic geochemistry can play in exploration and production, especially in the deep offshore. He analyses the research axes to be developed to make geochemistry a quantitative predictive tool.

Philippe Ungerer, who widely contributed to the development of basin modeling at *IFP,* proposes various aspects of geochemistry with studies in petroleum geochemistry that allows us to take into account secondary cracking, compositional kinetics and migration. He also tackles the problem of high pressure reservoirs and problems of oil-gas phase equilibrium through a thermodynamic approach. Finally, he opens up large perspectives towards molecular simulation as a predictive and quantitative tool in particular for the processing of H_2S rich gas.

Mireille Vandenbroucke reveals the extensive experimental process that was elaborated to characterize kerogens by chemical analysis and in particular through pyrolysis. She also demonstrates how the coupling of molecular modeling and a statistical approach gave clear insights on the very complex structure of kerogens and allowed us to visualise the various types of kerogen in sedimentary rocks.

Alain Prinzhofer and Anne Battani propose to complete the set of geochemical tools by a new approach based on the combined analysis of natural gas and noble gases isotopes. This new research direction offers promising results to improve constraints on the origin and the age of hydrocarbons.

In the final paper, Frédéric Schneider presents one of the most recent prototype software packages in basin modeling developed in *IFP*. It is a 2D tool simulating fluid flows (3 phases) in very complex structures such as thrust zones or salt diapir related structures. In particular, fault properties are taken into account during the basin evolution. This prototype is a first step toward the basin model of the future that will be able to work in four dimensions (the three space dimensions + time) and that could be used either in petroleum exploration or to evaluate and monitor the subterranean storage.

Bernard Colletta
Director of Geology-Geochemistry Department

Oil & Gas Science and Technology – Rev. IFP, Vol. 58 (2003), No. 2, pp. 171-178
Copyright © 2003, Éditions Technip

Introduction

La recherche des sources d'énergie utilisables aujourd'hui et susceptibles de le rester pendant des décennies voire pendant des siècles, est une des principales préoccupations de l'humanité.

Bernard Tissot a participé à cette recherche dans l'industrie pétrolière et la poursuit aujourd'hui au sein de l'industrie nucléaire. Il a d'abord contribué à élucider les processus physicochimiques qui conduisent de la matière organique vivante aux gisements d'hydrocarbures, des maturations dans les sédiments jusqu'au piègeage et à l'extraction.

Désormais, il se consacre à l'aval du cycle nucléaire. À la demande du gouvernement français, il préside et anime une commission chargée d'examiner et de proposer des voies d'enfouissement des déchets nucléaires en milieu géologique.

La science nucléaire élaborée pendant tout le XXe siècle s'est imposée de manière tragique à l'humanité, à l'occasion de l'explosion des premières bombes atomiques de Hiroshima et de Nagazaki qui mirent fin, en 1945, à la Deuxième Guerre mondiale. Pendant des décennies, marquées par les séquelles d'une guerre particulièrement meurtrière, le développement d'une industrie nucléaire militaire puis civile a été redouté par la population de la planète. Le souvenir de la guerre, l'abondance énergétique associée à la concurrence du pétrole et du charbon font de l'énergie nucléaire un objet de craintes.

Les accidents, spécialement dans les installations nucléaires, furent redoutés. Les persistances des désintégrations nucléaires dans les déchets les rendaient plus néfastes que ceux d'autres industries comme les industries chimiques. L'aval du cycle nucléaire devint une préoccupation majeure des scientifiques, des décideurs et des populations en général. L'entreposage de ces déchets ne pouvait être traité dans l'urgence et, fait exceptionnel dans l'industrie, le gouvernement dirigé par Michel Rocard mit en place, par la loi, une réflexion et un processus expérimental pour aborder scientifiquement et techniquement l'étude de l'enfouissement de ces déchets, du stockage de surface et de la transmutation, ceci avant toute décision politique.

Ainsi, la *Commission nationale d'évaluation des recherches pour la gestion des déchets radioactifs (CNE)* a été créée par la loi du 30 décembre 1991. Son évaluation porte sur les travaux de recherche définis par la loi :

– séparation et transmutation des déchets de haute activité et à vie longue ;
– stockage (permanent) de ces déchets en situation géologique ;
– la mise en conteneur et entreposage (temporaire) des déchets ;

ce qui représente l'ensemble des recherches sur la fin du cycle électronucléaire menées par les centres de recherche et l'industrie nucléaire française.

La Commission a été mise en place par le gouvernement en mai 1994. Elle comporte douze membres dont dix français et deux étrangers. Ils sont nommés pour moitié par le Parlement (c'est le cas de Bernard Tissot), et pour moitié par le gouvernement, sur proposition de l'Académie des sciences ou du Conseil supérieur de sécurité et d'information nucléaire. Bernard Tissot en fut nommé président en mai 1994. La Commission comporte actuellement cinq confrères de l'Académie des sciences (trois membres : MM. Bernard Tissot, Robert Dautray, Olivier Pironneau et deux membres-correspondants, MM. Robert Guillaumont, Ghislain de Marsily).

L'expérience géologique, géophysique et tout particulièrement géochimique de Bernard Tissot, et sa formation à l'*École des Mines de Paris* en génie minier étaient ainsi mises à profit pour aborder les problèmes de l'aval du cycle. Un autre élément favorable était la connaissance du domaine de l'énergie, dont le nucléaire est l'une des sources, importante et discutée. La question du stockage en situation géologique fait appel, pour maints aspects, aux mêmes disciplines scientifiques que celles nécessaires pour comprendre la formation et l'exploitation des gisements de pétrole et de gaz. De leur côté, les méthodes de séparation des actinides et des produits de fission sont proches de celles utilisées en géochimie minérale ; les problèmes de corrosion des conteneurs présentent des points communs avec ceux rencontrés en génie chimique.

La *CNE* a grandement contribué, en liaison avec le ministère chargé de la recherche, à l'établissement d'un programme synthétique des recherches que devaient conduire les acteurs de la loi (centres de recherches et industriels), ainsi qu'à la cohérence de ces recherches. La *CNE* fournit chaque année au gouvernement et au parlement un rapport d'évaluation, ainsi que, à la demande du gouvernement, des avis sur des points cruciaux comme l'inventaire des déchets, la réversibilité des stockages et le choix du site pour un laboratoire souterrain.

L'un des premiers problèmes rencontrés fut l'exploration de trois sites pour y creuser un ou des laboratoires souterrains. Le granite de la Vienne était très en faveur, à la suite en particulier des travaux réalisés en Suède et au Canada. Cependant, sur ce site français, la cartographie géologique montre l'existence d'un champ de failles dans la région proposée : c'est sur l'une d'entre elles qu'est situé l'épicentre d'un séisme ressenti, début juin 2001, jusque dans la ville de Nantes. En outre, les carottages comportent un grand nombre de fissures ou diaclases. En ce qui concerne la circulation des fluides, en l'occurrence la circulation de l'eau qui est l'ennemi de tout stockage souterrain, on se trouve dans une situation souvent rencontrée par les géologues pétroliers : une matrice rocheuse très peu perméable où les déplacements sont très lents et reposent principalement sur la diffusion et un ensemble de fissures où les transferts de matière sont plus rapides et liés au gradient de pression ou à celui de température. Les déchets sont une source de chaleur, propre à amorcer la circulation des fluides et la Commission fut donc amenée à donner un avis défavorable sur cette proposition.

Au contraire, le site proposé pour accueillir un laboratoire dans les argiles du Callovo-Oxfordien de l'est du Bassin de Paris présentait de nombreuses qualités : argile déjà fortement compactée, très grande régularité du dépôt, absence de faille décelable en sismique au niveau de la couche hôte, ou de sismicité notable. Tous ces éléments justifiaient pleinement l'implantation d'un laboratoire à Bure dans cet environnement géologique.

Enfin, le troisième site proposé près de Marcoule, dans le Gard, posait divers problèmes, notamment l'éventualité de variations latérales d'épaisseur des sédiments dont l'interprétation n'était pas claire. Sur demande de la *CNE,* des travaux complémentaires ont montré l'absence de discontinuités, en particulier de failles, ainsi que le caractère spécifique de la roche magasin, différente de celle de Bure. Il s'agit majoritairement d'un silt, c'est-à-dire une roche dont les éléments ont une granulométrie comparable à celle des argiles, mais qui sont constitués de silice, et non de silicoaluminates feuilletés. Le site de Marcoule offre donc une autre possibilité, dans un environnement différent de celui de l'est de la France.

Un grand nombre des travaux scientifiques de Bernard Tissot trouvent ainsi leur application dans le domaine du stockage souterrain des déchets nucléaires. Le problème central de tout ouvrage de ce type est, en effet, celui de la migration, c'est-à-dire la propagation éventuelle des radionucléides dans les argiles. On y retrouve le problème de la nature des transferts de radionucléides (diffusion, écoulement darcéen), le rôle de la mécanique des fluides dans un milieu poreux de très faible perméabilité, celui des failles, la remise en cause du coefficient de partage entre phase adsorbée et phase liquide, les liens entre spéciation et solubilité, etc.

Pour les radionucléides les plus critiques pour la sûreté, l'iode et césium, le facteur contrôlant la migration à partir du stockage sera la diffusion, processus très lent, et non un écoulement darcéen. Pour l'uranium et d'autres actinides, le facteur contrôlant sera la présence ou non de matière organique et d'un milieu réducteur : l'uranium U4+ est pratiquement insoluble dans ces conditions et restera piégé dans la matière organique des argiles encaissantes ; si au contraire un régime oxydant s'établit, U6+ est soluble et l'uranium pourrait migrer dans les formations géologiques. Les réacteurs naturels précambriens de la région d'Oklo au Gabon montrent que, dans ce type de situation, la migration est limitée à l'échelle métrique.

Plus généralement, le problème de fond du retour éventuel des radionucléides vers la biosphère est celui du cycle des éléments, tant géochimique que biogéochimique.

Plusieurs problèmes ont amené la *CNE* à déplorer, dans ses rapports au gouvernement et au parlement, des retards importants ou des insuffisances patentes. C'est notamment le cas de l'inventaire des déchets, des conteneurs d'entreposage et de stockage, et des dépôts provisoires (entreposages).

Les inventaires prévisionnels (en fin d'exploitation) établis par les divers producteurs de déchets en 1996 et 1998 et présentés successivement à la *CNE* ne présentaient pas de cohérence entre eux et ne permettaient aucun recoupement. Il apparut clairement que les producteurs avaient établi un inventaire prévisionnel prenant en compte les résultats espérés de recherches en cours qui auraient permis de déclasser certains déchets ou d'en réduire le volume de façon importante. Le gouvernement chargea le président de l'*ANDRA (Agence nationale des déchets radioactifs)* de mettre au point une procédure rigoureuse d'inventaire, qui fut soumise à la *CNE* et approuvée par elle. L'*ANDRA* reçut alors la mission d'établir, sur ces bases, un inventaire non critiquable des déchets existants, et de ceux attendus en fin d'exploitation des centrales électronucléaires en particulier.

L'absence de conteneurs répondant aux exigences de l'entreposage et du stockage, et validés par les organismes en charge de ces deux opérations, place la France loin derrière les pays comme les États-Unis, le Canada, la Suède ou la Finlande. Une remise à plat de l'ensemble du problème par les divers acteurs de la loi fit porter la priorité sur les déchets de moyenne activité (déchets B) et les combustibles usés. Les premiers sont placés de longue date dans des fûts qui ont des contenus et des comportements très divers, nécessitant pour certains d'entre eux la reprise des déchets et de nouveaux conteneurs. Rien n'est actuellement prévu pour les combustibles usés non retraités, en dehors des "piscines", puisque cette catégorie n'était pas prévue dans les schémas de gestion. Cette situation amena le *CEA (Commissariat à l'Énergie atomique)* à prendre l'engagement de présenter des démonstrateurs fonctionnels fin 2002, ce qui fut fait, et des conteneurs prototypes en vraie grandeur dès 2004.

L'existence d'un entreposage de fait pour certaines catégories de déchets ne répond pas nécessairement aux exigences de recherches demandées par la loi. Une étude dans ce sens portant sur des entreposages de durée séculaire a donc été promise par le *CEA* pour être présentée en 2004. On peut noter que la *CNE* avait affiché, plusieurs années avant le 11 septembre 2001, une nette préférence pour les ouvrages de subsurface (à flanc de colline) qui assurent une meilleure sûreté des entreposages (intrusion humaine, séisme, chute d'avion).

La modélisation numérique des stockages et du retour éventuel des radionucléides vers la biosphère est également un domaine suivi de près par la *CNE*. Beaucoup des problèmes traités sont proches de ceux que les recherches sur les hydrocarbures ont rencontrés dans les modèles de bassins sédimentaires au cours des deux dernières décennies.

Dans son rapport annuel, la *CNE* présente au gouvernement et au parlement l'avancement des travaux et les perspectives dans les domaines couverts par la loi de 1991. Son rapport n° 8 a été remis en septembre 2002. Au terme du délai prévu par la loi (décembre 2006), elle devra remettre un rapport final comportant le bilan des travaux et des recommandations sur les suites à donner aux recherches. Dans cette optique, elle s'efforce dès maintenant de présenter chaque année une étude critique des résultats obtenus et une perspective des résultats qu'on peut espérer atteindre d'ici 2006.

Pour conclure, je soulignerai le poids moral pris par la *CNE* dans la recherche scientifique sur l'aval du cycle nucléaire : la grande majorité de ses recommandations scientifiques ou technologiques est suivie par les acteurs de la loi et la *CNE* est réellement devenue une référence scientifique importante dans la recherche française sur ce domaine. La seule recommandation présentée avec persistance et toujours écartée concerne le souhait de voir accroître les recherches dans le domaine de la sociologie, pour mieux comprendre les fondements du refus ou de l'acceptation sociale. Divers évènements ont cependant montré à l'évidence qu'il s'agit là d'un élément important du processus décisionnel et ce domaine-là aussi est familier au chercheur pétrolier. La crainte de la pollution, du gaz carbonique rejoint celle de l'énergie nucléaire et elle s'étend désormais à d'autres sources d'énergie. Et pourtant, l'énergie est la base de notre civilisation.

Bernard Tissot consacre avec talent toute son attention à permettre que cette civilisation ne s'arrête pas ; beaucoup d'hommes et de femmes sur notre planète en bénéficieront pleinement.

Jean Dercourt

Professeur honoraire de l'université Pierre et Marie Curie
Secrétaire perpétuel de l'Académie des Sciences de l'Institut de France
Président du Conseil scientifique de l'IFP

Introduction

Search for today available sources of energy that could last for tens of years and eventually centuries, is one of the main concern of the humanity.

Bernard Tissot, who participated to this search in the oil industry continues it today in the nuclear industry. He first contributed in elucidating the physical and chemical processes that conduce from the living organic matter to hydrocarbon fields through maturation in the sediments, till the entrapment of hydrocarbons that are then exploited.

Nowadays his main concern is the downstream nuclear energy cycle. On request of the French Government he presides over and animates the commission in charge to examine and propose ways to store nuclear wastes in subterraneous geological places

The nuclear science elaborated during the XXth century stroke in a tragic way the humanity when, in 1945, the explosions of the first atomic bombs in Hiroshima and Nagasaki put an end to the Second World War. During tens of years marked by the aftermath of a particularly bloody war, the development of a military and then civilian nuclear industry was feared by the world's population. The memory of the war and the relative abundance of energy sources related to the competing carbon and petroleum industries, relegate the nuclear energy to a position of secondary importance and keep it in a fear context.

Accidents especially in nuclear plants were dread. The long lasting disintegration of nuclear matter caused the wastes more harmful than those issued of other industries like the chemical one. The downstream part of the nuclear cycle became a major concern for scientists, managers and populations in general. The disposal of these wastes could not be decided in the hurry and, exceptionally for the industry, the Michel Rocard Government installed by law a specific process to study scientifically and technically the storage of the nuclear wastes in the surface and in the subsurface and the study of the transmutation, before any political decision.

The *Commission nationale d'évaluation des recherches pour la gestion des déchets radioactifs (CNE)* was created by the 30 December 1991 law. Its role consists in evaluating the researches defined by the law:

- separation and transmutation of the long living and highly active wastes;
- storage (permanent) of these wastes in geological environment;
- conditioning and disposal (temporary) of the wastes;

which represents all the studies of the ultimate part of the electronuclear cycle conduced by French research centres and nuclear industry.

The Commission was installed by the Governement on May 1994. It consists of twelve members, among whom ten French and two foreigners. Half of the commission is named by the Parliament (it is the case of Bernard Tissot) and the other half is named by the Government following proposals from the *Académie des sciences* or from the *Conseil supérieur de sécurité et d'information nucléaire*. Bernard Tissot was named president of this Commission on May 1994. The Commission is composed at present by five members of the *Académie des sciences* (three members: Messrs. Bernard Tissot, Robert Dautray, Olivier Pironneau and two corresponding-members, Messrs. Robert Guillaumont, Ghislan de Marsily).

The experience and knowledge in geology, geophysics and more specifically in geochemistry of Bernard Tissot, as well as his background in mine engineering acquired in the *École des Mines de Paris* were profitable to tackle the problem of the nuclear waste storage. Another favourable element was his perfect knowledge of the energy world, the nuclear sources being an important but controversial part of it. The waste storage in geological environment calls, in several aspects, for the same scientific disciplines than those necessary to understand the formation and the exploitation of oil and gas fields. The methodologies for the separation of the actinides and products resulting from fission are close to the one used in mineral geochemistry; the problems of the corrosion of containers have several similarities to the problems found in chemical engineering.

The *CNE* widely contributed, in relation with the ministry in charge of the scientific research, in establishing a synthetic programme of the studies that were supposed to be performed by the research centres and the industry and in controlling the coherency of theses studies. Each year, the *CNE* provides the Government and the Parliament with an evaluation report, as well as, on the request of the Government, with advises on crucial points such as the inventory of the produced wastes, the reversibility of the disposal and the choice of a site for subterraneous laboratory.

One of the first encountered problem was the exploration of three sites to dig one of the subterraneous laboratory. The Vienne granites site was particularly favoured, in regard of previous works done in Sweden and in Canada. However, in the French granites, the geological mapping showed the occurrence of a fault array in the proposed area: the June 2001 earthquake that was felt till the Nantes town was located on one of these faults. Moreover, the cores performed in the granites show a large number of joints and fractures. Concerning the fluids flow and more precisely the water flow which is generally the killing factor for all subterraneous storages, the situation is the same than the one encountered by the petroleum geologists: an almost impervious rock matrix were displacements are very slow and correspond to diffusion and a set of joints were the matter transfers are more rapid and related to the pressure or temperature gradient. Nuclear wastes generally produce heat that can initiate the fluid flow and the Commission was led to give a negative advice for this proposed site.

On the contrary, the site proposed to host the laboratory in the Callovo-Oxfordian shales in the eastern part of the Paris Basin displayed numerous favourable properties: shales already strongly compacted, wide continuity of the strata, lack of recognisable fault in the host-layers on the seismic reflection data and lack of notable earthquakes. All these elements fully justified the choice of the Bure laboratory in such a geological environment.

Finally the third proposed site, close to Marcoule, in the Gard department, was bearing several problems, in particular the possible lateral thickness variations of the host sediments, the interpretation of which was not clear. Upon the request of the *CNE*, complementary works revealed the lack of discontinuities, such as faults, as well as the specificity of the; host sediments different of the Bure strata. Sediments are mainly composed of silt, that means a rock composed of grains the size of which is comparable to the one of the shales but with a mineralogy of silica and not of silico-aluminates. The Marcoule site corresponds to another possible laboratory emplacement, in a geological environment distinct from the Bure one in the eastern part of France.

A large number of the scientific works of Bernard Tissot could be applied to the domain of the subterraneous storage of nuclear wastes. The central problem of all these storages is in fact the one of the fluid migration, *i.e.,* the transfers of radionuclear elements in the shales. One finds the problems of the type of transfer of the radioelements (diffusion, transport by Darcy's flow), of the fluid mechanics in a very low permeability porous media, the role of faults, of the questioning about the sharing factor between the adsorbed and liquid phase, of the links between speciation and solubility, etc.

For the more safety critical radionuclear elements, the iodine and caesium, the controlling factors for migration will be the diffusion, very slow process and not a Darcy's type flow. For the uranium and other actinides, the controlling factor will be the presence of organic matter and reducing media : uranium U4+ is practically insoluble in those conditions and will stay trapped in the organic matter of the shales; if on the contrary, an oxidising regime prevails, U6+ becomes soluble and the uranium could migrate in the geological units. The Precambrian natural reactors in the area of Oklo in Gabon have shown that in this kind of situation; the migration is limited to few metres.

In a more general way, the fundamental problem of the possible transfer of radionuclear elements in the biosphere is the one of the cycle of all the geochemical and biogeochemical elements.

Several problems led the *CNE* to complain, in its reports to the Government and the Parliament about the large delay or obvious inadequacies. It was the case of the inventory of the wastes, containers, planned disposals and storage areas.

Provisional inventory (at the end of the exploitation) established by various waste producers in 1996 and 1998 and successively presented to the *CNE* were not self consistent and did not allow any cross-checking. It was obvious that the producers established an inventory taking into account the expected results of the ongoing research which could have permitted to discard some waste and reduce considerably their volume. The Government asked the *ANDRA (Agence nationale des déchets radioactifs)* president to elaborate a rigorous inventory procedure that was submitted to the *CNE* and approved by this commission. The *ANDRA* was then asked to establish , on these bases, a non questionable inventory of the existing wastes and of those expected to be produced at the end of the exploitation of the electronuclear plants in particular.

The lack of containers corresponding to the disposal and storage requirement, and validated by the organisms in charge of both operations, leaves France far beyond countries like the United States Canada, Sweden or Finland. A general questioning by all the actors of the law about all the problems led to put the priority on wastes of medium activity (wastes B) and used combustibles. The first ones have been placed for a long time in containers which have various content and behaviour, needing for some of them the reconditioning of the wastes and new containers. Nothing is presently planned for the non conditioned used combustibles except the "pools" because this class of waste was not planned in the management scheme. This situation led the *CEA (Commissariat à l'Énergie atomique)* to take the agreement to present viable solutions at the end of 2002, that was done, and prototypes of new containers as soon as 2004.

The existing disposals for some classes of wastes do not necessarily correspond with the research criteria asked by the law. A study about the century disposal was agreed by *CEA* to be presented in 2004. One can notice that the *CNE* mentioned several year before 11 September 2001, a clear preference for subsurface storage (in the hill flanks) that insure a greater safety for the disposals (human intrusions, earthquakes, aircraft crashes).

Numerical modeling of the storage and of the possible transfer of the radioelements to the biosphere is also a topic of high interest for the *CNE*. A lot of the treated problems are close to the ones encountered by the hydrocarbons exploration in the modeling of sedimentary basins during the two last tens of years.

In its annual report, the *CNE* presents to the Government and the Parliament the progress of the works and the prospective results in the domains covered by the 1991 law. Report No. 8 was delivered in September 2002. At the end of the time laps planned by the law (December 2006), it will provide a final report containing the synthesis of the achieved works and recommendations on the future researches. In this scope, the Commission try to propose every year a critical review of the results already obtained and a perspective of the expected results in 2006.

To conclude, I would like to underline the moral weight taken by the CNE in the scientific research on the downstream part of the nuclear cycle: most of its scientific and technological recommendations is followed by the actors of the law and the *CNE* truly became a major scientific reference for the French research in this domain. The only recommendation insistently proposed and always withdrawn concerns the wish to increase the sociological research to better understand the bases of the social acceptance or refusal. Various events demonstrated this is key element of the decisional process and it is also well known of the petroleum explorer. The fear of the contamination and of the carbon dioxide, meets up with the fear of the nuclear energy and is also transferred to other energy resources. Nevertheless, energy is fundamental in our civilisation.

Bernard Tissot dedicates with talent all his attention to allow this civilisation not to stop; a lot of men and women on our planet will fully benefit from this action.

Jean Dercourt
Emeritus Professor at the University Pierre et Marie Curie (Paris VI)
Permanent Secretary of the French Academy of Sciences
Chairman of IFP Scientific Board

Oil & Gas Science and Technology – Rev. IFP, Vol. 58 (2003), No. 2, pp. 179-182
Copyright © 2003, Éditions Technip

From the Field to the Model
Bernard Tissot's Path

L. Montadert[1]

1 Beicip-Franlab, 232, avenue Napoléon-Bonaparte, 92502 Rueil-Malmaison Cedex - France
e-mail: lucien.montadert@beicip.fr

This short article is not aimed to comment the long career of B. Tissot, but to single out certain periods of his professional life that were probably decisive. Of particular interest are the years in which he practiced the profession of oil explorer, and the transition period between these years and his entry into the profession of researcher into the scientific problems associated with the genesis and migration of oil.

One is first naturally inclined to try to understand the reasons for his attraction to Earth Science. After graduating from the *École des Mines de Paris*, he decided to take the Geology option at *École nationale supérieure du pétrole et des moteurs (ENSPM)*, and ended with a degree in Advanced Geology Studies at the *Université de Grenoble* in 1955. This was concerned with a "Geological study of the Massifs of Grand Galibier and Cerces" *(Fig. 1)*, hence with true alpine geology. It demanded to be in good form because when he lived in the chalet of the French Alpine Club of Monestier and only possessed a bicycle, it was a real grind to reach the study zone in the Galibier Massif. The answer to our initial question is simple: the quality of geology teaching at *École*

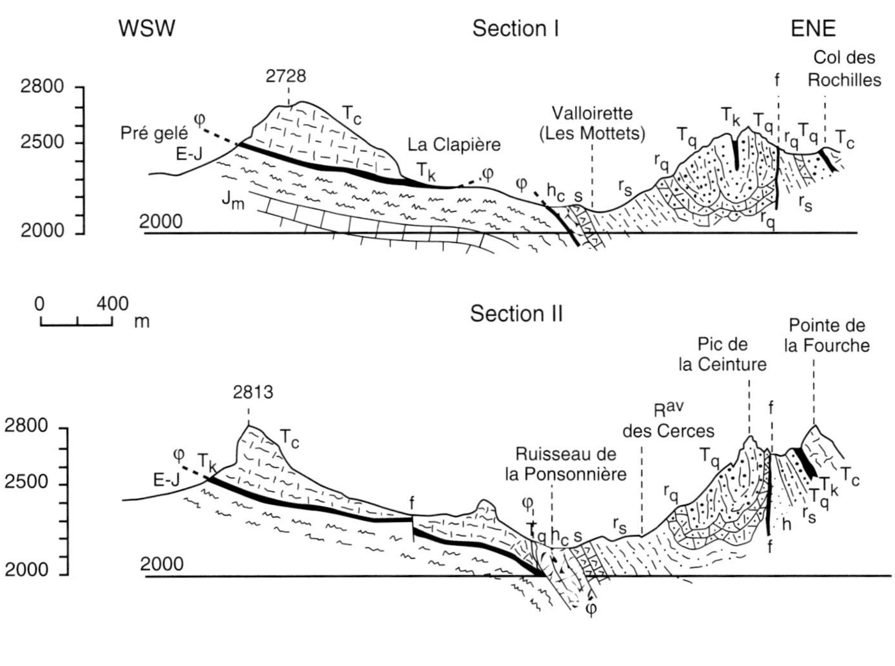

Figure 1

Two cross-sections issued from one of the first works published by B. Tissot about the geology of the Grand Galibier and Cerces Massifs in the "zone briançonnaise".

des mines de Paris by Professors J. Goguel and M. Lemoine was the direct cause of this leaning toward Earth Science.

THE ROUTE OF THE OIL EXPLORER

Following his period at *ENSPM*, B. Tissot joined *IFP* in 1955 in the *Bureau of Geological Studies (BEG)*, *IFP*'s arm for conducting studies in oil exploration worldwide for governmental agencies and oil companies. He started out in the Paris Basin with M. De Myans d'Archimbaut, working for *Gaz de France* on the Beynes gas storage facility, and for *Mobil* on the southwest of the Basin, an unpropitious area for the accumulation of oil and gas, but where *GDF* eventually built the world's biggest storage facility of this type, the *Chemery* facility.

Then came his first mission outside France. In 1956 and 1957, B. Tissot investigated the petroleum potential of New Caledonia. Outstanding field work followed with a detailed mapping in a region featuring highly complex tectonics. Only a good geologist could discern the big picture, identify the unseen thrusts and decide whether the sedimentary series observed at the outcrop were right side up or upside down. Naturally, the petroleum possibilities of the region appeared very mediocre, but the mission was the first that B. Tissot carried out from start to finish, as manager, coordinating the field work and the laboratory investigations which he had requested.

In 1958 and 1959, he did his military service, with an assignment to *Compagnie d'Exploration Pétrolière (CEP)* at Fort Platters: alongside other *BEG* colleagues he was mostly busy locating exploration boreholes.

On April 1st, 1960, he headed for Australia with his wife for a three-month stay, and returned after three years with a family, two daughters born in kangaroo land. This time, the task was to investigate the Australian sedimentary basins with a team of 22 to 23 persons. Following a short visit by M. Guillemot and M. Trümpy, B. Tissot was the Manager of this party, where laboratory studies, photogeology, petrography and interpretation of seismic lines went hand in hand. After the three-year spell, Dr Trümpy and B. Tissot presented a review of recent developments in prospecting in Australia, and proposed a list of the main onshore and offshore prospects. There isn't much to add today concerning this list.

These years dedicated to oil exploration were followed by some technical disappointment, because B. Tissot found that the geological data had been insufficiently used and that the boreholes were most often installed exclusively on a reflection seismic anomaly. Besides, the lack of scientific knowledge about the mode of formation of the oil and gas and their accumulation in the reservoirs, explained why judgments on the potential of the basins were essentially based on reasoning by analogy. The result was a low wildcat success ratio (1 to 10).

THE SCIENTIFIC JOURNEY

In 1964, B. Tissot began his career as a scientist, participating in a think tank on the research program of the *Geology Division* which became effective in June 1965, ultimately giving rise to a sweeping reorganization of this sector.

The genesis of oil and gas and the formation of reservoirs became priority topics of the division, *CEO R. Navarre* having demanded that these problems be resolved by *IFP* scientists. A number of positions taken in this program are worth recalling.

"The training of geologists, at least in France, is still primarily too naturalistic. This is manifested by the fact that more importance is ascribed to the description of the outward signs of phenomena (sense of observation) than on the search for their intimate mechanism. Given the development of geology, this results in the accumulation of an infinite number of observed facts, of no use to anyone because ignorance of the underlying processes makes it impossible to decide which of these facts is significant.

Thus in the specific field of petroleum geology, the existence of a relationship between the presence of organic matter and the occurrence of oil, the migration, and the accumulation of oil in preferential zones, are facts of observation. And yet until now, no one has derived a valid explanation of the formation of oil and gas reservoirs.

For knowledge to advance, it is critically important to stop focusing the effort on a naturalistic description of the basins and reservoirs, but rather on an understanding of the mechanisms involved. And this happens to be the only method for determining the general laws to open up new methodological perspectives for exploration.

For the geologist, this necessary progress in the fundamental knowledge of geological processes demands not merely the renunciation of his initial qualities as an observer, but the simultaneous development of knowledge in the field of the mathematical, physical and chemical sciences. It is only by a permanent back and forth process between the observation of nature and the critical analysis of phenomena in the laboratory, that one can succeed in formulating hypotheses reflecting the observed facts.

Furthermore, even with respect to purely analytical work, it is undeniable that an effort exploiting all the resources of physics and chemistry must be made to enhance the meaning and accuracy of the measurements. For example, at a time when chromatography and mass spectrometry are used to characterize components of an oil, it is curious to find that one can often adequately characterize an oil by a global property, such as a distillation curve, a density or a molecular weight.

A systematic effort must therefore be pursued, both to develop more objective methods of observation, and to acquire the basic knowledge of the processes involved."

This attitude perfectly expresses the approach pursued by B. Tissot to address research into the genesis of hydrocarbons.

It is interesting to note that upon the establishment of this new program, the *Geology Division* included the *Bureau of Geological Studies* and a "Research" section with a matrix organization.

The first step was a critical bibliography of the genesis of hydrocarbons. The aim was to estimate the possibilities of laboratory investigations and field observations of the specific conditions in which sedimentary organic matter is converted into hydrocarbons. Hence the first phase implied a critique of the hypotheses formulated and their replacement by new ones.

From autumn of 1965, with some intensification in 1966, the main conclusions were available to guide subsequent research. The central model was a thermocatalytic hypothesis (geothermal energy, inorganic catalysts). We can recall some of the conclusions of this analysis.

"If we refer to thermodynamic considerations, it appears that there is no reason to impose a time limit on the mechanisms examined. For each system of organic components and water, in a given set of physicochemical conditions, an equilibrium state exists, which the system probably does not reach, but towards which it will evolve insofar as the temperature is elevated and adequate catalysts are present. If the conditions change—especially the temperature and pressure with burial depth—the equilibrium position and the rate at which the system tends towards it, change, and the organic mixtures therefore evolve.

Subsequently, the age of the sediment in the formation of oil of the hydrocarbons does not appear to represent a restricted condition. One can very well agree that a sediment will remain for a long time at an insufficient depth and temperature, eventually reaching such conditions. Or even that having already supplied hydrocarbons and being subsequently placed at greater depth and high temperatures, it can supply even more. This would also help explain numerous reservoirs connected with an unconformity, where the rock rich in organic matter presumed to be the source of oil, is located under the unconformity and the erosion (Hassi-Messaoud, for example), which appears difficult in the case of an early genesis.

Similarly, it must be considered that wherever the oil is found, from the moment when it is mobile, the transformations can continue as attested by experiments on the evolution of crude oils under the influence of temperature, pressure and inorganic catalysts. In particular, it is unsure whether in the case of certain reservoirs, the migration has not concerned a product that is quite different from the present crude.

These remarks once again underscore the importance of the geological and physicochemical history of the formations, source rock and reservoir alike. On the completion of this study, if we single out the main steps of the genesis of petroleum accumulations, from the deposition of organic matter to the segregation of the oil in a reservoir, we find that some of the favorable factors are definitely associated with the specific composition of the source rock, while others, at least equally important (temperature, pressure, fluid expulsion law) are associated with the geological and physicochemical history of the fraction of basin considered."

The next step was particularly important, because as early as 1966, it helped establish the influence of temperature and pressure by observations on the Toarcian of the Paris Basin and by laboratory experiments. These results were presented by M. Louis and B. Tissot to the 7th World Petroleum Congress in Mexico in 1967.

"The choice of the Toarcian of the Paris Basin as a source rock makes it possible:

– to effectively investigate the influence of temperature and pressure because the other parameters vary only slightly (age, composition of the rock and depositional environment);
– through the ability to reconstruct the geological history of burial, to demonstrate the importance of the thermal history of the sediment *(Fig. 2)*;
– to clarify the role of different components (hydrocarbons, resins, asphaltenes, MAB extracts and insoluble organic matter).

The importance of these results for prospecting is immediately evident: for a formation where the organic matter and its mineral support have a given composition, a temperature exists above which the yield of the hydrocarbon formation reactions rapidly increases. Given the geothermal gradient and the reconstruction of the burial history, zones which have never been raised to this temperature are of no petroleum interest, even if they are rich in organic matter."

The results of the observations and experiments on the Toarcian of the Paris Basin enabled a decisive breakthrough, this time with the passage to mathematical modeling. In 1968 B. Tissot wrote a report on the "Preliminary Data on the Mechanisms and Kinetics of the Formation of Oil. Computer Simulation of a Reaction Flow Sheet", published in 1968 in *Revue de l'IFP*.

We can quote the conclusions here:

"It was therefore possible, starting with the results of observations and experiments on the Toarcian of the Paris Basin, to construct a simplified reaction flow sheet, and then simulate the conversion on a computer. The consistency of the overall results demonstrates that the reaction flowchart and the kinetics employed are compatible with reality. We thus had a model to represent natural phenomena and to contribute to their analysis.

The use of such a tool helps explicitly introduce the time and temperature parameters, in other words, the entire geological history of the sediment after its deposition, instead

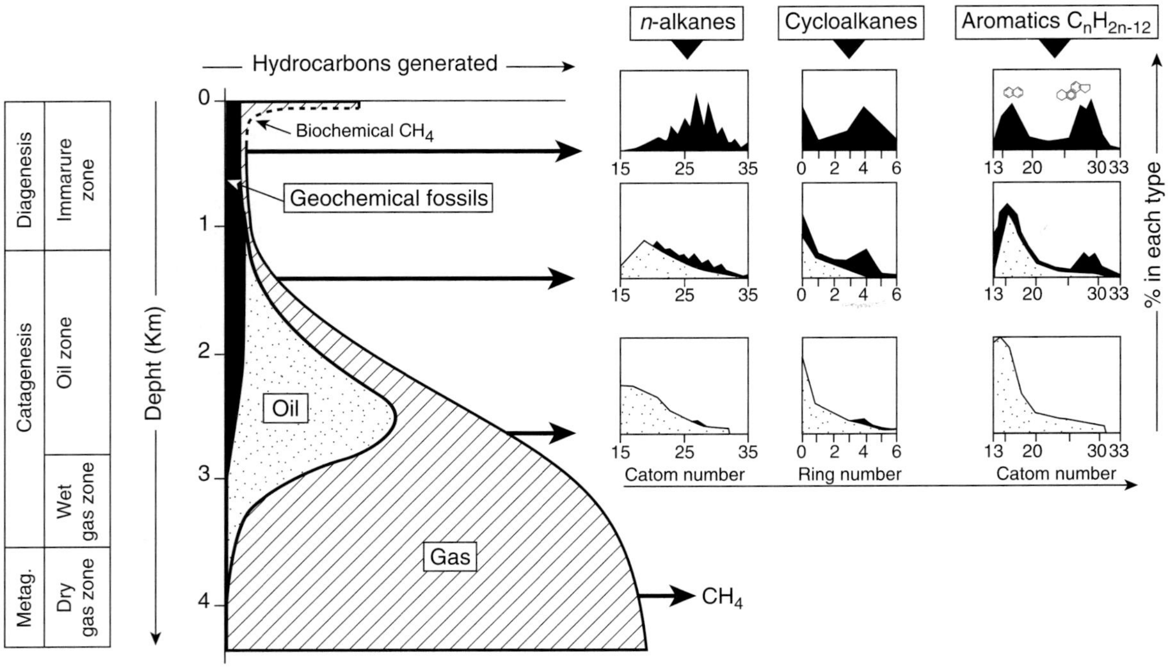

Figure 2

One of the most famous diagrams published by Bernard Tissot and co-authors illustrating the burial/maturation and hydrocarbons composition relationship. This concept was at the root of the further developments in petroleum organic geochemistry.

Tissot and Welte (1978) (2nd edition 1984) *Petroleum Formation and Occurence*, Springer Verlag.

of merely considering a specific episode (maximum burial depth or present depth). Besides, it becomes possible to separate the respective influence of time and temperature, which was hitherto impossible.

In terms of prospecting, this tool helps to simulate every point of the basin investigated, and with a good approximation, the quantities of oil formed since the time of deposition until today. We can thus determine the most favorable zones, in other words, those where the source rock has been the most productive. It is naturally easy to take account of the organic matter contents and the thickness of the formation.

We can also consider the generalization of this model to other formations, and to other sedimentary basins, to contribute to prospecting. This aspect is the subject of another publication. It suffices to have measurements on a small number of core samples (to adjust the constant specific to the formation concerned), a reconstruction of the geological history of the basin, and an evaluation of the geothermal gradient.

Finally, the use of the model supplies the preliminary data, still approximate, on the time and rate of the formation of oil. A comparison of these data with the age of the structural movements will perhaps, in certain basins, help to identify the most favorable traps."

It was therefore in four years, from 1965 to 1968, that the major steps were completed:

- choice of a mechanism based on a critical bibliography;
- demonstration of the influence of temperature and pressure by observations and experiments on a natural case: the Toarcian of the Paris Basin;
- mathematical modeling.

Investigations then naturally continued, particularly on the migration of oil and gas and the modeling of petroleum systems, first with one dimension (Genex) then with two dimensions (Temis 2D) and finally in three dimensions (Temis 3D).

B. Tissot, Head of the *Geochemistry Department* of *IFP* from 1965 to 1979, was the driving force of these investigations and the leader of an entire team. They spawned the development of a new discipline: organic geochemistry and its application to the exploration of oil and gas.

The work done at *IFP* had huge repercussions, not only in the scientific world but also in the oil companies, such as *Total, Exxon, Shell*, and *Chevron.*

This influence led B. Tissot to become Scientific Director of *IFP* in 1979, and then, from 1988 to 1995 Director of Research and Development, and finally Deputy Executive Officer. But that is another story.

Final manuscript received in March 27, 2003

Oil & Gas Science and Technology – Rev. IFP, Vol. 58 (2003), No. 2, pp. 183-202
Copyright © 2003, Éditions Technip

Preliminary Data on the Mechanisms and Kinetics of the Formation of Petroleum in Sediments. Computer Simulation of a Reaction Flowsheet

B. Tissot[1]

1 Institut français du pétrole, 1 et 4, avenue de Bois-Préau, 92852 Rueil-Malmaison Cedex - France

Résumé — Premières données sur les mécanismes et la cinétique de la formation du pétrole dans les sédiments. Simulation d'un schéma réactionnel sur ordinateur — Des travaux récents ont permis de suivre la formation des composés du pétrole dans les sédiments en fonction de l'enfouissement sur quelques cas géologiques : Toarcien du Bassin de Paris, Mio-Pliocène de Californie. Ces observations permettent de déterminer, pour la formation étudiée, les zones favorables où la matière organique a produit du pétrole.

Pour que ces résultats puissent être généralisés à d'autres formations et d'autres bassins, on a tenté de déduire des observations un schéma réactionnel synthétique, et d'écrire la cinétique correspondant à ce schéma. La simulation sur ordinateur de l'évolution de ce schéma réactionnel permet de calculer les quantités de pétrole formées à partir d'un gramme de matière organique primitive en tout point d'un bassin. Les données nécessaires sont une reconstitution géologique de l'histoire de l'enfouissement et un ensemble de constantes représentant la nature de la matière organique primitive.

Dans le cas du Toarcien du Bassin de Paris, les résultats montrent un excellent accord entre les valeurs ainsi calculées et celles effectivement observées.

On accède également au problème de l'époque et de la vitesse de formation du pétrole pour les comparer avec l'âge des mouvements structuraux qui déterminent la formation des pièges.

Abstract* — *Preliminary Data on the Mechanisms and Kinetics of the Formation of Petroleum in Sediments. Computer Simulation of a Reaction Flowsheet — *Recent studies investigated the formation of oil compounds in sediments, as a function of burial depth, in some geological settings of the Toarcian in the Paris Basin and the Mio-Pliocene in California. In the formations examined, these observations served to identify favorable zones where oil was generated by organic matter.*

To extend these results to other formations and other basins, an attempt was made to fit them into a synthetic reaction flowsheet and to determine the kinetics corresponding to this flowsheet. The computer simulation of the way in which the reaction flowsheet evolves helps to calculate the amount of oil formed from one gramme of original organic matter, at any point of a basin. The necessary data are the geological reconstruction of the burial history and a set of constants representing the type of original organic matter concerned.

In the case of the Toarcian in the Paris Basin, the results show close agreement between the values calculated in this manner and the values actually observed.

This also clarifies the problem of the period and the oil formation rate, allowing comparison with the age of the structural movements which determined the formation of the traps.

FOREWORD

This study is based on geological, geochemical and digital processing projects completed by Y. Califet-Debyser, G. Deroo, B. Durand, J. Espitalié, M. Louis, J.L. Oudin, R. Pelet and B. Tissot of the Geochemistry Laboratory of the *Institut français du pétrole*.

INTRODUCTION

The aim of the geochemical study of a sedimentary basin is to determine the zones in which the oil is formed, and, if possible, for each geological formation, to produce a map of the quantities of oil formed at any point. To do this, it is necessary first to identify the reaction mechanisms of the genesis of the oil and the corresponding kinetics.

The transformations of organic matter in the subsoil depend evidently on its initial composition and, like any chemical reaction, are governed by the temperature, pressure and any catalysts involved. Recent studies, based on observations in sedimentary series and confirmed by laboratory experiments, have demonstrated the important role played by burial related factors, and particularly the temperature. These studies include the work of Louis (1964) and Louis and Tissot (1967) on the Toarcian in the Paris Basin; Philippi (1965) on the Mio-Pliocene in California (Welte, 1966; McIver, 1967 and Vassoevich, 1967).

According to these results, the rise in temperature is the determining factor for the changes in organic matter. It promotes the formation of petroleum compounds, particularly hydrocarbons, from the insoluble organic matter (kerogen) in the sediments (*Fig. 1*). The hydrocarbon formation rate also varies with burial depth, since it is very low and becomes high below a certain depth, *i.e.* above a certain temperature threshold.

These observations make it possible to determine directly, for the formation investigated (Toarcian of the Paris Basin, Mio-Pliocene in California, etc.), the zones where organic matter succeeded in producing oil. However, any extension to other formations or other basins raises a difficulty: it becomes immediately clear that the numerical values, for example, the rate of conversion to hydrocarbons, are only valid for the formation where they have been observed. In fact, these values depend on the nature of the organic and inorganic fractions, which varies from one formation to another, and the temperature distribution as a function of time. These are in turn determined by the subsidence and geothermal gradient, *i.e.* by the geological history, which varies from one basin to another.

In order to generalize the results to other situations, it is therefore necessary to formalize them. This is the aim of the present work, carried out primarily with data of the Toarcian in the Paris Basin, and comprising the following steps:

- review of the data from observations and experiments;
- determination of a synthetic *reaction flowsheet*, which satisfies the mechanisms observed for the degradation of organic matter in the sediments;
- expression of the *kinetics* corresponding to this flowsheet;
- computer *simulation* of the evolution of this flowsheet.

This produces a *model* which, in certain conditions, serves to simulate the evolution of organic matter in other sedimentary units, or in other basins.

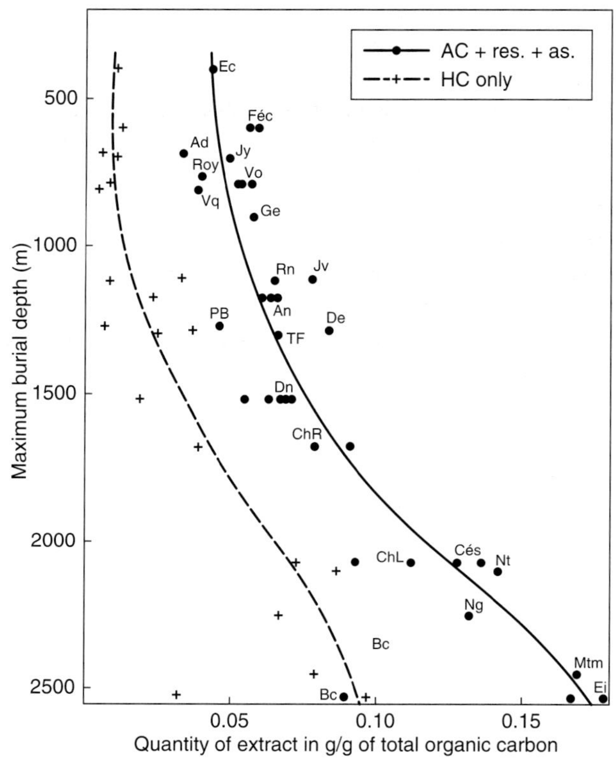

Figure 1

Quantity of hydrocarbons + resins + asphaltenes in the Lower Toarcian clays of the Paris Basin (in grams per gram of total organic carbon), as a function of maximum burial depth.

The study of the kinetics of oil formation brings up another problem, often raised in exploration: that of the "timing" of formation. This is an important point in determining the possibilities of hydrocarbon expulsion and trapping. In fact, we know that a conflict exists between the conditions favorable to the genesis of oil and to its expulsion (Tissot, 1966): oil formation only becomes significant beyond a certain depth threshold; by contrast, the potential for oil expulsion by compaction of the source rock decreases rapidly with increasing burial. The search for an optimum requires the knowledge of the respective rates of oil formation and fluid expulsion.

It is also important to consider the time when the oil was formed, to compare it with the age of the structural movements conditioning the formation of the traps.

1 DEFINITION OF COMPONENTS

The example provided by the Lower Toarcian clays in the Paris Basin is chiefly used for this study. The results of the analyses are given in Louis and Tissot (1967) and in a forthcoming publication that will provide further data. For insoluble organic matter, use is also made of the results on the Devonian and Cretaceous in Canada (McIver, 1967).

It is important to remark how difficult it is to compare the results of analyses conducted in various laboratories, on different bases, which deprives attempts at synthesis of a set of valuable data. It would in particular be necessary to have a protocol of rock extraction and fractionation that has a general value.

In this study, the terms employed to designate the different components of the organic matter are listed in Table 1: the components are separated by their differences of solubility in a scale of solvents. The finely ground rock samples are extracted in the Soxhlet apparatus first by chloroform, then by the triple mixture MAB (methanol 15, acetone 15, benzene 70), until completely depleted. These extracts are respectively called the $CHCl_3$ *extract* and *MAB extract*. Insoluble organic carbon is determined on the extracted rock.

The $CHCl_3$ extract is treated with hexane to precipitate a solid component called asphaltene here. Chromatography of the remaining liquid on alumina is used to separate the *hydrocarbons* and another fraction called *resins*. A portion of the asphaltenes is insoluble in carbon tetrachloride: thus two fractions are separated called CCl_4 *soluble asphaltenes* and CCl_4 *insoluble asphaltenes*. The term *kerogen* is applied to the insoluble organic residue after the two extractions. The kerogen is virtually inseparable from the inorganic matter except by destruction of the latter (hydrofluoric acid attack).

The fractionation thus carried out is justified, because it corresponds to a scale of chemical properties (condensation, oxygen content) and physical properties (molecular weights and polarity):

- Table 2 gives the elementary composition of the various heteroatomic compounds.
- Figure 2 shows the relation between condensation (measured by the C/H ratio) and the content of oxygen, the most abundant heteroatom; the resins, CCl_4 soluble asphaltenes and CCl_4 insoluble asphaltenes can be seen to form an orderly succession, in which condensation and oxygen content steadily increase. By contrast, the MAB extract is quite different, being at once richer in heteroatoms and the least condensed.
- The average molecular weights measured increase in the order of hydrocarbons, resins, asphaltenes.

TABLE 1

Separation of organic components of rocks and main analytical methods
(The methods of fractionation and detailed analysis of the hydrocarbons are not shown here)

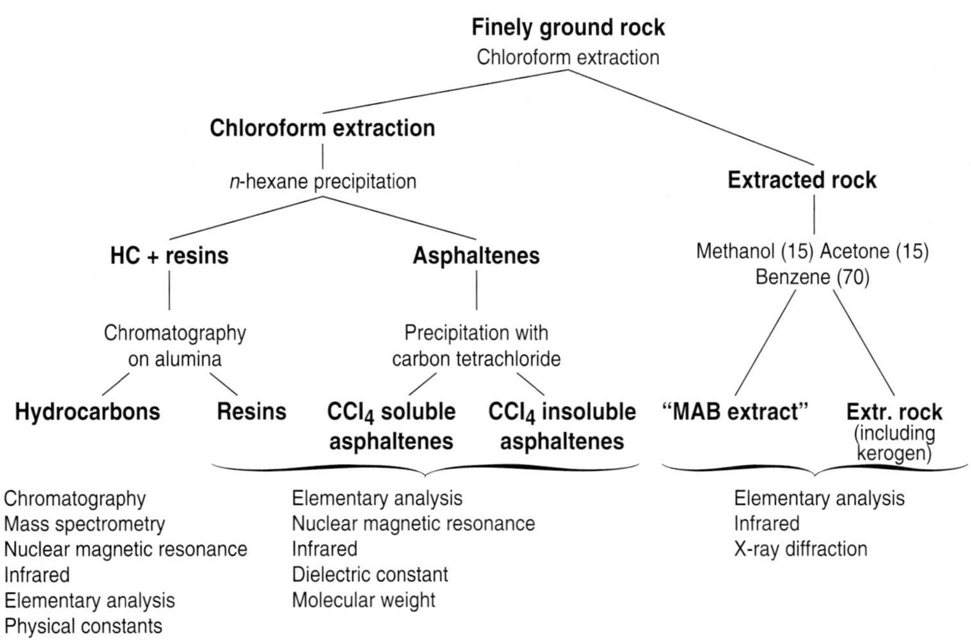

TABLE 2

Elementary composition of MAB extract and resins and asphaltenes extracted from Toarcian clays, compared with that of resins and asphaltenes of crude oils (average values)

Source	Component	C	H	C/H	O	O + N + S
Crude oil	Resins (average of 14 oils)	84.0	9.2	9.17		6.9*
	Asphaltenes (average of 18 oils)	85.6	8.2	10.39		6.2*
Toarcian extracts	Resins (average of 27 samples from 16 boreholes)	76.9	9.2	8.37	6.8	10.1
	CCl_4 soluble asphaltenes (average of 24 samples from 16 boreholes)	72.1	7.5	9.67	9.2	13.3
	CCl_4 insoluble asphaltenes (average of 22 samples from 16 boreholes)	69.8	6.8	10.37	13.9	17.5
	MAB extract (average of 28 samples from 20 boreholes)	70.1	8.4	8.40	16.0	20.3

* 100 – (C + H).

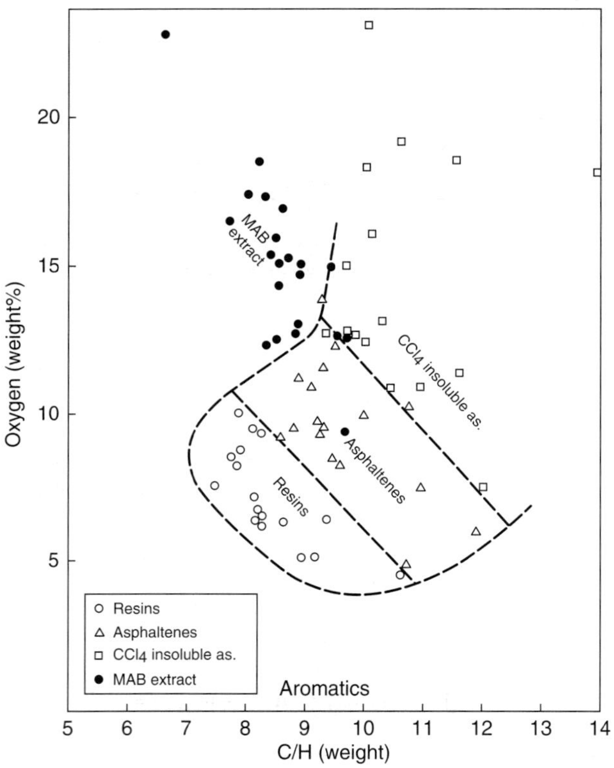

Figure 2

Carbon/hydrogen ratio and oxygen content of various components extracted from Toarcian clays from the Paris Basin.

– Table 3 shows the behavior of the dielectric constants for solutions in chloroform; they increase in the order of resins and CCl_4 soluble asphaltenes, CCl_4 insoluble asphaltenes. The MAB extracts are even more polar, cannot be extracted by chloroform, and demand highly polar solvents for extraction.

The various extractable components shown in Table 1 hence display clearly circumscribed ranges of physical and chemical properties. The same does not apply to kerogen, whose properties vary with the original organic matter, time and depth: elementary composition *(Table 4)*, and structure (see infrared diagrams, *Fig. 10*). This is not surprising because kerogen is defined as a residue in the above extraction protocol. To a lesser extent, the remark can be applied to MAB extract whose elementary composition can also vary *(Fig. 2)*.

TABLE 3

Dielectric constant of resins and asphaltenes
The dielectric constant ε was measured with a frequency of 1 MHz and at 10°C, for heavy products in solution in chloroform in a concentration of 20 g/l

Component	Number of samples	Mean ε	Standard deviation
Pure chloroform		5.2	
Resins	10	5.32	0.02
CCl_4 soluble asphaltenes	11 10	6.30 6.15	0.55 0.18
CCl_4 insoluble asphaltenes	6	6.82	0.20

After J.L. Oudin

TABLE 4

Elementary composition of insoluble organic matter (kerogen) of the Devonian and Cretaceous in Canada (mean values)

Depth (m)	Number of samples	C/H	O
< 1500	20	13.0	15.9
> 1500	7	18.6	6.3

Elementary composition of insoluble organic matter (kerogen) lower Toarcian in the Paris Basin (mean values)

Depth (m)	Number of samples	C/H	O
700 to 2500	10	11.8	12.7

After McIver, 1967

2 REVIEW OF DATA FROM OBSERVATIONS AND EXPERIMENTS ON THE GENERAL MECHANISM OF THE EVOLUTION OF ORGANIC MATTER

2.1 General: Method of Representation

The system consisting of the mass of rock containing 1 g of organic carbon (quantity easily accessible to measurement) is considered, in any form whatsoever; this quantity corresponds to about 1.3 g of soluble and insoluble total organic compounds. This definition assumes that the migration effect is ignored (which appears plausible in the case of the Toarcian in the Paris Basin) as well as the loss of light compounds (CO_2, H_2O and light hydrocarbons) which are entrained by the waters or lost in the laboratory during the extractions.

The two studies mentioned (Toarcian in the Paris Basin, Devonian and Cretaceous in Canada) revealed that the temperature rise is the determining factor for the evolution of organic matter. This temperature rise promotes the appearance of petroleum compounds *(Fig. 1)*, and hydrocarbons in particular, at the expense of kerogen, according to a flowsheet shown in Figure 3 (Louis and Tissot, 1967).

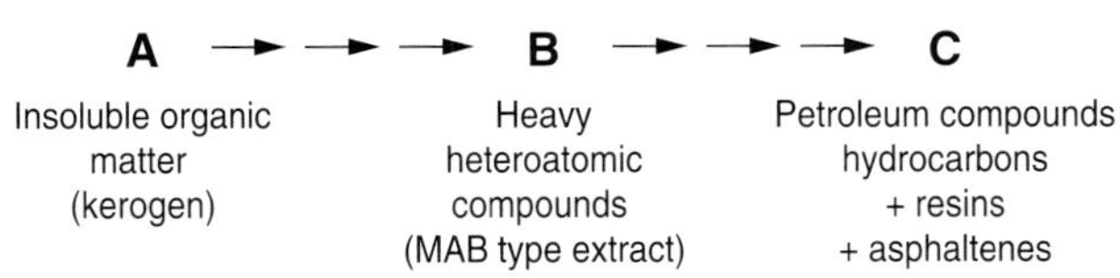

Figure 3

Synthetic reaction flowsheet inferred from observations and experiments on the Toarcian in the Paris Basin.

Each of the terms A and B actually represents a large number of individual compounds which are formed simultaneously or in succession.

The mutual relations of the different components in the example of the Paris Basin are clarified by Figures 4 and 5. These figures show, as a function of the maximum depth (*i.e.* the maximum temperatures) reached by the sample, the advancement of the transformation A → B → C *(Fig. 4)*, and the transformation B → C alone. These graphs resemble those normally employed to monitor the kinetics of a reaction, but are different in two aspects:

- on the *x*-axis, time is not plotted, but the maximum burial depth, which integrates the time and the subsidence rate;
- the *y*-axis does not show the concentrations, but the fraction of total carbon present in each component A, B, C, the latter being separated into hydrocarbons (AC), and resins and asphaltenes (RA).

For example, the fraction of total carbon present in hydrocarbons is defined as follows:

$$[C_{HC}] = \text{quantity of hydrocarbons (related to 1 g of } C_{\text{total organic}}) \times \text{HC carbon content (elementary analysis)}$$

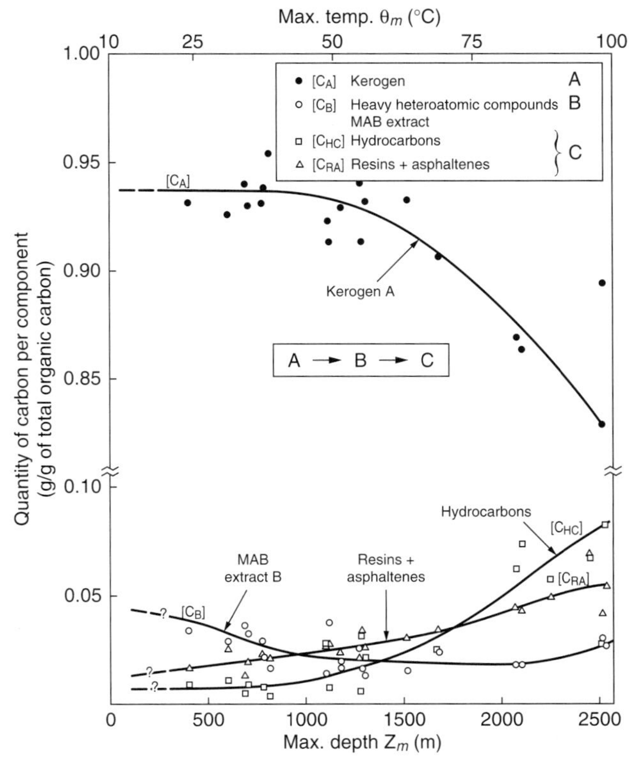

Figure 4

Evolution of the transformation A → B → C as a function of depth and maximum temperature reached (Lower Toarcian in the Paris Basin).

The advancement of the reaction is measured by the quantity of carbon present in the different components, the total being related to 1 g of total organic carbon. The quantities $[C_A]$, $[C_B]$, $[C_{HC}]$, $[C_{RA}]$ are defined as follows:

$$[C_{HC}] = \text{quantity of hydrocarbons (related to 1 g of total organic carbon)} \times \text{carbon content of hydrocarbons (elementary analysis)}$$

$$[C_A] + [C_B] + [C_{HC}] + [C_{RA}] + [C_{losses}] = 1$$

and similarly $[C_A]$, $[C_B]$, $[C_{RA}]$. We therefore have the equations:

- in Figure 4: $[C_A] + [C_B] + [C_{HC}] + [C_{RA}] + [C_{losses}] = 1$
- in Figure 5: $[C_B] + [C_{HC}] + [C_{RA}] + [C_{losses}] = 1$.

The value of using this parameter is that the quantity $[C_A]$ is the only one that is easy to measure in the kerogen; besides, it offers a better picture of the advancement of the reactions, in view of the remark on the variations composition of the kerogen and of the MAB extract made in Section 1: hence it is the only method that allows for a single parameter to represent the different contents.

Figures 4 and 5 explain why, in the flowsheet of Figure 3, the heavy heteroatomic compounds (MAB extract) are made to play the role of a necessary intermediate between

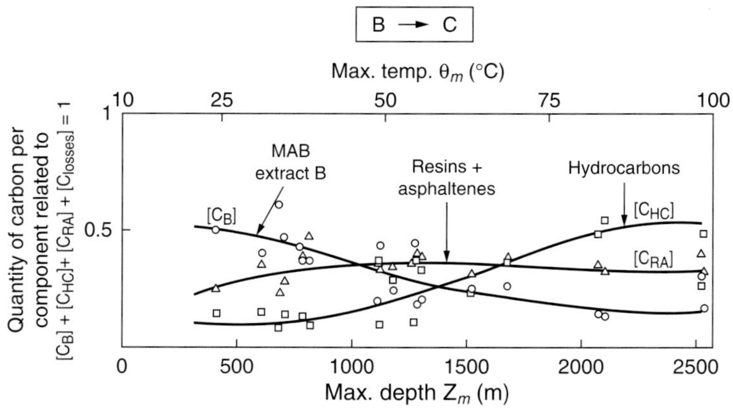

Figure 5

Evolution of the transformation B $\rightarrow$ C as a function of depth and maximum temperature reached (Lower Toarcian in the Paris Basin). The advancement of the reactions measured by the quantity of carbon present in the different components. The quantities $[C_B]$, $[C_{HC}]$, $[C_{RA}]$ are defined as in Figure 4. We obtain the equation:

$$[C_B] + [C_{HC}] + [C_{RA}] + [C_{losses}] = 1$$

For the meaning of the symbols, refer to Figure 4.

kerogen A and the compounds of oil C. This assumption is based on the rapid decrease of this extract *(Figs. 4 and 5)* in the domain where large quantities of hydrocarbons are formed. To confirm this, Louis and Tissot (1967) conducted an experiment with the same samples of the Toarcian from Paris Basin. The hydrocarbons, resins and asphaltenes were removed by solvent extraction. Only the kerogen and heavy heteroatomic compounds (MAB extract) remained in the rock. The samples were then raised in the laboratory to pressures and temperatures corresponding to those of a deep basin (180°C). The formation of new hydrocarbons, resins and asphaltenes was observed, but this formation was preceded by the formation of a new quantity of MAB extract.

Let us first examine the relations between the temperature and the contents observed on the natural examples, and then consider the problem of transformation rates.

2.2 Content-Temperature Relations

The observation made on the Toarcian of the Paris Basin revealed a very strong link between the quantity of compounds of oil C formed and the maximum depth, *i.e.* the maximum temperature θ_m to which the samples were raised *(Fig. 1)*. The graphs show that the same applies to each of the groups of reactions A $\rightarrow$ B and B $\rightarrow$ C, and, in the latter case, for various groups of products formed.

Figure 4 shows the quantity of A transformed, *i.e.* the advancement of the first group of reactions as a function of depth. The temperature rise strongly favors reactions of the type A $\rightarrow$ B, but this influence is only significant above 60-65°C. Below this temperature, the temperature dependence is slight and the quantities of B and C observed largely originate from the hydrocarbons and other compounds preexisting in the recent sediments.

A parallel can be drawn between this fact and the observations of McIver *(Fig. 6)*: in kerogen A, the degree of condensation marked by the C/H ratio varies first slightly with depth, and then increases sharply in a depth range comparable to the previous one.

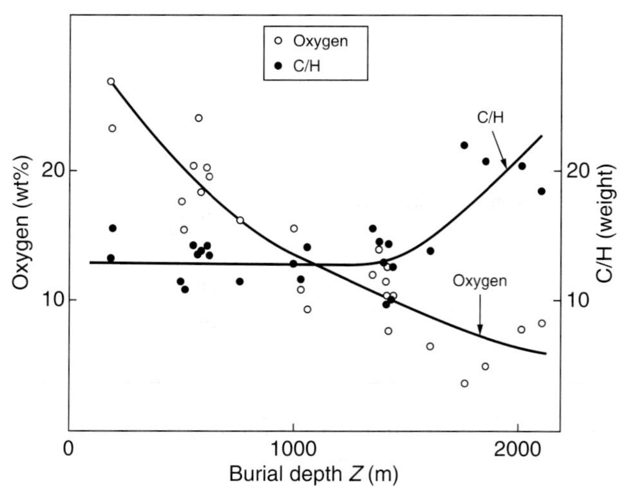

Figure 6

Variation in the elementary composition of kerogen of the Devonian and Cretaceous in Canada, as a function of burial depth (after the data of McIver, 1967).

In natural conditions, transformation A $\rightarrow$ B is far from complete. The fraction of total carbon remaining in the raw material A varies from 0.95 for shallow samples to 0.85 for samples raised to the highest temperatures (100°C) in the

TABLE 5

List of notations

1. Time, depth and temperature

t:	time (Ma)
z:	depth (m) a time t
θ:	temperature (°C)
z_m:	maximum burial depth
θ_m:	maximum temperature (corresponding to z_m)
T:	$\theta + 273$
λ:	subsidence coefficient (m/Ma)
G:	geothermal gradient (degrees/meter)
H:	mean temperature of soil

$$z = \lambda t$$
$$\theta = Gz + H = \lambda Gt + H$$
$$\theta_m = Gz_m + H$$

2. Subsidence law

In the following simple cases considered, it suffices to represent the example of the Toarcian: burial up to t_1, followed by erosion.

For $t \leq t_1$ $\quad \lambda_1 > 0 \quad Z = \lambda_1 t$

For $t > t_1$ $\quad \lambda_2 < 0 \quad Z = \lambda_1 t_1 + \lambda_2 (t - t_1) = Z_m + \lambda_2 (t - t_1)$

3. Extractable compounds

Type	Designation	Quantify for a system defined by 1 g of total organic carbon	
		At origin $t = 0$	At time t
Insoluble organic matter (kerogen)	A	x_0	x
Heavy heteroatomic compounds extracted with mixture MAB	B	y_0	y
Hydrocarbons + resins + asphaltenes	C	u_0	u
Hydrocarbons only	HC	v_0	v
Resins and asphaltenes	RA	w_0	w

4. Rate of formation of C

At constant temperature $\quad V_\theta = \left(\frac{du}{dt}\right)_\theta$

At constant subsidence $\quad V_\lambda = \left(\frac{du}{dt}\right)_\lambda$

case of the Toarcian of the Paris Basin. This part of organic matter can moreover be artificially decomposed: in the oil shale industry, at around 500°C, it can be converted to "shale oil" and gas, except for a cokefied fraction which represents about 20% of the total organic carbon.

Figure 5 shows that transformation B → C depends considerably on the temperature: this promotes the formation of the compounds of oil C, from the heavy heteroatomic compounds B. The total quantity of C (hydrocarbons + resins + asphaltenes) is observed to vary approximately exponentially with $1/T$ *(Fig. 7)*. T represents the absolute temperature ($\theta_m + 273$) to which each sample was raised in the basin.

Within this group of reactions B → C, the temperature rise favors certain products at the expense of others. Thus the ratio of hydrocarbons/resins + asphaltenes increases with temperature and also displays an exponential variation with $1/T$ *(Fig. 8)*. Among the saturated hydrocarbons, the formation of alkanes and cyclanes with a single ring is favored by the increase in temperature, while the polycyclic saturated hydrocarbons (cyclanes with 3, 4, 5 and 6 rings), and the optical activity associated with them decrease sharply (Louis and Tissot, 1967).

These remarks also confirmed a statistical observation, valid at world scale: the density of the crude oils, and their

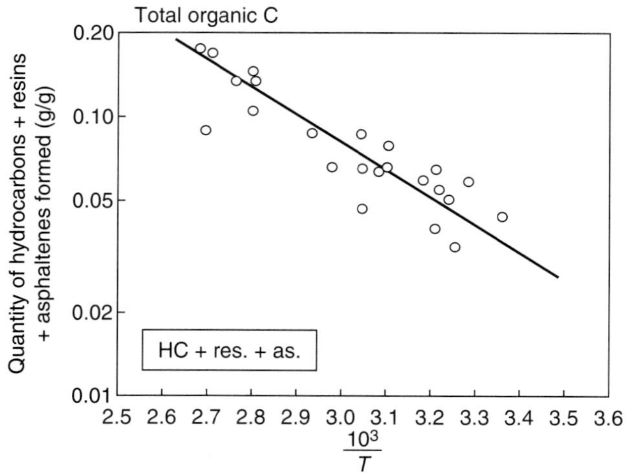

Figure 7

Variation in the quantity of hydrocarbons + resins + asphaltenes formed as a function of maximum temperature reached T in kelvin (Lower Toarcian in the Paris Basin).

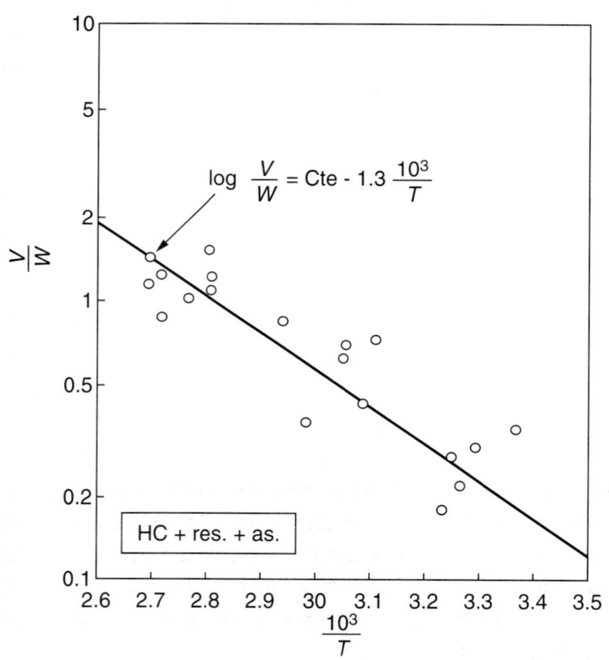

Figure 8

Variation in ratio $\frac{v}{w} = \frac{\text{hydrocarbons}}{\text{resins + asphaltenes}}$ as a function of maximum temperature T in kelvin (Lower Toarcian of the eastern half of the Paris Basin).

total cyclane content decreases with the depth of the reservoir, and hence with rising temperature.

It may be observed that Figures 7 and 8 imply a kinetics wherein the degree of advancement of the reactions varies with temperature according to an Arrhenius' type law.

2.3 Content-Time Relations

The depth and temperature of a sediment are both related to time by the subsidence λ, *i.e.* the burial rate measured in meters per million years. In each time interval where subsidence is constant, the depth and temperature are a linear function of time *(see Table 5 for notations meaning)*.

All the Toarcian samples analyzed were first buried (subsidence $\lambda_1 > 0$) up to time $t = t_1$, which reached 25 to $150 \cdot 10^6$ years depending on each case, followed by a rise to the surface ($\lambda_2 < 0$) up to the Present $t_2 = 180 \cdot 10^6$ years. For many of them, the mean value of λ_1 is close to 15 m/Ma (between 13 and 17).

If a sample raised to temperature θ_m is brought to the surface and cools, it can be assumed that the reaction is "frozen", especially if cooling is rapid (highly negative λ_2). Theoretical calculations from the equations given in Chapter 4 show that as a first approximation, this assumption is acceptable. In these conditions, to have an approximate value of the quantities of hydrocarbons, resins and asphaltenes formed at time t_1 for a subsidence 15 m/10^6 years, use is made of the values currently measured on the samples subjected to this subsidence up to t_1.

We finally obtain Figure 9, determined from Figure 1, by eliminating the samples whose subsidence is less than 13 or

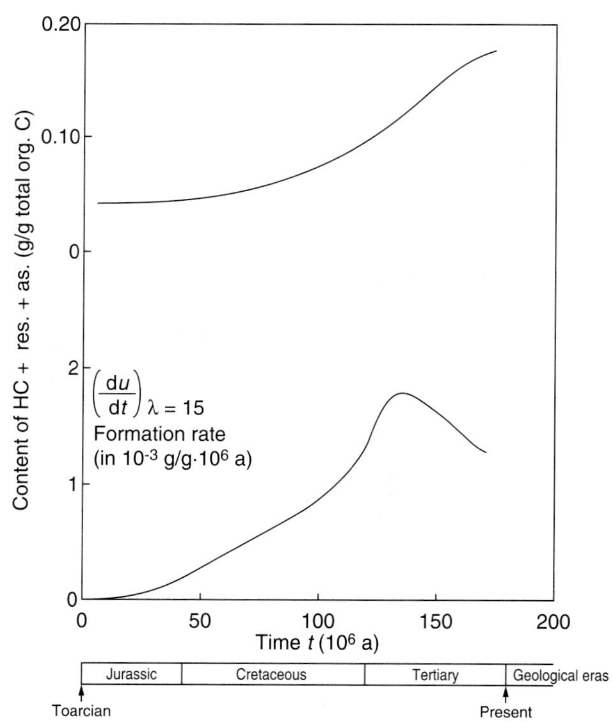

Figure 9

Reconstruction of the content of hydrocarbons + resins + asphaltenes and their rate of formation, as a function of time, for 1 g of total organic carbon and for a subsidence $\lambda = 15$ m/10^6 years (Lower Toarcian in the Paris Basin).

greater than 17 m/10^6 years, and by converting the maximum depths Z_m into subsidence time t_1 by the equation $Z_m = \lambda_1 t_1$.

It is easy to see that in Figure 9, the formation of hydrocarbons only becomes effective after a long induction period, since this is the time necessary for burial of the formation to a depth equal to that of the threshold already identified.

2.4 Rate-Time Relations

An attempt can also be made to evaluate the formation rates of the compounds of oil C and hydrocarbons at constant subsidence. Let u denote the quantity of C (hydrocarbons + resins + asphaltenes) at time t and per gram of total organic carbon; v is a quantity of hydrocarbons per gram of total organic carbon. The rates are:

$$\left(\frac{\mathrm{d}u}{\mathrm{d}t}\right)_{\lambda=\mathrm{Cte}} \qquad \left(\frac{\mathrm{d}v}{\mathrm{d}t}\right)_{\lambda=\mathrm{Cte}}$$

Using the same assumptions as above, we can start with Figure 1 which gives n and u as a function of z and constitutes the base document.

We have the general equation:

$$\mathrm{d}u = \left(\frac{\partial u}{\partial t}\right)_\theta \mathrm{d}t + \left(\frac{\partial u}{\partial \theta}\right)_t \mathrm{d}\theta$$

In a transformation at constant λ, we obtain:

$$\left(\frac{\mathrm{d}u}{\mathrm{d}t}\right)_\lambda = \left(\frac{\partial u}{\partial t}\right)_\theta + \left(\frac{\partial u}{\partial \theta}\right)_t \cdot \left(\frac{\mathrm{d}\theta}{\mathrm{d}t}\right)_\lambda = \left(\frac{\partial u}{\partial t}\right)_\theta + \lambda G \left(\frac{\partial u}{\partial \theta}\right)_t$$

and furthermore:

$$\left(\frac{\mathrm{d}u}{\mathrm{d}z}\right)_\lambda = \left(\frac{\partial u}{\partial t}\right)_\theta \cdot \left(\frac{\mathrm{d}t}{\mathrm{d}z}\right)_\lambda + \left(\frac{\partial u}{\partial \theta}\right)_t \cdot \left(\frac{\mathrm{d}\theta}{\mathrm{d}z}\right)_\lambda = \frac{1}{\lambda}\left(\frac{\partial u}{\partial t}\right)_\theta + G\left(\frac{\partial u}{\partial \theta}\right)_t$$

hence:

$$\left(\frac{\mathrm{d}u}{\mathrm{d}t}\right)_\lambda = \lambda\left(\frac{\mathrm{d}u}{\mathrm{d}z}\right)_\lambda$$

Hence only the points corresponding to λ close to fifteen are selected from Figure 1, $\left(\frac{\mathrm{d}u}{\mathrm{d}t}\right)_\lambda$ and $\left(\frac{\mathrm{d}v}{\mathrm{d}t}\right)_\lambda$ are calculated from $\frac{\mathrm{d}u}{\mathrm{d}z}$ and $\frac{\mathrm{d}v}{\mathrm{d}z}$. The results are shown in Figure 9 for the formation rate of all the oil compounds $\left(\frac{\mathrm{d}u}{\mathrm{d}t}\right)_\lambda$ and for a subsidence of 15 m/10^6 years.

This confirms that the formation rates are far from constant. In the case of the Toarcian in the Paris Basin, they only have a significant value from the Middle Cretaceous to the Present (between 75 and $180 \cdot 10^6$ years) peaking in the Early Tertiary.

3 ESTABLISHMENT OF A REACTION FLOWSHEET

3.1 The Original Organic Matter

We now leave behind the region of facts established from analytical data to enter the region of interpretation. This will rely on a number of simplifying assumptions, which can be justified if they culminate in relations between the parameters that are confirmed by actual data. In particular, given the large number of individual components, most of them are unknown, we shall have to treat as an overall transformation what is in fact a series of successive or parallel reactions. We obviously run the danger of advancing too far along this route.

Living organic matter is the original material from which all organic products found in the sediments are developed. In the free state, it can be fractionated and its analysis shows that it is nearly entirely composed of proteins, carbohydrates, lipids and lignin.

On the floor of the seas and lakes, the freshly deposited sediments are the seat of considerable microorganism activity (particularly bacteria). In aerobic medium, the organic matter is rapidly destroyed; by contrast, in impermeable rocks (clay, fine limestone mud), the medium rapidly becomes closed, the oxygen present is consumed, and an anaerobic regime is installed, so that part of the organic material is not destroyed.

The activity of the microorganisms decreases very rapidly: it is divided by 20 around 70 cm depth, and there is practically no active life beyond 10 to 15 m. At the same time, a very sharp decrease is observed in the simple organic compounds that can be separated and determined (sugars, amino acids, even fatty acids). Nearly all the remaining organic matter (more than 80%) consists of *kerogen*, a poorly defined compound, in commonly used solvents, and whose structure is poorly known.

It can be assumed that kerogen represents the residue inassimilable by the microorganisms. The disappearance of the microorganisms can be simply explained by the disappearance of the compounds necessary for their nourishment, as in case of a closed culture medium which peaks and declines over time. The action of the microorganisms, and perhaps of certain inorganic species, could explain the appearance of a macromolecular structure in very moderate temperature conditions.

In the rest of this discussion, we shall focus on the kerogen degradation reactions, leading to the formation of most of the oil; we shall no longer discuss hydrocarbons,

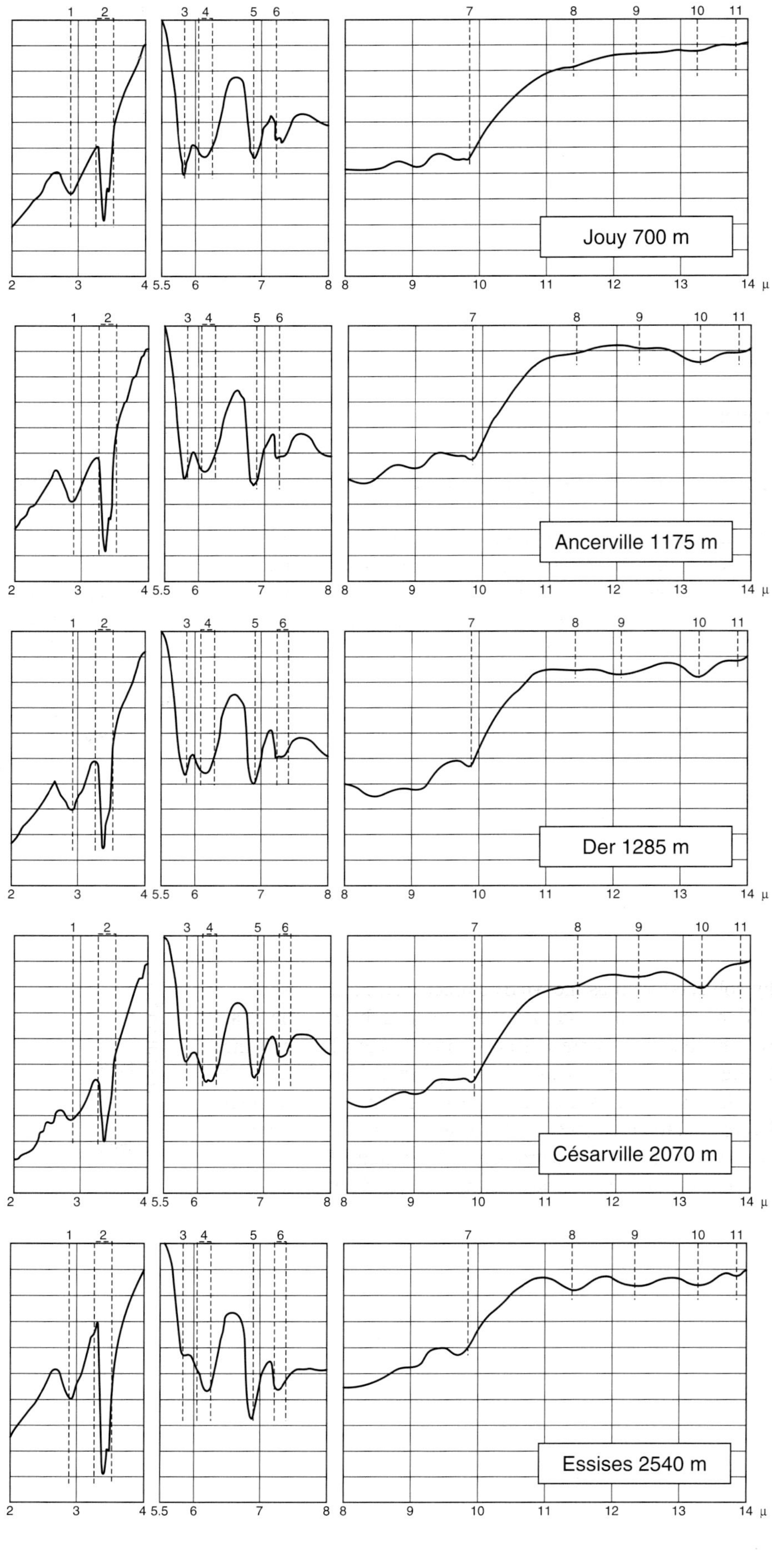

Figure 10

Infrared spectra of kerogen samples of the Toarcian in the Paris Basin, showing the evolution of the structure according to the burial depth (M. Pennequin).

Attribution of absorption bands:

1) 2.82 μ (3450 cm^{-1}): OH stretching
2) 3.25-3.50 μ (3080-2860 cm^{-1}): CH_2 and CH_3 stretching
3) 5.80 μ (1725 cm^{-1}): C = O
4) 6.05-6.25 μ (1650-1600 cm^{-1}): H_2O deformation and C = C aromatic
5) 6.82 μ (1466 cm^{-1}): asymmetric CH_3 bending
6) 7.20-7.30 μ (1389-1370 cm^{-1}): CH_2 bending and CH_3 asymmetric bending
7) 9.85 μ (1015 cm^{-1}): C–O–C aromatic?
8) 9, 10, 11.10 to 13.33 μ (900-750 cm^{-1}): protons adjacent to aromatic rings
11) 13.80 (725 cm^{-1}): $(CH_2)_n$, $n \geq 3$?

It may be observed that:

1) The C = O band at 1725 cm^{-1} decreases with increasing depth *(see also Fig. 11)*.
2) Bands 8, 9, 10 grow with depth. The attributions made after Erdman (1965).

The aromatic C = C band 4 also increases.

lipids, amino-acids and other compounds which exist in the recent sediment alongside the kerogen. However, they cannot be ignored: they could be the source of the compounds found at rather shallow depths (less than 1000 m) in the Toarcian in the Paris Basin, and in particular:

- polycyclic cyclanes possessing optical activity (Louis and Tissot, 1967), from steroids, triterpenoids, etc.;
- porphyrines, from chlorophyls; pristine, phytane etc., from phytyl chains;
- predominantly odd numbered *n*-alkanes, from fatty acids.

Although these compounds subsequently account for a minor part of the oil, because diluted among the more abundant compounds formed at the higher temperature, they remain an invaluable indication of the original organic matter.

3.2 Structure and Transformation of Kerogen

In the absence of a satisfactory definition and effective analytical method, the structure of kerogen is poorly known. We can nonetheless try to represent it with a few data obtained by thermal degradation, X-ray diffraction and infrared, and by analogy with the structures proposed for coals. It is a macromolecule, a sort of giant polymer, probably three-dimensional. The more condensed nuclei consist of naphtheno-aromatic structures; the bonds between these nuclei, or with other less condensed structures are mainly of the heteroatomic type.

Insofar as burial and temperature increase, these bonds (such as C–S, C–O, sometimes C–C) break in succession and the polycondensation intensifies (visible in infrared). The splitting of these bonds expels a part of the oxygen and the other heteroatoms in the form of light products (H_2O, CO_2 etc.), and liberate structures smaller than kerogen (extractable by solvents) such as the MAB extracts. These structures are still heteroatomic and probably have a high molecular weight, but they represent an important step between the kerogen and the petroleum compounds. However, the residual kerogen becomes more condensed: the carbon-hydrogen ratio increases, while oxygen and nitrogen decrease *(Fig. 6)*; in infrared spectrometry (Pennequin, *Fig. 10*), the carbonyl band decreases with the production of the petroleum compounds *(Fig. 11)* and the band corresponding to the aromatic group increases; at the same time, the patterned organic matter, visible under the microscope (spores, pollens, etc.), displays increasing carbonization, going from light yellow to brown and then to black (Correia, 1967 and *Fig. 12*). At extreme depths, the kerogen evolves towards the structure of graphite (appearance of graphite 0002 peak under X-ray diffraction) perhaps by development of graphitic islands within the kerogen (Long *et al.*, 1968).

Similarly, the heavy heteroatomic compounds (MAB extract type) undergo a progressive rearrangement, with a loss of bonding heteroatoms and disproportionation to form a range of products extending from hydrocarbons to the heavy and highly condensed structures. At extreme depth (very high

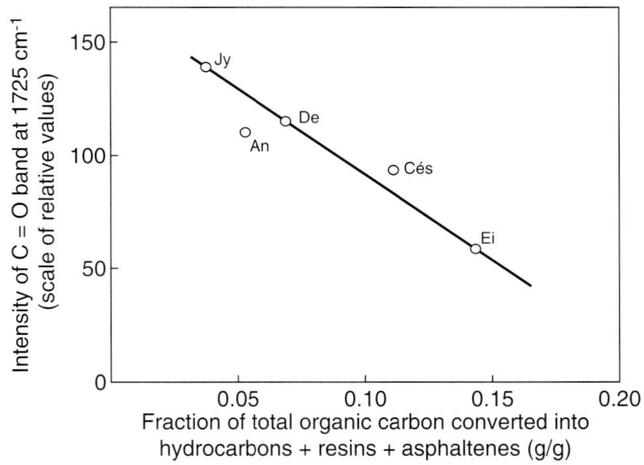

Figure 11

Relation between the formation of petroleum compounds and the decrease of the carbonyl function on kerogen (after the infrared spectra of M. Pennequin).

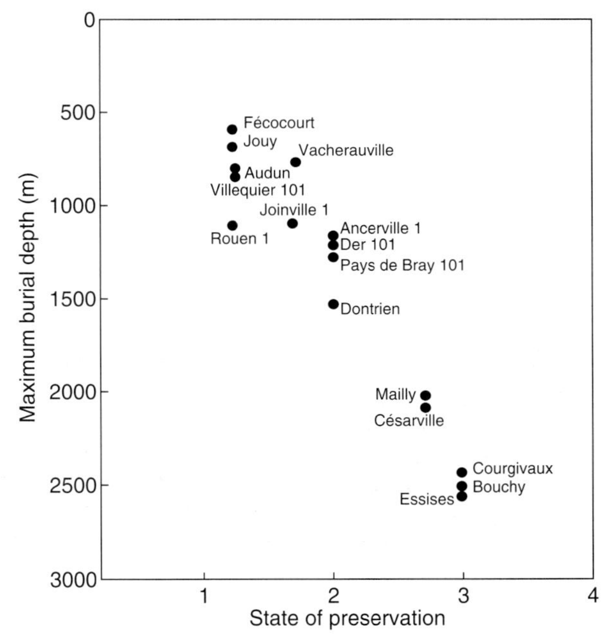

Figure 12

Variation in state of preservation of patterned elements of organic matter as a function of maximum burial depth. (Lower Toarcian in the Paris Basin) (M. Correia).

The spores, pollen grains and other fragments making up the patterned elements of the organic matter visible under the microscope, depending on their variable coloration from yellow to black, help to identify different states of preservation.

temperatures), this evolution continues towards methane and towards graphite.

All of these transformations can be summarized by the reaction flowsheet in Figure 13a. The elementary composition (C, H, O) of the kerogen and extractable compounds shown in Figure 14 helps to observe that the first group of reactions mainly corresponds to reduction (elimination of oxygenated bonds and functions); the second includes a new elimination of oxygen and a disproportionation resulting in compounds richer in hydrogen (hydrocarbons for which C/H is around 6 to 9) and more condensed compounds, poor in hydrogen.

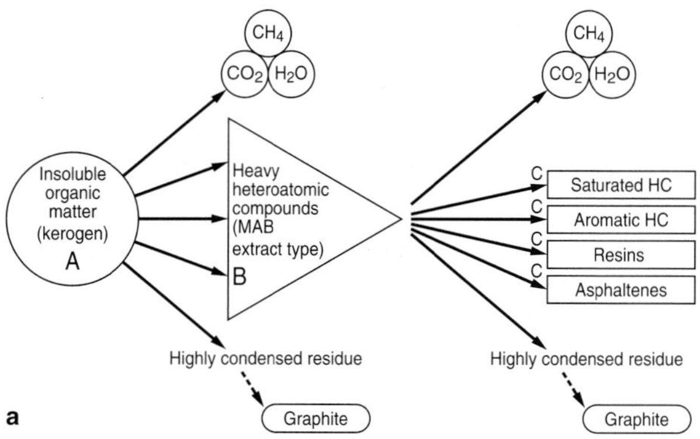

Figure 13a

Kerogen degradation flowsheet.

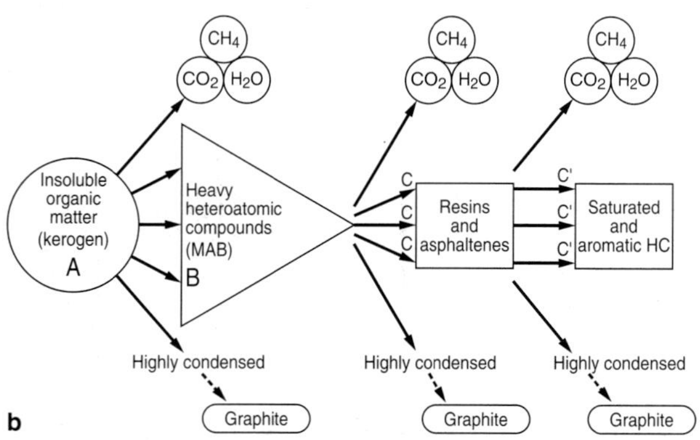

Figure 13b

Variant of above flowsheet.

Note that the highly condensed residues which evolve towards graphite cannot be separated from kerogen A with common fractionation methods.

3.3 Remarks

The reaction flowsheet proposed calls for the following remarks:

- The arrows actually represent a group of parallel or successive reactions whereof the intermediate steps are inaccessible to observation. A steady state is assumed to be established and these intermediates are ignored, just as only states with a long half-life are retained in the disintegration equations.
- An examination of Figures 5 and 14 indicates a variant of the previous reaction flowsheet: the resins and asphaltenes could represent an intermediate before culminating in hydrocarbons. This variant is shown in Figure 13b. It has not been used in the calculations in Chapter 4, to avoid needlessly complicating these calculations.

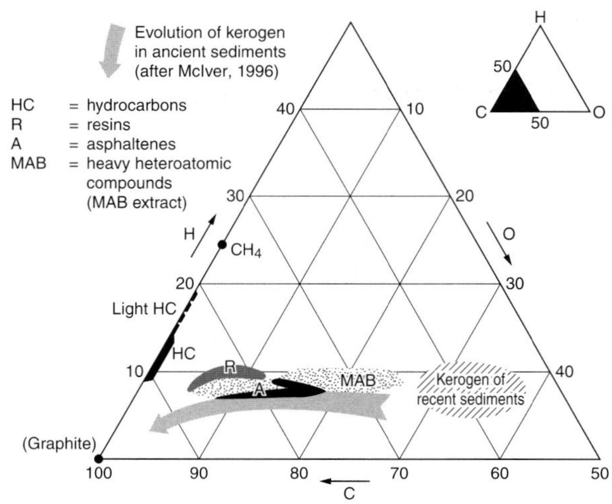

Figure 14

Elementary composition of the various organic components of sediments and evolution of kerogen.

- On the whole, the rest of the reactions cannot be expected to be reversible. If, for example, the system is raised to 80°C, a part of the organic raw material is decomposed but, if we return to the temperature and pressure conditions of the superficial environment of the deposit (by erosion), the original organic matter is not reformed. Besides, the probably very abundant formation of CO_2, H_2O, and CH_4, which are more mobile and disappear, also makes the reactions irreversible.
- In natural conditions, most of the kerogen A remains in the macromolecular state and does not undergo the conversion A → B. For example, in the case of the Toarcian in the Paris Basin, the organic carbon thus immobilized, depending on the temperatures reached, represents 0.85 to 0.95 of the total organic carbon. We know, however, that other bonds can be broken—and new extractable products formed—by raising the sediment to higher temperatures: 180 to 300°C in the laboratory experiments, up to 500°C in the oil shale industry. This accordingly leads to the concept of the distribution of activation energies for the elimination of the different

functional bonds, a distribution such that, for each temperature range, a new category of bonds occurs, whereas those corresponding to the higher energies are practically untouched.

4 EXPRESSION AND COMPUTER SIMULATION OF A KINETIC MODEL

4.1 Expression of Equations

It would be misleading to try to monitor the transformations of one compound or another in detail to establish specific laws which govern an individual transformation. The construction of a kinetic model makes it possible to simulate the reaction flowsheet in Figure 13 on a computer, and thereby to analyze the functioning of the complex transformation.

One essential advantage of the model is the introduction of time and temperature, *i.e.* of the geological history of the sediment after its deposition. The model thereby helps to calculate, for a given formation and at each point of the basin, the result of the transformation of the organic matter as a function of its geological history. After having written it, we can therefore calculate the quantities of oil formed in the Toarcian in the Paris Basin up to the Present, at each point where samples are available (borehole); the necessary numerical constants are evaluated from the observations and/or adjusted by comparison with the observations and laboratory experiments. The consistency of the results obtained justifies the validity of the model for simulating natural phenomena.

We therefore start with the reaction flowsheet in Figure 13a, which is symbolized by:

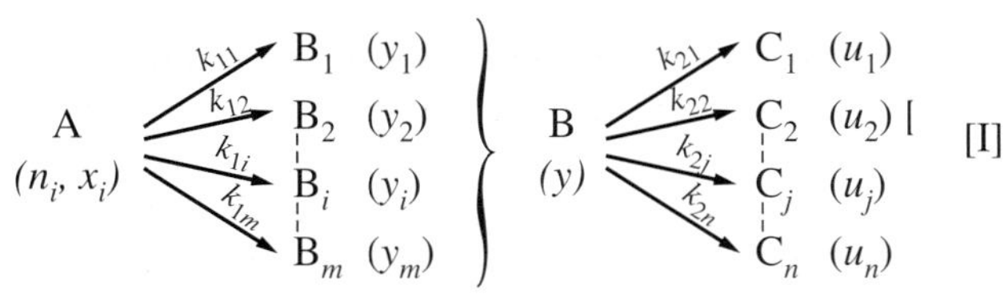

At time t, n_i represents a number of bonds of a given type i, x_i, the quality of matter involved in the ith reaction (splitting of the i bond), y_i and u_j the quantity of compounds B_i and C_j. The quantities resulting from the observation of natural samples are x (kerogen), y (heavy heteroatomic compounds from MAB extract) and u_1, u_2, … (saturated aromatic hydrocarbons, resins, asphaltenes).

As stated earlier, the $A \rightarrow B$ group represents a set of parallel reactions (*i.e.* the probability of splitting of a given bond is independent of the presence of the other bonds) and/or successive. We assume that the probability of the splitting of a bond is independent of the time elapsed; the breakage of the bond accordingly obeys a Poisson's ratio, which is widely encountered in nature (radioactivity), *i.e.*:

$$-\frac{dn_i}{n_i} = k_{1i}\, dt$$

where k is a constant at a given temperature.

If we assume that the density of the bonds is uniform throughout the volume (homogeneous structure), and if P_i is the frequency of the i bond in the original organic matter at time $t = 0$, we have:

$$x_{i0} = x_0 P_i \qquad \text{(parallel reaction)}$$

$$x_i = \mu n_i \qquad \text{(homogeneous material)}$$

we therefore obtain:

$$-\frac{dx_i}{dt} = k_{1i}\, x_i$$

the subscript 0 representing the quantity at time $t = 0$ and μ being a constant.

For each of the i reactions, we therefore obtain equations identical to those of the first order reactions. We also assume that the liberated structure of polymer A is independent of the nature of i. We can therefore consider B_i as statistically identical and for the second transformation only consider $B = \sum_i B_i$. We can then similarly write, and for the same reasons, the differential equations of the $B \rightarrow C$ transformations.

The quantities x_i, and u_j are then solutions of a system of first order differential equations with variable coefficients since the k depend on the temperature, hence on time, via the subsidence:

$$\left.\begin{array}{ll} -\dfrac{dx_i}{dt} = k_{1i}\, x_i & (1) \\ \dfrac{du_j}{dt} = k_{2j}\, y & (2) \\ y = \sum_i y_i & (3) \\ \sum_i x_{i0} + \sum_i y_{i0} + \sum_j u_{j0} = \sum_i x_i + \sum_i y_i + \sum_j u_j & (4) \end{array}\right\} \quad [II]$$

It can be observed immediately that the differential equations of the system [II] are not integrable with the help of simple functions if we assume that k vary with temperature θ, hence with time t, according to a law like the Arrhenius' law. We therefore, first in Section 4.2, carry out the theoretical calculation of the transformation at constant temperature (for example the maximum temperature or the mean temperature), to have an idea of the general form of the results. We shall then examine, in Section 4.3, an actual case at constant subsidence. In fact, although the equations are not integrable, they can be simulated on a computer.

4.2 Theoretical Study of Transformation at Constant Temperature

We shall only consider bonds of a single type (i) of frequency $P_i = 1$, and reactions $B \rightarrow C_i$ of two types:

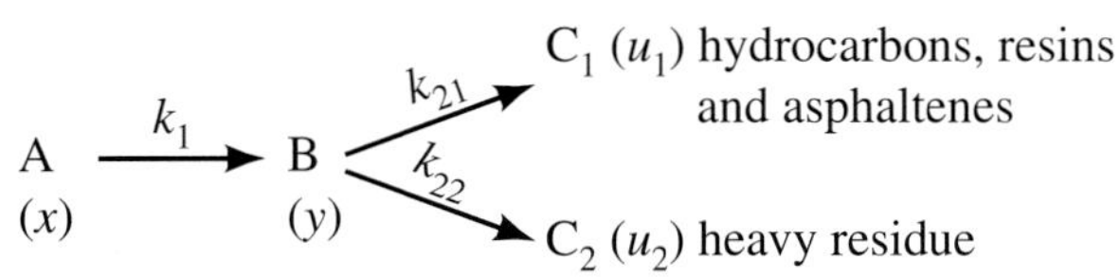

We also assume that the values of the origin y_0, u_{10}, u_{20} are zero.

The calculation is performed in Appendix I. The following values are obtained for y, u_1 and u_2.

$$y = \frac{x_0 k_1}{k_1 - (k_{21} + k_{22})}\left[e^{-(k_{21}+k_{22})t} - e^{-k_1 t}\right]$$

$$u_1 = S\frac{k_{21}}{k_{21} + k_{22}} \qquad u_2 = S\frac{k_{22}}{k_{21} + k_{22}}$$

by setting:

$$S = u_1 + u_2 = x_0\left[1 + \frac{(k_{21} + k_{22})\, e^{-k_1 t} - k_1 e^{-(k_{21}+k_{22})t}}{k_1 - (k_{21} + k_{22})}\right]$$

The shape of the curves given x, y, u_1, u_2 as a function of t is shown in Figure 15.

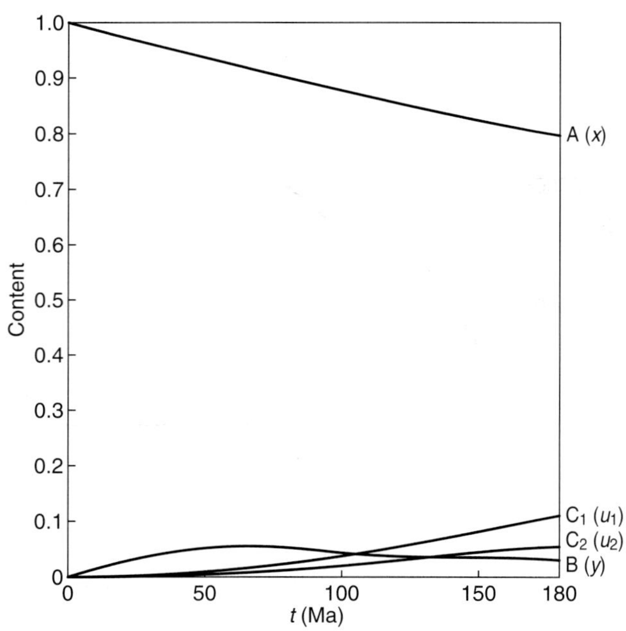

Figure 15

Example of evolution of contents as a function of time for a transformation at constant temperature.

4.3 Computer Simulation of an Actual Case

In the first step, we only considered two values of i and two values of j. A second more complete version is being constructed.

The actual case of the burial of a sediment can be treated as a succession of periods in which the subsidence is constant. This is in fact the form in which the burial history is reconstructed by the geologist. During each period, the temperature is accordingly a linear function of time $T = \mu t + \nu$.

For the variation of k_{1i} and k_{2j} with temperature, we adopt a function of the Arrhenius' type:

$$k_{1i} = A_{1i}\, e^{-\frac{E_{1i}}{RT}}$$

In fact, the examination of natural phenomena *(Figs. 7 and 8)* implied that the degree of advancement of the reaction varies with temperature according to Arrehnius' law. Appendix II shows that this is effectively acceptable for a group of products, and that some of the E_{2j} can be evaluated.

The integration of system [II] is no longer feasible using simple functions; a computer simulation was therefore carried out. In fact, a numerical integration by the finite differences method is possible, because the functions involved in [II] are continuous throughout as well as their primitives.

Let Δt be the integration step. If x_i, y_i, u_i are the values of the valuables at time t, they are $x_i + \Delta x_i$, $y_i + \Delta y_i$, $u_j + \Delta_{uj}$ in the following step $t + \Delta t$. By replacing k_{1i} by its value, from Equation (I) of the system [II], we obtain that Δ_{xi} from x_i:

$$\Delta x_i = -A_{1i}\, e^{-\frac{E_{1i}}{R(\mu t + \nu)}} x_i \,\Delta t \tag{5}$$

At time $t + \Delta t$, the value thus calculated for x_i is $x_i + \Delta x_i$. If the concentrations x_{i0} at time 0 are known, we can therefore calculate the successive approximate values of x_i, as well as $x = \Sigma x_i$ by summating the results obtained for the i equations, at each step.

From Equation (4), we obtain:

$$y = x_0 + y_0 + u_0 - \sum_i x_i - \sum_j u_j \quad \text{with: } x_0 = \Sigma x_{i\,0},\ u_0 = \Sigma_{j\,0} \tag{6}$$

By setting this result in (2), considering that the increase Δt corresponds to the increase Δu_j of u_j, and by replacing k_{2j} by its value, we obtain for the jth equation:

$$\Delta u_j = A_{2j}\, e^{-\frac{E_{2j}}{R(\mu t + \nu)}}\left(x_0 + y_0 + u_0 - \sum_i x_i - \sum_j u_j\right)\Delta t \tag{7}$$

If x_0, y_0, u_0, are known, the approximate values of u_j can be calculated by successive steps by setting in (7) at each step

$\sum_j u_j$ and $\sum_i x_i$ determined in the previous step; at time $t = 0$, we give to Σx_i and Σu_j the values x_0 and u_0; y is simultaneously obtained using (6).

A discrete series of approximate values of x_i, u_j and y is therefore easily calculable as a function of time.

The difference between the curves thus constructed and the integral curves decreases as the time step Δt is decreased. The approximation obtained is considered sufficient when, for two successive values of a step, the difference between the two calculated values of the same variable corresponding to the same time t is lower than a fixed value (in the present case, a ratio of this difference to the calculated value of the variable less than 10^{-3} at the final time is adequate). The calculation shows that this condition is satisfied in most cases for a step of 0.25 Ma.

The x_0, y_0, u_0, as well as the A_{1i}, E_{1i}, A_{2j} and E_{2j} are the free parameters of the system. The values at the origin of time are selected from the values observed in the present sediments, particularly those of the Black Sea where the depositional conditions appear similar (Gorskaya, in: Vassoevich, 1958). The A and E are determined in two steps. The order of magnitude of the E_{2j} and A_{2j} is derived from our own observations and experiments *(Appendix II)*; the optimal adjustment is then carried out by iteration between the calculation results and the observed values.

The calculations presented below were performed using the following values[1], and the time units are millions of years:

$E_{11} = 2$ kcal/mol $A_{11} = 1$
$E_{12} = 17$ kcal/mol $A_{12} = 3 \cdot 10^8$
$E_{21} = 15$ kcal/mol $A_{21} = 10^8$ (HC + res. + as.)
$E_{22} = 7$ kcal/mol $A_{22} = 3.3 \cdot 10^2$ (heavy residue)

All the calculations were performed on a CDC 6400 computer.

4.4 Results Obtained by Simulation of the Actual Case

By way of example, Figure 16 shows the curve of the formation of petroleum components (hydrocarbons + resins + asphaltenes) from one gram of organic carbon, calculated by the model for the Lower Toarcian of the Essises well. The lower part of the figure shows the corresponding formation rate, also as a function of time, in millions of years. The angular shape of the rate curve is explained by that of the burial curve, hence of the temperature as a function of time.

(1) If the laboratory experiments at high temperature are also to be represented, it is necessary to introduce other higher values of E_1: this yields a distribution of E_{1i} where the mean is about 30 kcal/mol.

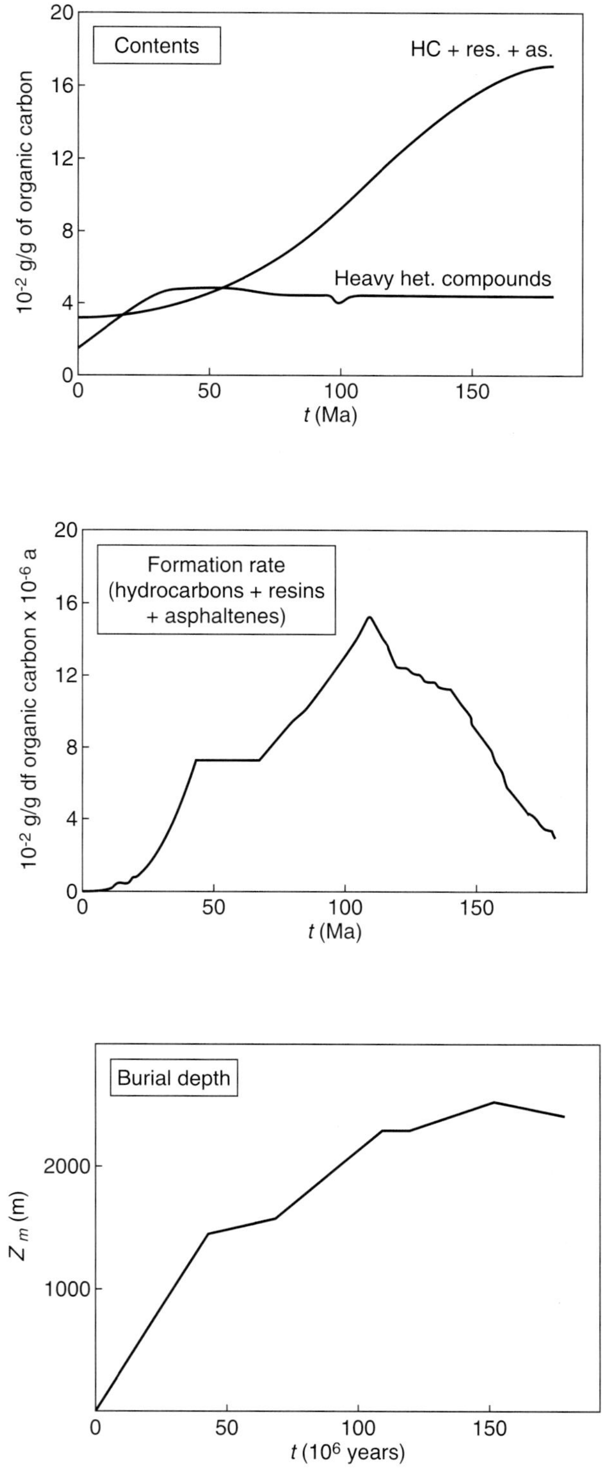

Figure 16

Content of petroleum compounds (hydrocarbons + resins + asphaltenes) in heavy heteroatomic compounds, and formation rate as a function of time for the Toarcian of the Essises well.

Given at the bottom is the burial curve reconstructed by the geologist, which serves to determine the temperature distribution as a function of time.

The results for all the twenty-three wells show excellent agreement between the contents of the petroleum components thus calculated and those observed on the samples *(Fig. 17)*. The correlation factor is 0.90; it reaches 0.96 if we ignore one of the wells whose behavior on natural samples appears to be anomalous with respect to its neighbors. The mean square error between calculated and observed values is about 10^{-2}.

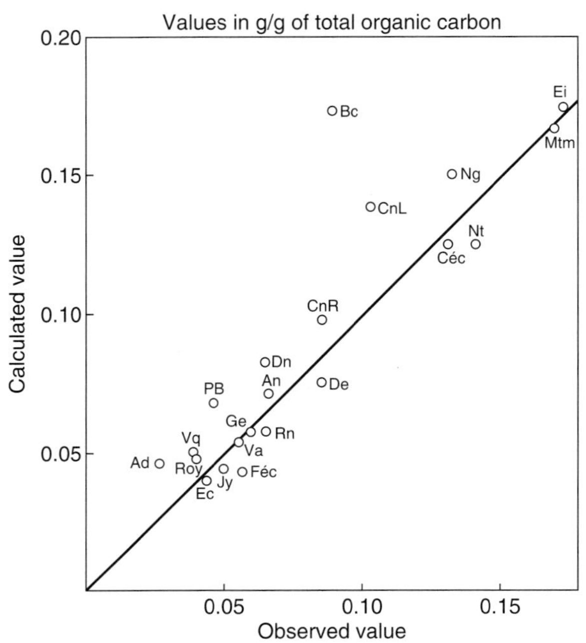

Figure 17

Correlations between quantities of hydrocarbons + resins + asphaltenes measured on the clay samples of the Toarcian in the Paris Basin, and the quantities calculated by computer simulation of the kinetic flowsheet.

CONCLUSIONS

Using the results of observations and experiments on the Toarcian in the Paris Basin, it was accordingly possible to construct a simplified reaction flowsheet, and then simulate the transformation on a computer. The consistency of the overall results shows that the reaction flowsheet and the kinetics implied are compatible with reality. We therefore have a *model* to represent the natural phenomena and contribute to their study.

The use of this tool helps explicitly to introduce the *time* and *temperature* parameters, *i.e.* the overall geological history of the sediment after its deposition, instead of only considering a specific episode (maximum burial depth or present depth). Besides, this makes it possible to separate the respective influence of time and temperature, which was not hitherto possible.

In terms of prospecting, this tool helps to simulate the quantities of oil formed at any point of the basin investigations, and with a good approximation, from the time of deposition to the present. We can thus determine the most favorable zones, *i.e.* those in which the source rock has been the most productive. It is obviously easy to take account of the organic matter contents and the thickness of the formation.

We can also consider extending this model to other formations, and then to other sedimentary basins, to contribute to prospecting. This aspect is the subject of another publication[2]. It suffices to have the measurements on a small number of core samples (to adjust the constants specific to the formation concerned), a reconstruction of the geological history of the basin, and an evaluation of the geothermal gradient.

Finally, the use of the model provides the first data, still approximate, on the timing and the formation rate of the oil; the comparison of these data with the age of the structural movements will perhaps make it possible, in some basins, to select the most favorable traps.

REFERENCES

Correia, M. (1967) Relations possibles entre l'état de conservation des éléments figurés de la matière organique (microfossiles palynoplanctologiques) et l'existence de gisements d'hydrocarbures. *Rev. Inst. franç. du pétrole*, **22**, 9, 1285-1306.

Deroo, G., Durand, B., Espitalié, J., Pelet, R. and Tissot, B. (1968) Possibilités d'application des modèles mathématiques de formation du pétrole à la prospection dans les bassins sédimentaires. *4th International Meeting on Organic Geochemistry*, Amsterdam, Sept. 1968.

Erdman, J.G. (1965) The Molecular Complex Comprising Heavy Molecular Fractions. *Hydrocarbon Analysis, ASTM, STP*, **389**.

Long, G., Neglia, S. and Favretto, L. (1968) The Metamorphism of the Kerogen from Triassic Black Shales, Southeast Sicily. *Geochim. et Cosmochim. Acta*, **32**, 647-656.

Louis, M. (1964) Études géochimiques sur les « Schistes cartons » du Toarcien du Bassin de Paris. In : *Advances in Organic Geochemistry*, G.D. Hobson et M.C. Louis (eds), Pergamon Press, Oxford, 85-94.

Louis, M. and Tissot, B. (1967) Influence de la température et de la pression sur la formation des hydrocarbures dans les argiles à kérogène. *7th World Petrol. Congress*, Mexico, **2**, 47-60.

McIver, R.D. (1967) Composition of Kerogen. Clue to its Role in the Origin of Petroleum. *7th World Petrol. Congress*, Mexico, **2**, 25-36.

(2) Deroo, G., Durand, B, Espitalié, J., Pelet, R and Tissot, B (1968) Possibility of Application of Mathematical Models of Petroleum Formation in Prospecting in Sedimentary Basins. *4th International Meeting on Organic Geochemistry*, Amsterdam, Sept. 1968.

Philippi, G.T. (1965) On the Depth, Time and Mechanism of Petroleum Generation. *Geochim. et Cosmochim. Acta*, **29**, 1021-1049.

Tissot, B. (1966) Problèmes géochimiques de la genèse et de la migration du pétrole. *Rev. Inst. franç. du pétrole*, **21**, 2, 1621-1671.

Vassoevich, N.B. (1958) Problèmes de la formation du pétrole. (En russe) *Trudy VNIGRI, Public.*, **128**, *Gostoptekhizdat.*

Vassoevich, N.B., Visotski, I.V., Guseva, A.N. et Olenin, V.B. (1967) Hydrocarbons in the Sedimentary Mantle of the Earth. *7th World Petrol. Congress*, Mexico, **2**, 37-45.

Welte, D.H. (1966) Kohlenwasserstoffgenese in Sedimentgesteinen : Unterserchungen über den thermischen Abbau von Kerogen unter besonderer Berücksichtigung der *n*-Paraffinbildung. *Geol. Rundschau*, **55-I**, 131-144.

Paper translated from Tissot, B. (1969) « Premières données sur les mécanismes et la cinétique de la formation du pétrole dans les sédiments. Simulation d'un schéma réactionnel sur ordinateur », Revue de l'Institut français du pétrole, ***24**, 4, 470-501.*

APPENDIX I

THEORETICAL STUDY OF TRANSFORMATION AT CONSTANT TEMPERATURE

We consider the flowsheet:

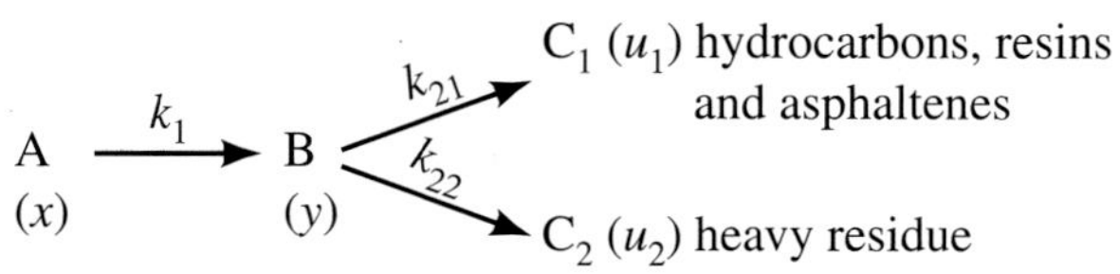

In system [II], we accordingly have a single value of i with $P_i = 1$ and two values of j: 1 and 2. y_0, u_{10}, u_{20} are zero.

The system becomes:

$$-\frac{dx}{dt} = k_1 x$$

$$\frac{du_1}{dt} = k_{21} y$$

$$\frac{du_2}{dt} = k_{22} y$$

$$x_0 = x + y + u_1 + u_2$$

Setting $S = u_1 + u_2$; we obtain:

$$x = x_0 e^{-k_1 t}$$

$$\frac{ds}{dt} = (k_{21} + k_{22})\, y = (k_{21} + k_{22})\,(x_0 - x - s)$$

$$\frac{ds}{dt} = (k_{21} + k_{22})\, S = (k_{21} + k_{22})\, x_0(1 - e^{-k_1 t})$$

By integrating the equation without a second member, we have:

$$S = M \cdot e^{-(k_{21} + k_{22})t}$$

$$\frac{ds}{dt} = e^{-(k_{21} + k_{22})t} \left[\frac{dM}{dt} - M(k_{21} + k_{22}) \right]$$

Let us set in the complete differential equation:

$$\frac{dM}{dt} e^{-(k_{21} + k_{22})t} = (k_{21} + k_{22}) x_0 (1 - e^{-k_1 t})$$

$$\frac{dM}{dt} = (k_{21} + k_{22})\, x_0 \left[e^{(k_{21} + k_{22})t} - e^{-(k_1 - k_{21} - k_{22})t} \right]$$

$$M = x_0 e^{(k_{21} + k_{22})t} \left[1 + \frac{k_{21} + k_{22}}{k_1 - k_{21} - k_{22}} e^{-k_1 t} + \mathrm{Cte} \cdot e^{-(k_{21} + k_{22})t} \right]$$

$$S = x_0 \left[1 + \frac{k_{21} + k_{22}}{k_1 - k_{21} - k_{22}} e^{-k_1 t} + \mathrm{Cte} \cdot e^{-(k_{21} + k_{22})t} \right]$$

In fact, for $t = 0$, $s = 0$, hence:

$$\mathrm{Cte} = -\frac{k_1}{k_1 - (k_{21} + k_{22})}$$

This gives:

$$S = x_0 \left[1 + \frac{(k_{21} + k_{22})\, e^{-k_1 t} - k_1 e^{-(k_{21} + k_{22})t}}{k_1 - (k_{21} + k_{22})} \right]$$

and since:

$$\frac{u_1}{u_2} = \frac{k_{21}}{k_{22}}$$

$$u_1 = S \frac{k_{21}}{k_{21} + k_{22}} \qquad u_2 = S \frac{k_{22}}{k_{21} + k_{22}}$$

$$y = \frac{x_0 k_1}{k_1 - (k_{21} + k_{22})} \left[e^{-(k_{21} + k_{22})t} - e^{-k_1 t} \right]$$

APPENDIX II

ANALYSIS OF TRANSFORMATION RATES EVALUATION OF PSEUDO-ACTIVATION ENERGIES

1) We consider here the formation of compounds of petroleum C in transformations at constant subsidence λ, *i.e.* where the temperature θ is a linear function of time t. In the beginning *(Fig. 4)*, when t and θ are low, reaction $A \to B$ practically does not occur and everything takes place as if C was formed from reactant B available in quantity y at the start of the reaction. We can therefore try to evaluate the constants of reaction $B \to C$. This is no longer valid beyond about 60°C, since reaction $A \to B$ occurs and therefore modifies the concentration and rates.

2) If we consider the rates $V = \left(\frac{du}{dt}\right)_\lambda$ (calculated in Section 2.3 from the rates observed $\left(\frac{du}{dz}\right)_\lambda$, we find in Figure 18 that log $1/y\ V_\lambda$ varies more or less linearly with $1/T$ at moderate temperatures (up to about 60°C). The alignment is good and encourages seeking an interpretation with an Arrhenius' type equation. We shall do this by introducing various assumptions, but we first face two difficulties:

   a) the rate observed concerns a set of reaction products, characterized either by a property (solubility in $CHCl_3$), or by their structure (hydrocarbons), to the exclusion of the other reaction products;

   b) kinetic interpretations classically use the relation between the formation rate V_θ (at constant temperature) and $1/T$; it is therefore necessary to analyze the relation between V_λ (at constant subsidence) and V_θ;

   these two points are considered in 3) and 4).

3) We consider a system in which a reactant B is converted into a large number n of compounds C (miscellaneous hydrocarbons, resins, asphaltenes):

$$\beta B \to \gamma_1 C_1 + \gamma_2 C_2 + \gamma_3 C_3 + \ldots \gamma_n C_n$$

where y and u_i represent the quantity of B and C_i.

If the nature of the reaction does not change in the temperature range considered we have:

$$-dy = \frac{\beta'}{\gamma'_1} du_1 = \frac{\beta'}{\gamma'_2} du_2 = \ldots = \frac{\beta' \sum_i^p du_i}{\sum_i^p \gamma' i}$$

where β' and γ'_i represent the masses of β molecules B, γ_i, molecules C_i, etc.

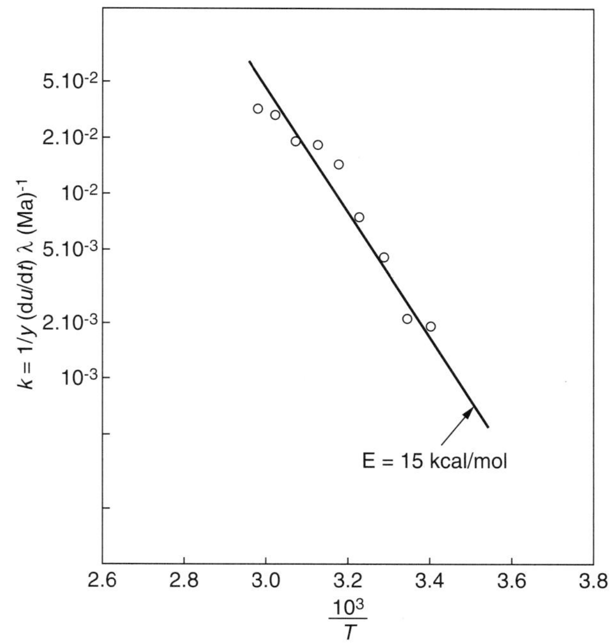

Figure 18

Relation between rate constant and temperature for the formation of petroleum compounds.

Hence if $u = \sum_i^p$ is the sum of a set of small p components, with $p \le n$, we have:

$$-dy = \xi du$$

where ξ is a constant that only depends on the type of reaction.

In the present case, we denote by u all the hydrocarbons, resins and asphaltenes, extractable by common solvent (chloroform) knowing that n–p components remain, which are not extractable, or disappear (such as CO_2 entrained by water and light hydrocarbons lost during extraction). The reaction rate is therefore:

$$-\frac{dy}{dt} = +\xi \frac{du}{dt}$$

4) If, as in Section 2.3, we consider the general equation:

$$du = \left(\frac{\partial u}{\partial t}\right)_\theta dt + \left(\frac{\partial u}{\partial \theta}\right)_t d\theta$$

it becomes in a transformation at constant subsidence:

$$\left(\frac{du}{dt}\right)_\lambda = \left(\frac{\partial u}{\partial t}\right)_\theta + \lambda G \left(\frac{\partial u}{\partial \theta}\right)_t \qquad V_\lambda = V^\theta + \lambda G \left(\frac{\partial u}{\partial \theta}\right)_t$$

Even in the case of the simple system that we have considered, the calculation of the difference $V_\lambda - V_\theta$ cannot be made accurately. However, we can calculate V_λ

and V_θ using the model constructed on a computer and which is presented above. We therefore see that in the range of transformations considered, this difference can be ignored (less than 10%) of the V_λ). This is not surprising, since the temperature rises about a half a degree per million years.

5) To analyze the variation in the rate constant k of reaction B → C, we have plotted in Figure 18, $\log k = \log\left(\frac{1}{y}\frac{du}{dt}\right)$ as a function of $1/T$ in the interval limited to 60°C.
The agreement observed with a linear distribution in $1/T$ shows that an Arrhenius' type law is acceptable, and helps to calculate a pseudo-activation energy E (for the *total reaction* considered, with all the reservations this implies, but not for an individual reaction of the formation of a compound). The table below shows the pseudo-activation energies for a set of compounds C, and for the various families they contain:

Formation of C (hydrocarbons + resins + asphaltenes)	15 000 cal/mol
Formation of hydrocarbons	20 000 cal/mol
Formation of resins and asphaltenes	14 000 cal/mol.

6) We observe that the pseudo-activation energy for hydrocarbons is higher than that for all petroleum compounds, and in particular than that of resins and asphaltenes. This hence implies that the more the temperature increases, hence the burial, the more the hydrocarbons (particularly saturated hydrocarbons) are favored over the heavy products.
In the case of the Toarcian in the Paris Basin, Figure 4 shows that this is certainly the case since the percentage of hydrocarbons in the transformation products increases with depth. This is not an ordinary observation from early results since the absolute quantities of the different components per g of organic carbon all increase in the same way. At the world scale, we know that shallow oil reservoirs are statistically heavier and richer in resins than asphaltenes.
If, to simplify, we consider two twin reactions:

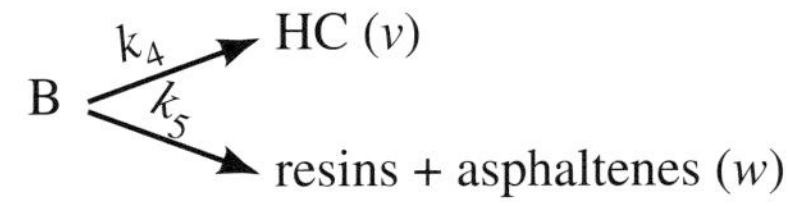

and if the reactions are of the same order, we have:

$$\frac{dv}{dw} = \frac{k_4}{k_5}$$

If we assume as a first approximation that the transformation occurs at constant temperature (for example the maximum temperature or the mean temperature), we have at any time:

$$\frac{v}{w} = \frac{k_4}{k_5}$$

In fact:

$$\log\frac{k_4}{k_5} = \log C - \frac{E_4 - E_5}{RT}$$

where log C is a constant. Using the above values (20 and 14 kcal/mol for E_4 and E_5, we obtain:

$$\log_{10}\frac{k_4}{k_5} = \log_{10} C - \frac{6}{4.57}\cdot\frac{10^3}{T} = \log_{10} C - 1.3\cdot\frac{10^3}{T}$$

The graph giving log v/w as a function of $1/T$ *(Fig. 8)* for the Toarcian samples, and the line with slope $1.3{\cdot}10^3$, shows that a law of this type fairly accurately represents the real mean distribution, although the differences are sometimes significant.

Oil & Gas Science and Technology – Rev. IFP, Vol. 58 (2003), No. 2, pp. 203-231
Copyright © 2003, Éditions Technip

A History of Organic Geochemistry

B. Durand[1]

1 Bernard Durand, 2, rue des Blés-d'Or, Dizée, 17530 Arvert - france
e-mail: bledor@wanadoo.fr

Résumé — Histoire de la géochimie organique — La géochimie organique est née des interrogations sur l'origine du pétrole. Son développement a pour l'instant été lié à celui de l'exploration pétrolière. Elle ne s'est constituée en science autonome qu'un peu après 1960. Les années 1965-1985 furent particulièrement productives : pendant cette période les mécanismes de la formation des gisements de pétrole et de gaz naturel furent explicités et de nombreux biomarqueurs, témoins de l'origine organique du pétrole furent identifiés. La connaissance des kérogènes, matière première de la formation du pétrole, fit également des progrès décisifs et des outils d'aide à l'exploration pétrolière virent le jour, comme les méthodes de corrélation huiles-roches mères, le Rock-Eval, et les modèles mathématiques de simulation de la formation et de la migration du pétrole. En même temps eut lieu un rapprochement avec la science des charbons et une extension de la géochimie organique à des domaines variés, tels que la sédimentologie organique, la microbiologie des sédiments ou la formation des gîtes minéraux.

La période actuelle se caractérise par une application croissante des connaissances acquises à des domaines autres que l'exploration pétrolière et par son intégration à l'ensemble des disciplines géologiques. Cette intégration a lieu, entre autres, par sa contribution au développement des « simulateurs de bassin », qui simulent mathématiquement l'évolution des bassins sédimentaires au cours du temps et la formation des gisements de pétrole et de gaz naturel qui en résulte.

Dans les années qui viennent, la géochimie organique aura encore un rôle important à jouer dans l'exploration et la production du pétrole et plus généralement des combustibles fossiles. Mais son avenir à plus long terme est probablement dans l'étude des interactions des produits de l'activité humaine, en particulier les polluants organiques et les gaz à effet de serre, avec la géosphère, en relation avec la recherche d'un développement durable et la compréhension des mécanismes des changements climatiques.

Le développement de la géochimie organique moderne doit beaucoup aux instituts professionnels de recherche et en particulier à deux d'entre eux, l'*Institut français du pétrole (IFP)* en France et le *Kernforschungsanlage (KFA)* en Allemagne : il faut en créditer les créateurs et premiers animateurs des départements de géochimie organique de ces deux instituts, B. Tissot et D. Welte.

Abstract* — *A History of Organic Geochemistry — *Organic geochemistry is the fruit of questions about the origin of petroleum, and for the time being, its development has been closely linked to that of petroleum exploration. It only became an autonomous science shortly after 1960. The years 1965-1985 were extremely productive: during this period, the mechanisms of the formation of oil and natural gas fields were clarified, and many biomarkers, testimony of the organic origin of oil, were identified. The knowledge of kerogens, the raw material for the formation of oil, also made decisive progress, and tools to aid petroleum exploration were created, such as oil-source-rock correlation methods, Rock-Eval, and mathematical models simulating the formation and migration of oil. A rapprochement took place at the same time with coal science and an extension of organic geochemistry to various fields such as organic sedimentology, the microbiology of sediments and the formation of orebodies.*

The current period is characterized by the growing application of the knowledge gained to fields other than petroleum exploration, and its integration with all geological disciplines. This integration has, among other factors, taken place through its contribution to the development of "basin simulators", which mathematically simulate the evolution of the sedimentary basins over time and the resulting formation of oil and natural gas fields.

In the coming years, organic geochemistry will continue to play an important role in the exploration and production of oil, and of fossil fuels in general, but its long term future probably lies in the study of interactions of the products of man's activity, in particular organic pollutants and greenhouse gases, with the geosphere, in relation with the search for sustainable development and the understanding of mechanisms of climatic changes.

The development of modern organic geochemistry owes a great deal to the professional research institutions and particularly to two of them, Institut français du pétrole (IFP) *in France and* Kernforschungsanlage (KFA) *in Germany: credit is certainly due to the creators and the early leaders of the organic geochemistry departments of these two institutions, B. Tissot and D. Welte.*

1 A SCIENCE BORN OF PETROLEUM

Organic geochemistry is the study of organic substances in rocks. It attempts to understand their composition, their origin, formation mechanisms, modes of deposition and distribution, as well as their relations with one another and with the mineral world. It is a young science, most of it having been acquired in less than 40 years. The beginnings of organic geochemistry are closely linked to research on the origin of petroleum.

Let us first point out that for an organic chemist, petroleum is a mixture of saturated and aromatic hydrocarbons which account for about 85% of its weight, and molecules containing "heteroelements" alongside carbon and hydrogen: mainly sulfur, oxygen and nitrogen. The large majority of heteroatomic molecules are high molecular weight molecules which are called resins and asphaltenes.

The petroleum family includes bitumens, also called asphalts, which are highly viscous and sometimes solid oils mainly containing resins and asphaltenes, as well as natural gas, which contains gaseous hydrocarbons in normal temperature and pressure conditions, and a variable proportion of nonhydrocarbon gases, chiefly carbon dioxide, hydrogen sulfide and nitrogen.

Oil and natural gas are found in sedimentary basins at depths ranging from the surface to about 10 km, in the pores of permeable rocks of variable extension called reservoir rocks. These are generally sandstones or coarse-grained or fractured limestones. Bitumens and asphalts are found in the same type of rock, but mainly at or near the surface. Bitumens and asphalts are in fact products of the alteration of petroleum by surface waters.

1.1 Inorganic Theory *versus* Organic Theory: the Scientists of the Late 19th Century

The understanding the origin of petroleum was for a long time merely an object of curiosity, because no economic importance was attached to it. Petroleum found at the earth's surface, chiefly in the form of bitumen outcrops (the famous Judea bitumen for example) and small surface flows supplied the only uses.

There was hardly any theory on this subject that could be qualified as scientific before the 19th century. Emblematic scholars of the late 19th century delved into the question: for Berthelot (1866) petroleum was the product of the action of water on inorganic carbides such as iron carbide and aluminum carbide, presumed to be present in abundance in the earth's crust. For Mendeleev (1878), who spent some time in his early career analyzing the petroleums of Pennsylvania and the Caucasus, petroleum is formed at depths where there is no longer any question of organisms. This so-called inorganic theory predominated for a very long time. L. Pasteur is said to have supported a biogenic theory, according to which petroleum was the result of bacterial action on organic debris.

The present so-called organic theory, according to which petroleum originates in the organic matter present in sedimentary rocks, took a long time to be established. Yet it already had partisans in the early 19th century and even before: the Swiss geologist A. Jaccard, a defender of this theory, published a work in 1895 where he dates it in 1801, to L. de Buch, a scholar attached to the King of Prussia, yet the premisses of the theory can already be found in a memorandum presented in 1770 in France to the Royal Academy of Sciences by M. Fougeroux de Bondaroy. A. Jaccard and L. Buch were partisans of an origin in animal debris, while M. Fougeroux de Bondaroy proposed a plant origin. The bibliography compiled by Jaccard identifies also the work of the French geologist M. Lesquereux, who explained in a letter sent in 1865 to J. von Liebig the reasons for believing in an origin in marine plants, such as fucus.

1.2 The Progress of Organic Theory: Hunt, White, Engler, Verdnasky, Treibs

It is with the North American petroleum and coal geologists of the late 19th and early 20th century that the first truly comprehensive approach can be found to the mechanisms of petroleum formation, at a time when petroleum demand was growing significantly. In 1863, the Canadian petroleum geologist T.S. Hunt is also one of the fathers of the concept by which petroleum accumulates preferably in the anticlinal zones of the earth's crust; this concept was for decades and still remains one of the principal guides to petroleum exploration declared that petroleum was formed from plant and animal debris, particularly microscopic marine forms, which are present in sedimentary rocks. According to him, the transformation by which the organic matter present in the sediments becomes bitumen and petroleum is similar to that which produces so-called "bituminous" coals from peat and lignite. In 1915, the American coal geologist D. White showed that a relation existed in Pennsylvania between the location of the petroleum reservoirs and the map of "coalification" stages of the very abundant coals in the region.

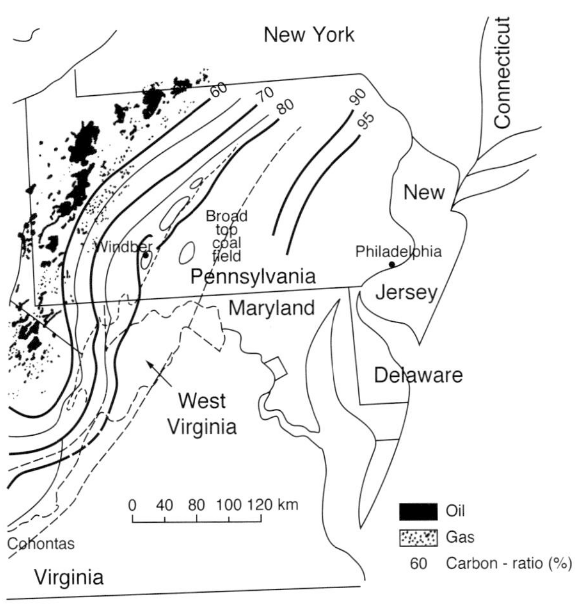

Figure 1

Location of oil and gas fields in Pennsylvania compared with isocontent curves of carbon in coals (carbon ratio), according to D. White, 1915. After Robert (1985).

Coalification is the natural mechanism of transformation that makes peat, the original organic matter of coal, as it goes through its very progressive burial in the depths of the sedimentary basins, pass through successive stages of lignite, bituminous coal, anthracite and metaanthracite. It is accompanied by physical and chemical transformations in the coals and at the time, these different stages of transformation were identified by the organic carbon content of the coal (carbon ratio), which increased with coalification. White observed that petroleum fields were only found in areas where this content was lower than a value of about 60%; in the same way, natural gas fields disappeared when the carbon ratio rose above 70% (Fig. 1). *White thereby concluded, as already done by T.S. Hunt, that the formation of petroleum and natural gas reservoirs and coalification were two aspects of the same mechanism of progressive transformation of organic matter in the sediments during their burial. He also stated that the intensity of this transformation could be calibrated with a simple parameter, the carbon content of the coals found in the sedimentary series.*

In the late 19th century, an industry developed in the United States and Europe producing a mineral oil similar to petroleum, by pyrolysis of rocks rich in organic matter called "oil shales" at temperatures of 400 to 500°C. The oil produced was called "shale oil". It was therefore tempting to suggest by analogy that in nature, the factor responsible for the formation of petroleum from the sedimentary organic matter was a thermal cracking of this organic matter under the effect of increasing temperature with depth. This increase in temperature was well known to all who had drilled boreholes or conducted mining operations. The German chemist C. Engler (1888, 1912), a partisan of animal origin, tried to prove this experimentally and showed that substances similar to petroleum could be produced by pyrolyzing a variety of animal substances in the laboratory (fish, molluscs, whale oil). Investigations were also conducted at the same time with sedimentary rocks by the American geologists A.W. McCoy and P. Trager in 1919 and, according to Bordenave (1993) by the English geologists MM. Kennaby and May of the *Anglo-Iranian Oil Company* in 1931. Later on, the American geologist and chemist, W.E. Pratt (1943), observing variations in the chemical composition of petroleums according to their geological age, concluded that they were the result of a "progressive cracking in nature's laboratory".

Yet the temperatures necessary to produce shale oil in the oil shale industry were several hundred degrees, while such temperatures are only encountered in the surface areas of the earth's crust in volcanic regions. Besides, comparisons made between petroleums and Engler's pyrolysates show that even with the makeshift analytical resources of the time, significant differences were observed. The experiments of Engler and his imitators were therefore deemed undemonstrative to most scientists at the time.

The idea that petroleum originated in sediments containing organic matter was nonetheless well established among North American petroleum geologists in the 1930s. The concept of "source rock" was born, illustrated inter alia by the work of D. Trask (Trask and Wu, 1930; Trask *et al.*, 1932) for whom the origin of petroleum in source rocks was positively established, and who wondered whether the petroleum was formed in the sediments at the time of their deposition. In 1926, the *American Petroleum Institute (API)* and the *US Geological Survey (USGS)* launched a vast program to analyze borehole cuttings and cores in order to find criteria for characterizing source rocks, a program that culminated with the publication by Trask and Patnode (1942) of the analysis of 35 000 samples. These petroleum geologists, however, still failed to agree on the mechanisms of petroleum formation.

Very similar observations and investigations also took place in the early 20th century in the Soviet Union. In 1924, the Russian geologist V.I. Verdnasky synthesized them in a work where he defended the theory of the origin of petroleum in the thermal transformation of organic matter present in sedimentary rocks. Yet he went far beyond simple observation, and firmly put petroleum formation into the more general framework of a description of the geochemical cycle of elements and the interactions between the biosphere and geosphere. His work, translated in 1934 into English with the title "Outlines of Geochemistry", contained what could be called the first treatise of organic geochemistry. Verdnasky is therefore considered in Eastern Europe as the founder of organic geochemistry.

In 1934, the German chemist A. Treibs revealed the close similarity of chemical structure between porphyrins, organic molecules found in petroleums and in sediments, and chlorophyll pigments (Fig. 2). *This relationship is considered to be the first and really undeniable proof of the organic origin of petroleum. And it is to A. Treibs that the paternity of organic geochemistry is attributed in Western Europe.*

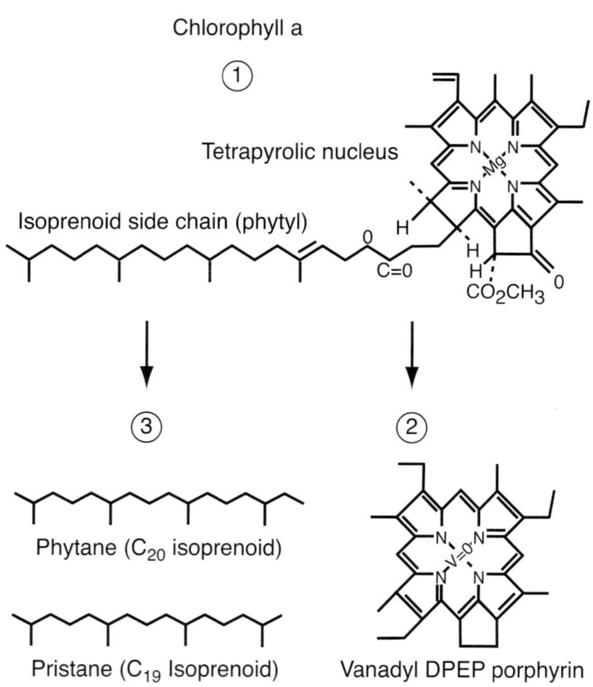

Figure 2

Chemical structure of chlorophyll a and its derivatives in sediments: the chlorophyll found in plants (1) and vanadyl DPEP porphyrin (2) found in oils possess the same tetrapyrolic nucleus. This observation led Treibs (1934) to infer that petroleum was of biological origin.

The alteration of chlorophyll in sediments by rupture of the phytyl side chain and saturation of its fragments, also leads to the formation of isoprenoid hydrocarbons (3). After Bordenave *et al.* (1993).

With permission of Éditions Technip.

1.3 The Theory of Direct Accumulation

Later, the observation of the existence of traces of hydrocarbons in recent sediments, in particularly by V. Smith (1954), gave rise to the theory of direct accumulation (Baker, 1959; Meinschein, 1959). According to this theory, the mechanisms of the formation of petroleum accumulations is expulsion followed by concentration of these hydrocarbons by compaction and deformation of the sediments during their burial. This theory does not contradict the organic theory: it is rather a theory on the mechanisms of its migration, in other words, its displacement from the source rock, recent sediments in this case, up to its reservoir.

This theory was soon contradicted by the finding of wide differences in chemical structure between substances extracted from recent sediments and those found in petroleums (Stevens *et al.*, 1956; Erdman *et al.*, 1958). Besides, volumetric calculations demonstrated that the quantities of hydrocarbons found in the recent sediments were too small to produce large fields, and centrifuge experiments designed to simulate compaction only succeeded in extracting an infinitesimal quantity of these hydrocarbons, showing that they could not be expelled from the sediments by compaction at the same time as water (Peake and Hodgson, 1965).

1.4 Biogenic and Radiogenic Theories

While the organic origin of hydrocarbons found in recent sediments had not been challenged, the mechanisms of their formation had not been clarified. The American microbiologist E.C. ZoBell and his colleagues (ZoBell, 1947; Stone and ZoBell, 1952), who investigated the action of microorganisms on petroleum hydrocarbons at the time, suggested that these hydrocarbons were produced by bacteria and that the mechanism of petroleum formation was to be found there, as Pasteur himself had believed. Yet this research direction ultimately failed, because while certain bacteria were known for a long time to produce methane, all the laboratory efforts to cause bacteria to produce other hydrocarbons from sedimentary organic matter failed. However, the idea was not completely sterile, since, as we shall see further on "bacterial" methane proved to be the source of large fields of natural gas. It was also observed that some bacteria were capable of degrading petroleum in its reservoirs.

Also at the time appeared the theory of the formation of petroleum by the action of the radioelements. According to this "radiogenic" theory, the radiation produced by the radioelements present in sedimentary rocks caused an alteration of the organic matter and the formation of hydrocarbons (Lind, 1938). Yet this theory soon proved to be difficult to credit for quantitative and qualitative reasons. While an alteration of organic matter present in certain sediments in contact with grains of radioactive minerals is effectively sometimes observed under the microscope, as in the case of Alum shales for example, shales of Silurian age rich in radioactive minerals found in Scandinavia, the mechanism was too rare for generalization, and the analyses of the products formed showed that their composition was very far from that of common petroleums (Whitehead, 1951). Yet radioactive decay is probably the cause of a large part of the helium and argon found in certain natural gas reservoirs.

1.5 Growing Petroleum Demand and the Creation of EAOG

1960 was the year of entry into the petroleum era, when petroleum became the main source of energy for humanity. Petroleum demand started growing rapidly, while most of the shallow reservoirs had been discovered, usually next to surface leaks (the oil "shows"). No method of direct detection in depth had been imagined. The exploration managers in the oil companies then realized that they increased their chances of finding oil if they clearly understood the mechanisms of formation and accumulation in the reservoirs. Specialized laboratories therefore increasingly appeared in these companies and large professional institutes, led by teams of geologists and chemists who cooperated with the university laboratories interested in these matters. Publications on the origin of petroleum began to proliferate. On this subject, the bibliographic studies of Stevens (1956) and Debyser and Deroo (1969) are enlightening.

It is in Europe that the developments were the most significant:

The leading European laboratories formed the structure of the European Association of Organic Geochemistry (EAOG), *which organized its first congress at Milan in 1962, and then organized the International Meeting on Organic Geochemistry every two years, the main rendezvous of organic geochemists. The proceedings of this congress are the series of* Advances in Organic Geochemistry. *The* EAOG *also struck the Treibs medal to reward the most significant efforts in organic geochemistry.*

The first Gordon Research Conference in Organic Geochemistry was held in 1963 in the United States, at the Tilton School in New Hampshire, bringing together American and European organic geochemists. This conference, alongside the EAOG *conference, marks the activity of this discipline today.*

1963 also saw the publication by I. Breger et al. *of the first book with the title* Organic Geochemistry. *Organic geochemistry had therefore acquired all the attributes of an autonomous science. Yet the participants at the first Gordon Research Conference in Organic Geochemistry were divided by a finding of uncertainty not about the origin of the petroleum, because there was too much evidence of its origin in organic matter in the sediments, but about the precise mechanisms of its formation: many theories had been compared, but none was considered to offer positive proof.*

2 A PARALLEL HISTORY: COAL SCIENCE

Coals are sedimentary rocks whose organic matter accounts for at least half of their weight. In their overwhelming majority, they were formed from debris of higher plants, particularly in the marine and lacustrine deltas of rivers crossing areas of abundant vegetation at the time of their deposition. This is why they are called humic coals (from humus, a soil formed of plant debris), as opposed to more occasional coals, coals of algae mainly formed of debris of unicellular algae and coals of spores formed chiefly of debris of plant spores. Humic coals are associated, in the sedimentary series where they are found, called "coal series", with rocks containing a lesser abundance of the same type of organic matter. The coal series may be several kilometers thick. The coal levels are rarely more than one meter thick, but can extend over several kilometers.

Coals were considered in ancient times as products of volcanic phenomena or the condensation of asphaltic materials. Their origin in the debris of higher plants was nonetheless recognized clearly from the early 19th century. Coal geology made tremendous progress in the late 19th century and early 20th: the depositional modes and eras of coal formation were investigated: the quantitative importance of coals formed in the Carboniferous, 300 to 350 Ma ago was recognized in particular. Noteworthy among the great coal geologists of the time were C.E. Bertrand and B. Renault in France, H. Potonié in Germany and D. White in the United States. In 1913, D. White, together with the American petrographer R. Thiessen, published a work that stands as a reference *The origin of coal.*

Since the origin of coal was clear and coal was abundant on the earth's surface, coal science which began to develop at the time did not focus on the problems of its origin and exploration, as the case of organic geochemistry with petroleum, but on problems of a technological nature: in the variety of coals, what criteria could be used to recognize the fittest for their intended use: home heating, steam generation, manufacture of cast iron and steel? The aim was also to improve the industrial processes of coal utilization. Two disciplines emerged, providing each other with mutual support: coal petrography and coal physical chemistry.

Shortly after 1830 the English geologist Hutton (not to be confused with J. Hutton (1726-1797), one of the creators of modern geology), using a transmission microscope to examine thin slices of coal, discovered remains of spores and plant cells. In doing so, he founded coal petrography, a discipline that then grew considerably. In 1913, H. Winter in Germany created the technique for examination by reflection microscope on a polished block, perfected in 1927 by E. Stach with the use of a lens immersed in oil (van Krevelen, 1981). In 1943, the German petrographer M. Schochardt introduced the technique of illumination with ultraviolet light, which made fluorescent some of the composants examined under the microscope.

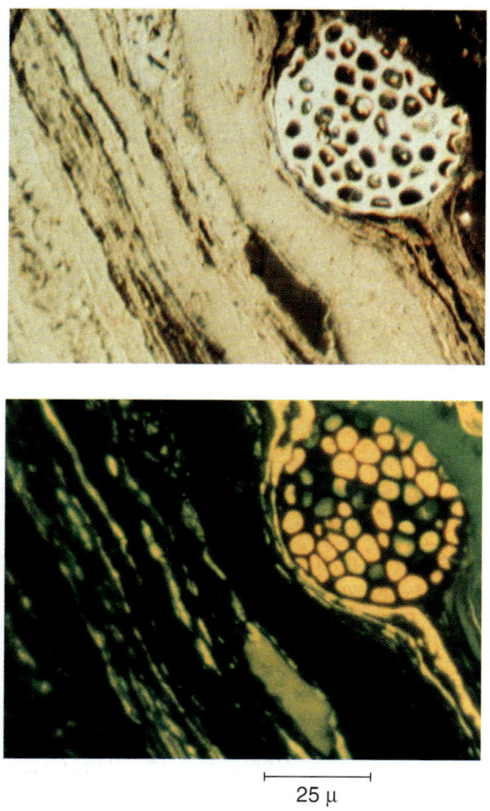

Figure 3

Microphotographs in reflected light of coal fragments bound in a resin. At top, illumination with light at 546 nm. The medium reflectance particles are vitrinite, and the low reflectance particles, which are dark, liptinite, and the high reflectance particles inertinite (a mushroom stalk here). At bottom, ultraviolet illumination causes fluorescence of the liptinite particles. Photographs B. Alpern.

Coal petrography witnessed a conceptual advance alongside its technical progress: the objects observed under microscope were classed according to their biological origin. Particularly noteworthy was the work in France of H. Fayol and in England the work of C.A. Seyler and M. Stopes. Stopes named these objects maceral by analogy with rock minerals and, in 1919, proposed the so-called Heerlen classification which, in 1935, became the base of the one used today. This classification was improved later on by the German petrographers E. Stach, M.Th. Mackowsky and M. Teichmüller (Stach *et al.,* 1982), the Dutch petrographer P. Jongmans and the French petrographer B. Alpern, among others.

Macerals belong to three main families *(Fig. 3)*, classed primarily by their reflectance under the reflecting microscope. Their biological origin can be identified easily in most cases.

The vitrinite family, with medium reflectance, accounts for about 70% of the coal on average. The macerals of this family are gels formed in sedimentation environments from lignocellulose debris (wood) of higher plants.

The liptinite family, with low reflectance, contains exinites, formed from the "outer parts of plants": spores, pollens, leaf cuticles, the alginites from unicellular algae initially living in the sedimentation environment, and miscellaneous amorphous substances. The macerals of this family display the feature of becoming fluorescent under ultraviolet light.

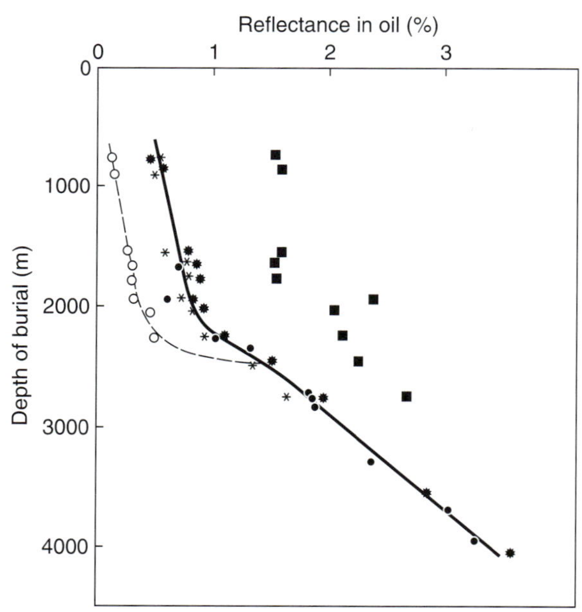

Figure 4

Example of variation reflectance of coal macerals with depth (and generally type 3 kerogens): the Logbaba (Cameroon) boreholes: vitrinite in solid lines and liptinite in dotted lines. The black squares represent inertinite. After Durand and Espitalié (1976).

The inertinite family, with high reflectance, mainly consists of particles that were oxidized before or during sedimentation, for example under the effect of forest fires or residence or prolonged transport in aerobic media.

Petrographic examination helps to observe the wide variety of coals and connect it with the variety of depositional environments. It also helps to anticipate the technological properties and identify the stages of coalification. *The vitrinite reflectance measured in standard conditions (light of wave length 546 nm, normal incidence to a perfectly polished section, lens with immersion in oil of index 1.517), which increases steadily during coalification, from about 0.2% in the peat stage to 3% or more in the anthracite and metaanthracite stage* (Fig. 4) *is the most widely used for this latter use.* The best summary of the petrographic research of the time is no doubt the work by Stach (1935), *Lehrbuch der Kohlenpetrographie.*

Coal physical chemistry developed at the same time as petrography. The pioneers, most of whom worked in institutes dedicated to these studies, include the following: concerning the more chemical aspects, in England W.A. Bone at the *Imperial College of London* and R.W. Wheeler in Sheffield, and in Germany, F. Fischer at Mülheim and F. Bergius who had the rare distinction at the time of financing his own researches—concerning the more technological aspects, E. Audibert in Montluçon in France, F.S. Sinnatt in Greenwich in England and A.C. Fieldner in Pittsburgh in the United States. Research institutions employing many researchers were then created in many countries: Belgium, Spain, Holland, Poland, Yugoslavia, USSR, China, India, and Australia. All the efforts were aimed at a better understanding of the composition and chemical structure of coals, their mechanical and rheological properties, surface properties etc., to improve the industrial processes in which they were used and to discover new uses. One of the major problems to be solved was to improve the quality of metallurgical coke, which is a reducing agent and an energy supply in the production of cast iron and steel.

The best summary of the work done at the time in physical chemistry of coals is no doubt the work by D.W. van Krevelen *Coal, Typology, Physics, Chemistry, Constitution* (1961, reprinted in 1981). In 1950, this researcher also proposed a rectangular diagram for use in determining the properties of coals and their macerals, plotting the O/C atomic ratios on the x-axis and the H/C on the y-axis. *This "van Krevelen's diagram" proved to be invaluable in monitoring the coalification process. It is therefore an excellent tool for synthesis between physical chemistry and petrography. It went on to be very popular with petroleum geochemists.*

However, coal science and organic geochemistry were completely unrelated at the time. One exception was the work of M. Teichmüller, who published a still classic article in 1958 on the location of oil and gas reservoirs around the Bramsche Massiv in Germany, which is a volcanic intrusion in sedimentary formations containing levels with organic matter. The temperature elevation of the sediments resulting from this intrusion caused the formation of oil and gas fields from the organic matter; the zones where they are found are bounded by vitrinite isoreflectance curves, indicating a relation between coalification stages and oil and gas formation stages. This work is very similar in principle to that of D. White on the oil fields in Pennsylvania.

3 1965-1985: THE YEARS OF EXPANSION

3.1 The Successors of Hunt and Verdnasky

It is important to consider those who demonstrated the mechanisms of the formation and accumulation of oil by primarily relying on geological observations along the path initiated by T.S. Hunt and V.I. Verdnasky. In the western world, the most significant pioneering work was no doubt done by Philippi (1965) at *Shell,* Louis and Tissot (1967) at the *Institut français du pétrole,* and Albrecht and Ourisson (1969) at the *University of Strasbourg. These researchers used organic solvents (chloroform, methanol, benzene) to extract series of sedimentary rock samples rich in organic matter taken by petroleum boreholes at different depths in homogeneous sedimentary series. The extracted fraction, called the soluble extract or bitumen, also contained hydrocarbons. This extraction left an insoluble fraction in the sediments, quantitatively the largest fraction, called kerogen by Forsman and Hunt (1958). The series originated in different sedimentary basins with considerable variations in the type of sediment, age, geothermal gradient and burial rate, but they all had one common feature: at the time the sediments were deposited,* i.e. *in the recent sediment stage, the samples of each series had very similar mineralogical and organic compositions. In consequence, the analysis of samples taken at increasing depths revealed the effects of burial on a material that was initially the same, all other things remaining equal.*

All the observations led to the same conclusion: the progressive burial of the sedimentary organic matter increased the formation of hydrocarbons, first liquid, and then gas. It was also observed that the depth at which the formation of oil began was also shallower as the average geothermic gradient of the basin was lower and/or the age of the basin was higher, suggesting that the formation of oil obeyed rules of chemical kinetics (Tissot *et al.*, 1975). Moreover, the hydrocarbons extracted from shallow sediments were very different from those found in the oilfields, but the resemblance grew progressively with burial: a minimum depth of about 1 to 2 km was necessary for a mixture of hydrocarbons to be formed in the sediments, with a composition close to those found in the reservoirs.

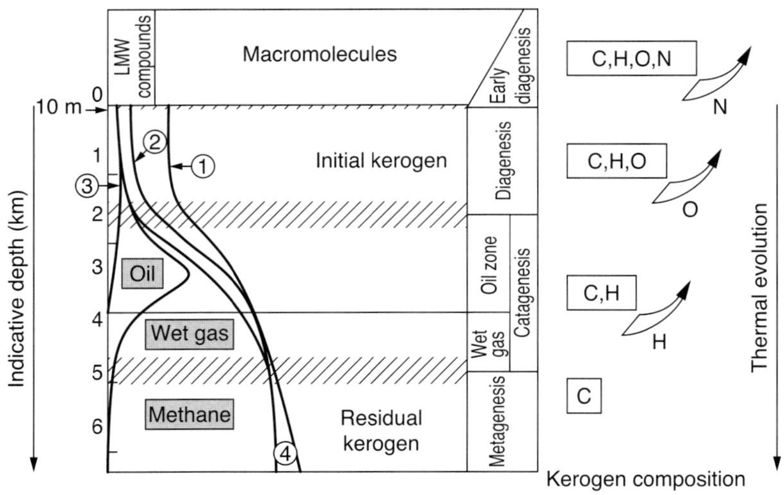

Figure 5

Flowsheet of the formation of oil and gas from kerogen as a function of depth. 1) low molecular weight oxygenated compounds; 2) resins and asphaltenes; 3) biomarkers; 4) insoluble pyrobitumen (coke) formed from oil as a consequence of gas formation. After Tissot and Welte (1978), Durand (1980) and Vandenbroucke *et al.* (1993) in Bordenave *et al.* (1993).

With permission of Éditions Technip.

Identical conclusions were drawn at approximately the same time by Russian researchers, particularly A. Kartsev, V. Sokolov and N. Vassoievich (Vassoievich *et al.*, 1969; Kartsev *et al.*, 1971), who had samples and documents taken from the immense variety of sedimentary basins of the Soviet Union. Since the only variable parameters in each of the series investigated was the temperature, pressure and time associated with burial, it was necessary to conclude that these variants were responsible for the formation of oil in the sedimentary basins. Since pressure could hardly cause the radical chemical transformations observed and would rather tend to delay them, temperature and time were the best candidates. These observations finally led to the establishment of a flowsheet for the formation of oil and natural gas as shown in Figure 5, which is commented on here.

In recent sediments, only a very small part of the organic matter present is extractable in organic solvents. Kerogen hence forms most of the sedimentary organic matter at this stage. The extractable fraction, bitumen, mainly contains high molecular weight and strongly functionalized molecules, belonging to categories called resins and asphaltenes, which cannot be analyzed molecule by molecule. The molecules that could be separated and analyzed were mainly geochemical fossils, i.e. *molecules undeniably deriving from biological molecules. There were very few hydrocarbons among them.*

During their burial, sediments containing organic matter are exposed to increasing temperatures. Three successive stages of a transformation called maturation are then observed; they were called diagenesis, catagenesis and metagenesis by Tissot and Welte (1978). Maturation corresponds to a thermal cracking of the kerogen under the effect of the rise in temperature accompanying the burial of the sediments.

During diagenesis, very few hydrocarbons are formed. This is why diagenesis is also called the immature stage by petroleum geochemists. Mobile oxygenated products, particularly water and carbon dioxide, as well as resins and asphaltenes, are liberated at this stage by the thermal degradation of the kerogen. Catagenesis, the second stage, is that of the oil, and then of the "wet" thermal gas, so called because it is a mixture of gaseous and liquid hydrocarbons. This thermal gas is formed from the remaining kerogen but also from the oil already formed, which explains why the formation of oil passes through a peak in the figure. During catagenesis, the gas/oil ratio steadily increases and the geochemical fossils, which are fragile molecules, are progressively destroyed. At the end of catagenesis, the kerogen has practically lost its capacity to produce hydrocarbons. The third stage, metagenesis, is that of the dry thermal gas, so called because the hydrocarbons it contains are gaseous hydrocarbons. This gas is formed from the remaining kerogen, but mainly from the oil and wet gas previously formed.

Finally, a general description of the formation of oil and gas and their accumulation in the reservoirs materializes. It is illustrated in Figure 6.

After their formation during burial, oil and natural gas are expelled from their source rock, in a process called primary migration, and then travel in permeable drains towards traps where they accumulate, in a process called secondary migration. These traps consist of a permeable "reservoir rock" and a permeability barrier which stops the hydrocarbons from advancing. They are usually located in the upper zones of the geological structures, the anticlines. The oil and gas can also escape from the traps to form surface shows or secondary reservoirs. This process is called tertiary migration.

The formation of the oxygenated molecules followed by oil and gas necessarily also results in a progressive modification of the physicochemical properties of the kerogen, the insoluble fraction of sedimentary organic matter (McIver, 1967; Durand and Espitalié, 1973; Tissot *et al.*, 1974). The observation of these changes by a variety of physicochemical

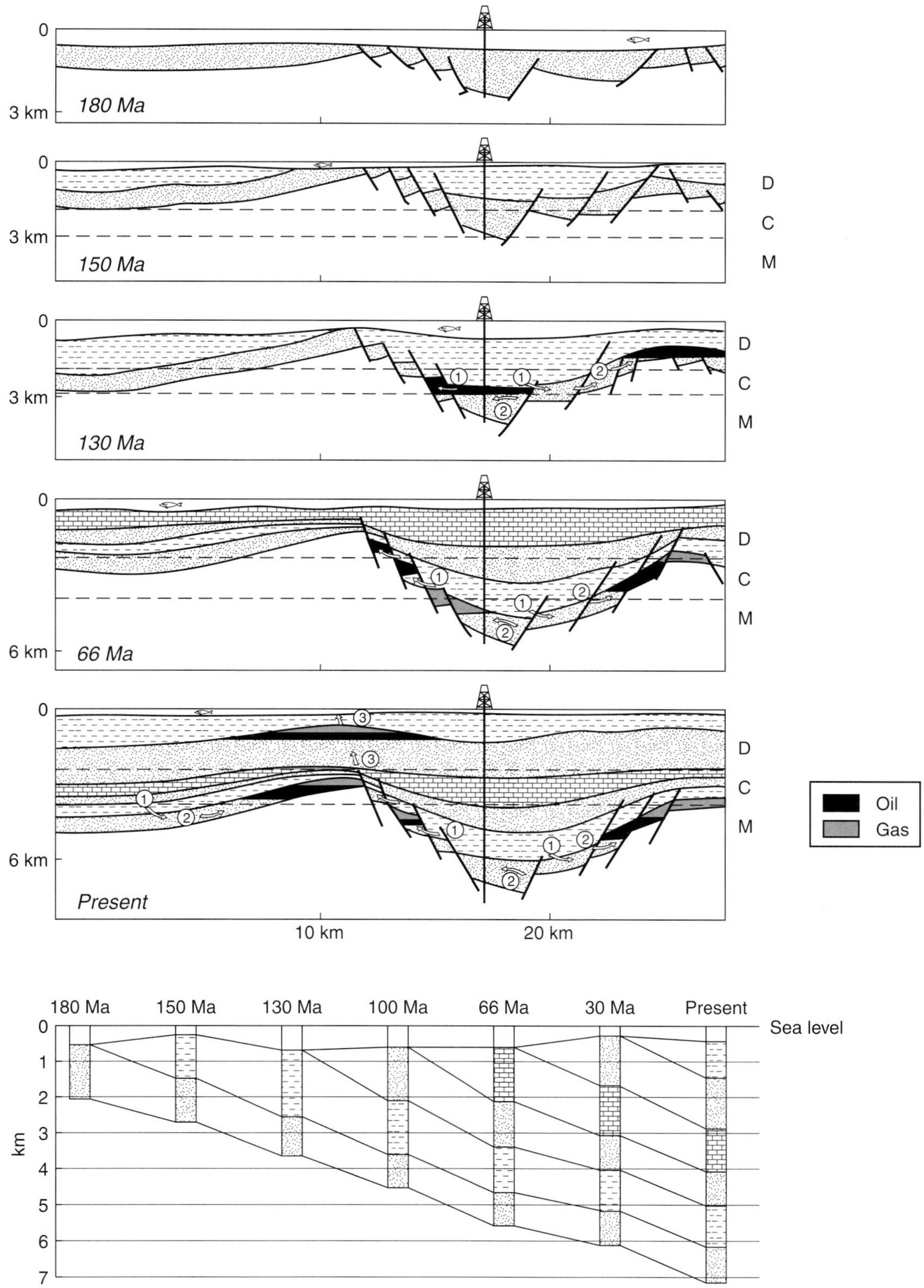

Figure 6

Flowsheet of the formation of oil and gas reservoirs in a sedimentary basin during its burial, based on the example of the Viking graben in the North Sea. Upper part of the figure represents the successive steps in the evolution of a basin of a rock rich in organic matter and buried to increasing depths and temperatures. The oil and gas formed in this source rock at stages C (catagenesis) and M (metagenesis) have been expelled (arrows 1) towards drains where they flow (arrows 2) to accumulate in traps and form fields. They could ultimately escape from these traps (arrows 3) to form secondary fields higher up. Lower part of the figure shows the evolution, over geological time, of the disposition and burial of levels identified at the fictitious well at the center of the basin.

methods was mainly the achievement of Durand *et al.* (1980), who were also among the earliest to develop relations between organic geochemistry and coal science.

Kerogens are formed mainly of carbon, hydrogen and oxygen. They also contain nitrogen, but much less than the living matters in which they originate, as well as sulfur, more on average than living matter. A convenient method to track the evolution of kerogens as a function of burial of the sediments and to characterize their variety is hence a C, H, O diagram. Durand *et al.* selected the rectangular diagram already used by the Dutch researcher van Krevelen (1950) to track the physicochemical changes in coals, a diagram in which the O/C atomic ratios are plotted on the x-axis and the H/C atomic ratios on the y-axis *(Fig. 7)*.

In this diagram, the kerogens isolated from series of sediments in which the organic matter was approximately the same at the time of deposition, are grouped together in narrow zones called evolution or maturation tracks by Durand and Espitalié (1973). Kerogens of the shallower sediments have high O/C *and* H/C *atomic ratios. Diagenesis results in a greater decrease in* O/C *than* H/C, *corresponding to the liberation of carbon dioxide and water. Catagenesis results in a sharp decrease in* H/C, *corresponding to the liberation of hydrocarbons. Metagenesis does not result in any substantial change, because at this stage, the kerogens have lost most of their oxygen and hydrogen, and have become carbonaceous residues, but the dark-field electron microscopy shows that this step corresponds to the development of an organization of the carbon into small "graphitoid" domains measuring some tens to hundreds of microns, formed of stacks of roughly parallel polyaromatic sheets (Oberlin* et al., *1974).*

The kerogen maturation tracks of the different series analyzed are also not merged in the diagram and are distinguished according to the sedimentology of the series examined. Thus Tissot et al. *(1974) distinguished between three types of reference of sedimentary organic matter:*

- *type 1, from a planktonic biomass thoroughly reworked by bacteria in lacustrine or evaporitic sediments;*
- *type 2, from a marine planktonic biomass;*
- *type 3, from debris of higher terrestrial plants* (Fig. 10).

Their maturation paths converge towards the carbon pole of the diagram with increasing depth. Kerogens initially with the highest H/C *ratios are those that produce the most oil per unit weight during maturation. The composition of the oils formed differs according to the type and evolves as a function of maturation.*

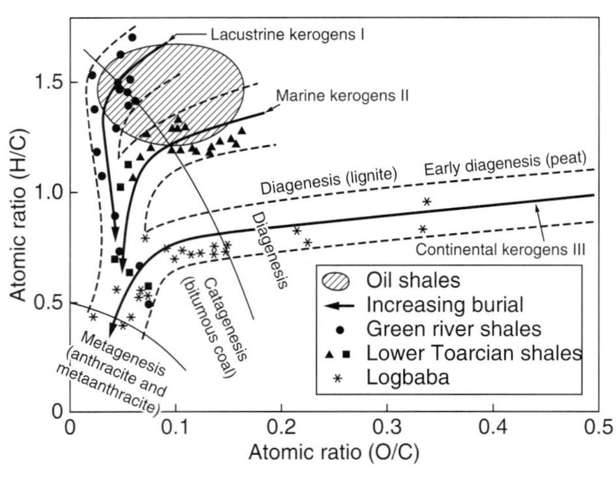

Figure 7

Evolution (maturation) pathways of kerogens of the three reference types in a van Krevelen's diagram. Also indicated are the position in this diagram of the kerogens present in oil shales and the correspondence between the main steps in the formation of oil and gas with those of coalification. After Durand and Espitalié (1973), Tissot *et al.* (1974) and Durand (1987).

The diagram allows an easy classification of fossil fuels. These belong to three categories.

- Oils and allied substances which are asphalts or bitumens and natural gas: as we have seen, these are by-products of thermal cracking of the kerogens during their maturation. Natural gas is produced at the end of maturation, while asphalts and bitumens are the degradation product of oil in the reservoirs.
- Coals: these are in fact sediments very rich in organic matter. In the vast majority of cases, coals are "humic" coals, *i.e.* where the organic matter derives from higher terrestrial plants. Hence it is a special case of type 3 kerogens, and the coalification of coals is part of the more general process of the maturation of sedimentary organic matter, as claimed by D. White: the peat stage corresponds to the recent sediment stage, the lignite stage to diagenesis, the bituminous coal stage to catagenesis, and the anthracite and metaanthracite stages to metagenesis.
- Oil shales, which, as we have seen, are rocks rich in organic matter, from which a petroleum-like oil is produced by pyrolysis at 400 to 500°C called shale oil: oil shales are in fact sediments that have undergone relatively shallow burial and consequently contain a still immature kerogen. This kerogen is rich in hydrogen, hence of type 1 or 2, and therefore has a high potential for hydrocarbon production per unit weight. Temperature causes pyrolysis of the kerogen in these rocks, in other words, an artificial maturation, comparable to the natural maturation that burial would produce.

3.2 Treibs' Successors

An enormous task was achieved from 1965 along the path initiated by A. Treibs. This was considerably facilitated by the rapid development of the technology of analytical instruments: since the invention in 1906 by M. Zswett of paper chromatography, much progress had been achieved in the separation of organic mixtures into broad chemical classes, but from 1960, the development of gas chromatography and mass spectrometry, followed in 1970 by the coupling of these two instruments, combined with major advances in chemical synthesis methods, helped make tremendous headway in the separation and identification of molecules. Rigorous analytical and comparison protocols of the components of oils and extracts of sediments by organophilic solvents were developed. This led to the identification in sediments and oils of thousands of molecules undeniably derived from biological molecules. These molecules were called geochemical fossils or biomarkers. Many of them were optically active explaining the observation made several years earlier that many oils had a rotary power over light. Among the researchers focusing on "molecular geochemistry", the most active were G. Ourisson, P. Albrecht *et al.* (Albrecht and Ourisson, 1971) at the *University of Strasbourg*, G. Eglinton, J. Maxwell *et al.* at the *University of Bristol*, P. Schenck, De Leeuw *et al.* (De Leeuw, 1986) at the *University of Delft*, J. Connan of *Société nationale des pétroles d'Aquitaine* in Pau, W. Seifert and M. Moldowan of *Chevron* in the United States, M. Radke (Radke, 1987) and J. Rulkötter at *Kernforschungs-anlage (KFA)* in Germany. A. Petrov (Petrov, 1984) at the *Institute of Geology and Exploitation of Fossil Fuels* in Moscow, and E. Whitehead and A. Mackenzie of *British Petroleum* in England.

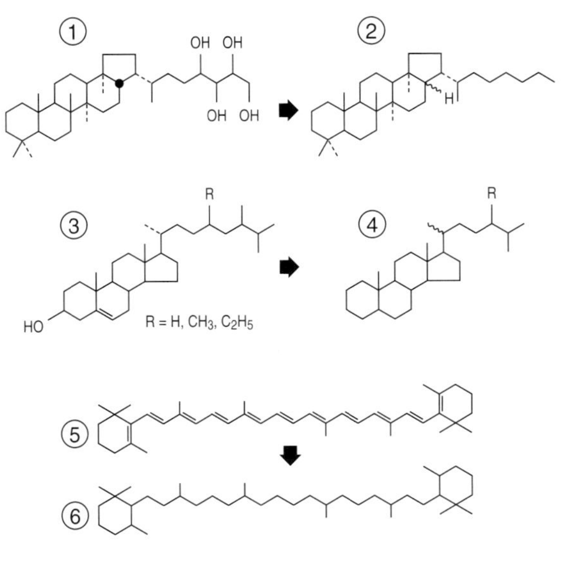

Figure 8

Structures of frequent biomarkers in sediments and oils, and biological molecules from which they derive: the bacterio-hopane tetrol (1) produces hopanes (2) sterols (3) produce steranes (4) and β-carotene (5) produce β-carotane (6). In Bordenave *et al.* (1993).

With permission of Éditions Technip.

These molecules are for example porphyrins discovered by Treibs, as well as long chain n-*alkanes derived from the cuticular waxes of higher plants or fatty acids of marine or lacustrine plankton, isoprenoids derived from the side chain of chlorophyll pigments, as well as bacterial lipids, steroids and triterpenoids derived from the cell membranes of microorganisms, and carotenoids derived from algae and bacteria* (Fig. 8). *The triterpenoids of the hopane family (so named after the English botanist hope), characterized by a pentacyclic carbon skeleton, are found everywhere. According to Ourisson* et al. *(1979), the precursors of these molecules are present in the cell membranes of prokaryotic organisms (bacteria, blue algae) where they play the rigidifying role played by steroids in eukaryotic organisms.*

Following the work of Craig, who in 1953 in the United States developed a method to measure the isotopic composition of carbon in rocks, the development occurred of the isotopic analysis of oils and organic matter in sediments, championed among others by W. Sackett (Sackett, 1975) and S.R. Silverman (Silverman and Epstein, 1958) in the United States, E.M. Galimov in the Soviet Union (Galimov, 1980), W. Stahl (Stahl, 1977), M. Schoell (Schoell, 1983) *et al.* in BGR in Germany. These researchers demonstrated the low C_{13} content of carbon in sedimentary organic matter and its derivatives, compared with that of carbonates, and helped to establish the isotopic signatures of these organic matters and their different fractions, from the isotopic composition of carbon, nitrogen and sulfur they contained.

All these results undeniably confirmed the organic origin of oil. They were used for a closer understanding of the composition of the sedimentary organic matter and oils and, above all, in the path opened up by the American geochemists Bray and Evans, who in 1961 proposed a criterion called the Carbon Preference Index (CPI) based on a measurement of the unevenness of the n-*alkanes of the* C_{25} *to* C_{35} *range, to establish what petroleum geochemists call oils/source rock correlations,* i.e. *the identification of the sedimentary rocks from which the oils found in the basins are derived, by comparison of their "fingerprints" provided by the distributions of the main markers—see inter alia Welte* et al., *1975; Deroo (1976) and Connan (1984, 1987). They were also used to establish criteria of maturation of the source rocks as well as oils. In fact, oils, just like kerogens, find their composition evolve with the burial of their reservoir under the effect of variations in thermodynamic conditions. The example in Figure 9, after Mackenzie (1984), shows an example of the evolution of the ratio of the two enantiomer forms of a sterane with burial of the sediments. The measurement of this ratio, in principle, helps calibrate the maturation of a source rock or an oil.*

However, the biomarkers are destroyed at relatively shallow depths, thereby limiting their potential in such applications to relatively immature oils.

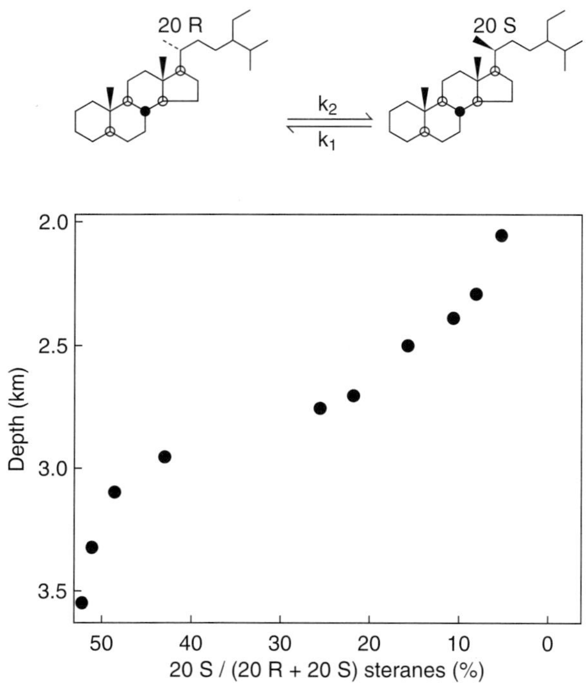

Figure 9

Example of evolution of the 20 S/(20 R+20 S) steranes ratio as a function of depth. After Mackenzie (1984).

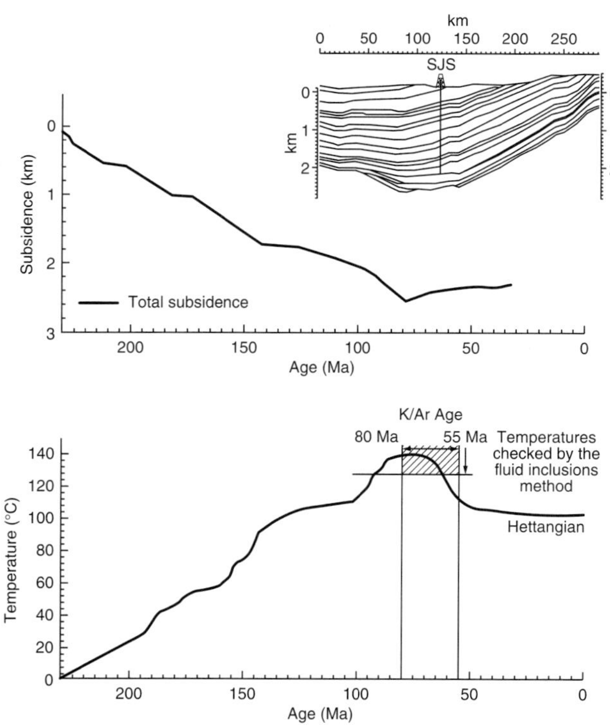

Figure 10

Evolution of burial and temperature of the Hettangian of the Paris Basin over geological time, at a well located in the central part of the basin, as indicated in the inset. After Burrus (1997).

3.3 The Thermal Regime of the Sedimentary Basins and the Modeling of Oil Formation

In a sedimentary basin, the temperature varies as a function of depth (on average 30°C/km, but in a range of 10 to 50°C/km depending on the basin and the historical periods of a given basin) which itself varies as a function of time: every sedimentary layer therefore has a different thermal history *(Fig. 10)*.

However, the temperatures reached in sedimentary basins rarely exceed 200°C, and in some cases, oil is already found to be formed at temperatures of around 50°C. Besides, oils also contain optically active compounds which are not stable above 200°C, indicating that most oils are formed at lower temperatures. In fact, much higher temperatures are necessary to produce shale oil from oil shales or to form oil-like products in the laboratory in Engler's experiments. This factor was much more disturbing for chemists than geochemists: it appeared impossible for some of them to produce thermal cracking reactions at temperatures as low as those. This is no doubt why they had imagined a theory on the structure of kerogens according to which they contained hydrocarbons in the occluded or adsorbed state. This only demanded relatively low temperatures, comparable to those existing in the sedimentary basins, to desorb these hydrocarbons and produce oil. Yet the question arose why extractions by solvents as powerful as methanol were incapable of extracting these hydrocarbons and why they were not liberated by heating in the laboratory at moderate temperature, as they were supposed to be during burial. Another of their ideas was that the thermal cracking reactions were catalyzed by the clay minerals, as done at the time in refineries for crude oil cracking. Yet the clay minerals are poorly crystallized and in the hydrated state in the sedimentary rocks. In these conditions, they cannot be catalysts for cracking (Durand *et al.*, 1972).

This made it necessary to conclude that oil was formed as a result of uncatalyzed thermal cracking of kerogen at low temperature, at a very low rate, but producing very significant results at geological timescales. Moreover, the shape of the oil formation curves as a function of depth suggested that this formation obeyed rules of chemical kinetics. The chemists connected with the oil shale industry, particularly those investigating the Green River shales in the United States, the biggest accumulation now known in the world, for decades used chemical kinetics to describe the production of oil in the pyrolysis furnaces as a function of temperature and time. The work of these oil shale specialists culminated in a mathematical modeling of the production of shale oil as a function of pyrolysis temperature and time (Allred, 1966; Fausett, 1964; Braun and Rothman, 1975). *One of them, P. Abelson, showed in 1963 that the kinetic parameters employed could be extrapolated to temperatures and times corresponding to the formation of oil in the sedimentary basins.*

On their side, the coal geologists and physical chemists who tried to understand the process of coalification had for long recognized the role of temperature and geological time via burial, and developed pragmatic methods and charts to describe its advancement as a function of these two parameters. A kinetics of coalification, based on a first order reaction and an Arrhenius' formalism to calculate the reaction rate had been proposed in particular by Huck and Karweil (1955) and Karweil (1955). Lopatin, a geochemist and student of Vassoevitch, imagined a simple method in 1971 to describe the formation of oil from kerogens, which was then popularized by D. Waples in 1980 under the name of the Time/Temperature Index (TTI). This method consists in assigning to each 10°C increment of the temperature rise during burial, a temperature factor, called the γ factor, which is multiplied by 2 on entry into the upper 10°C increment. Each of these coefficients is multiplied by the residence time of the sediment, in millions of years, in the temperature interval concerned. This produces a series of increasing numbers, each corresponding to a temperature interval of 10°C. Their sum is the TTI which is compared to a preprepared chart, in which a value of TTI corresponds to a stage of advancement of the formation of oil. For the basin investigated, the geological sections and maps can be provided with iso-TTI curves in order to delimit the most favorable zones and eras for the formation of oil.

But it was Tissot (1969, 1973) who had the merit of being the first to formalize the natural production of oil and natural gas by thermal cracking of kerogens, in the form of a mathematical model simulable by computer. B. Tissot illustrated the chemical structure of the kerogens in the form of immobile organic fabrics of a dimension such that they are not soluble in organic solvents, and in which the atoms are linked by chemical bonds with different rupture energies. The rupture of these bonds by an increase in temperature produces smaller structures and molecules that can then be soluble in the solvents or mobile in the sediment, such as hydrocarbons.

The thermal cracking of kerogens is accordingly described as two successive sets of independent first order reactions (parallel): the cracking of the bonds of a given type i, *whereof the quantity in a type of kerogen is characterized by its frequency* xi *and occurs at a rate characterized by a frequency factor* A_i *and an activation energy* E_i *(Fig. 11). The temperature depends on the burial depth and hence time for the equation* $T = T_0 + g(z) \cdot z(t)$, *where* T_0 *is the soil temperature,* z(t) *the depth at time* t *and* g(z) *the geothermal gradient at depth* z. *The model can also be used to simulate pyrolysis in the laboratory.*

This model contains "free" kinetic parameters xi, A_i, *and* E_i, *which must be calibrated for use in practice. This calibration was initially done by fitting to the measurements of quantities of hydrocarbons formed as a function of the depth in thoroughly investigated basins like the Paris Basin, and subsequently by laboratory experiments as described below. Naturally, a computer was necessary to perform this task, and only the development of data processing from 1965 allowed this to happen. It was also necessary to accurately reconstruct the temperature variations as a function of time in the basins analyzed, for all the sedimentary levels rich in organic matter and consequently forming potential oil source rocks.*

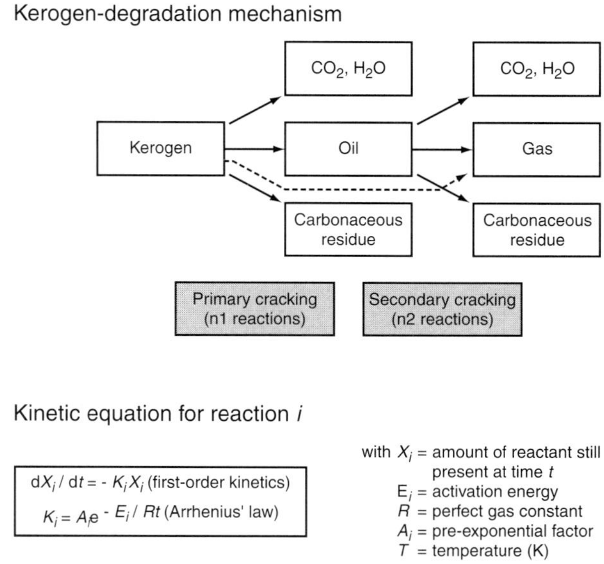

Figure 11

Principles of the formation model of oil and natural gas created by B. Tissot. In Tissot and Welte (1984).
With permission of Éditions Technip.

All the models used today by the oil companies to predict the location of oil reservoirs in the basins are based on this first model. They serve to calculate the quantities of oil and gas formed in the basins explored over geological time, as well as the changes in the locations of this formation.

3.4 Engler's Successors

As a technique for simulating oil formation, pyrolysis earned many disciples following the pioneering work of Engler. The first experiments were conducted in an autoclave with immature sediments rich in organic matter. The products formed were extracted by organic solvents and analyzed by the protocols employed to analyze products found in natural sediments. Subsequently, with the development of gas chromatography, "on-line" analytical techniques were created for products formed from very small samples of sediments during pyrolyses in open or closed media in very small capsules, with or without intermediate condensation in a cold trap, thereby considerably increasing the speed of the operations and varying the experimental conditions.

These experiments revealed the relationship between laboratory pyrolysis and the formation of oil in the basins. The similarity was so great that an experienced eye could recognize the type and stage of maturation of a kerogen by a simple examination of the distribution of its pyrolysis products.

However, differences also existed: far more resins and asphaltenes were formed in the laboratory experiments than in natural sediments, as well as unsaturated hydrocarbons that were not found in oils. The kerogen also lost its hydrogen too early compared with natural maturation. Pyrolyses in the presence of water produced pyrolysates closer to oils than in the absence of water (Lewan *et al.*, 1979). This led to modeling the reactions, but the interest for pyrolysis as an experimental simulation of oil formation gradually declined.

Yet three fields of application of laboratory pyrolysis became extremely productive: the characterization of the nature and stage of maturation of the source rocks of oil, the study of the structure of kerogens, the calibration of mathematical models of oil formation. Hence the precise simulation of oil formation by pyrolysis was abandoned, as this was impossible, given the very different temperature and time ranges in which the two processes take place, to focus on the plotting of correspondence charts between the natural processes of oil formation and the parameters obtained by pyrolysis.

- The first route included the creation of the "fixed carbon" method by Gransch and Eisma (1970) in the *Shell* laboratory in Holland, to characterize the stage reached by a kerogen in the oil formation process: the idea was to measure the carbon present in the kerogen analyzed before and after pyrolysis in standard conditions. Clearly the more a kerogen has produced oil in nature the less these two values differ from each other. The ratio of the two values hence decreases towards 1 as the oil is formed, and this ratio can be used to calibrate the advancement of this formation in a basin. This is a more direct method than the method of D. White, which consisted in using the carbon contents of the neighboring coal levels, which do not necessarily exist in the basin.

  In 1967, at the Belgian company *Pétrofina,* Leplat developed a method in which a flame ionization detector was used to measure the quantities of hydrocarbon compounds produced by isothermal pyrolysis in open medium of very small samples of sediments containing organic matter. The quantities of products formed and the form of the pyrograms provided information about the type and quantity of organic matter present and their stage of maturation. Similar researches were also conducted at the *USGS* by Claypool and Reed (1976).

  Bordenave *et al.* at *Compagnie française des pétroles (CFP)* in 1970, followed by Larter *et al.* in 1977 at the *Organic Geochemistry Laboratory* in Newcastle led by A. Douglas, similarly used a pyrolysis-gas chromatography coupling.

  Yet the most complete form of the characterization of the source rocks by pyrolysis resides in the creation at Institut français du pétrole *of the Rock-Eval® (Durand and Espitalié, 1975; Espitalié* et al., *1977), an instrument that uses pyrolysis at programmed temperature and in open medium of very small samples of ground sediments: the liberation of pyrolysis products is recorded in real time by two detectors, a flame ionization detector which records the production of hydrocarbon compounds, and a thermal conductivity detector which records the production of oxygenated compounds. This makes it possible to rapidly measure the characteristic parameters of the quantity of organic matter present in the samples, its type and its stage of maturation, and thereby to plot logs using rock samples taken by oil boreholes, and yield a measurement of the petroleum potential of the rocks analyzed,* i.e. *the quantities of oil and gas that they can produce during their burial. It is also possible to transfer the results to geological sections and maps. The automation of this instrument, followed by its coupling with a computer equipped with mathematical models of oil formation, designed to use the parameters gathered and, based on them and the geological data like the history of burial and the thermal history of the samples, allows the historic and geographic reconstruction of oil formation in the basins investigated, and rapidly supplied an indispensable tool for oil exploration. The development and success of the Rock-Eval® were due particularly to Espitalié* et al. *at* IFP.

- The second route was a closer understanding of the chemical structure of kerogens. Thanks to this technique, which supplemented the structural analyses carried out by physical processes, or by the study of extracts of sediments and selected chemical degradation products, the representations of the average structure of the kerogens as a function of their type and degree of maturation were clarified (Behar and Vandenbroucke, 1987). However, these average structures were unpopular with the specialists in molecular geochemistry and the petrographers, because kerogens are not formed of a single entity: they accordingly undertook several projects to identify, in the pyrolysis products, the mark of biological entities participating in the formation of the kerogens. Contrary to coals, in which many biological forms are preserved, these entities are invisible to the optical microscope in most kerogens, which consist essentially of amorphous particles. In some cases, however, bacterial membranes are nonetheless visible to the electron microscope (Largeau *et al.*, 1990). A review of many of these investigations can be found in Rullkötter and Michaelis (1990). These efforts, while they demonstrated the variety of biological entities at the source of the kerogens, failed to establish a quantitative inventory.

- In the third route, the concern was basically to determine the "free" parameters xi, A_i, and E_i of the mathematical models of oil formation: in fact, during the exploration of a basin, the data necessary for calibration of the models on the observation of the quantities of oil formed in the sediments are only available late, and these data are largely uncertain. Pyrolysis helps to achieve this calibration by extrapolating the measurements of these parameters made in the laboratory at elevated temperature and with short times to the geological conditions of the formation of oil, *i.e.* low temperatures and very long times (Ungerer and Pelet, 1987).

3.5 The Avatars of the Inorganic Theory of Oil Formation

Despite all these proofs that had been provided of the organic origin of oil, and its confirmation in practice by every day petroleum exploration, the diehards of the inorganic theory persisted. They were generally chemists with very little geological knowledge. The best known belonged to the school of Porfiriev (Sokolov, 1974) in Ukraine: they saw evidence of their theory in the fact that some oil reservoirs exist in nonsedimentary rocks: granitic, metamorphic or porous volcanic rocks. While this was perfectly true, all the studies conducted by geologists had demonstrated that these rocks served as a reservoir for an oil from a sedimentary source rock located not far from there. These "inorganicists" received unexpected support from the American astronomer T. Gold, one of the men responsible for the discovery of quasars. He claimed (Gold, 1979) that natural gas reservoirs resulted from leaks of a primordial methane, *i.e.* formed at the time of the construction of the earth's first atmosphere and trapped at the time in the depths of the earth's crust, or produced from early bitumens buried at the time. These leaks were favored by earthquakes. The gas could be "polymerized" to form oil as it rose from the depths: in brief, the North Sea oil was formed from gas leaks from the Scandinavian shield. This theory met with some success among mineral geochemists, volcanologists and, in general, with geologists unfamiliar with petroleum. But it had even more success in the media, due to the fantastic nature of this theory, and Gold's relentless proselytism. Petroleum geochemists were taxed with scientific conservatism and placed on the accused bench. They received humiliating memoranda from the administrations, embassies and gas companies. In Europe, a joint meeting was specially organized in Brussels. This was an instructive story on the conditions in which the professions of scientific research are sometimes conducted at the present time.

Finally, T. Gold succeeded in securing funds for a borehole in Sweden in the Lake Siljan region. The choice of this site was dictated by the fact that the lake was installed in an extremely ancient meteoritic crater: the impact of the meteorite was presumed to have fractured the rocks at depth, producing a reservoir with gas rising from the depths. The borehole was naturally a failure (Jeffrey and Kaplan, 1988).

3.6 The Mysteries of Primary Oil Migration

While the mechanisms of oil formation were now well understood and modeled, the same did not apply to the mechanisms oil accumulation in reservoirs. Since T.S. Hunt, it was known that these accumulations were preferably found in the anticlinal zones of the sedimentary basins. This was primarily explained by the immiscibility of oil with water: for this reason, in the subsoil, oil was at a separate phase from water which, as we know, is omnipresent in the pores of sedimentary rocks because they are deposited in an aqueous environment. This phase having a specific gravity generally ranging from 0.7 to 0.9, it underwent a buoyancy force which pushed it towards the higher points. The anticlinal structures played the role of a collector and accumulator, as done by a lake for surface runoff. Where the sediment water was in motion, it could sometimes undergo a hydrodynamic thrust in the direction of the water movement. A movement of the oil was opposed by the capillary pressure applying between fluid phases in the porous media, a pressure which is greater as the pores are smaller. In consequence, an oil phase could circulate in the sediments if the thrust exerted on this phase was greater than the capillary pressure. This explained why rocks with very small pores such as shales, had the capacity of stopping the oil in its general upward movement, and why its circulation in the subsoil could only take place in coarse grained rocks, hence exhibiting high permeability, or in fractures and faults. All this had been formalized in 1953 by the American hydrogeologist M.K. Hubbert and updated in 1973 by the French hydrogeologist A. Chiarelli at *SNEA*, among others. The travel of the oil from its source rock to its reservoir rock, sometimes several hundred kilometers away, was therefore understood and modelizable.

The expulsion of the oil from its source rocks, called primary migration, as opposed to secondary migration, which is the flow of the oil in the drains, nonetheless remained a mysterious occurrence: it was found that oil source rocks were extremely fine-grained rocks, chiefly shales and marls, because the sedimentary organic matter, as we shall show further on, can only accumulate significantly in this type of rock. Hence the oil could not be formed in the coarse grained rocks in which it flowed and accumulated. But in this case, how could the oil formed in a very fine-grained rock be expelled from it since this rock was presumed to immobilize it due to the enormous capillary pressures presumed to reign between water and oil? Many tests were conducted to try to demonstrate that the oil was sufficiently soluble in the water to be expelled at the same time as the water during compaction of the sediments, or have the ability to diffuse in the water of their pores. These attempts all failed. Solubility measurements of hydrocarbons in water (McAuliffe, 1966, 1980) and diffusion coefficients in water saturated clays (Leythaeuser *et al.*, 1982; Krooss, 1986) finally showed that only methane, ethane and benzene could be expelled this way

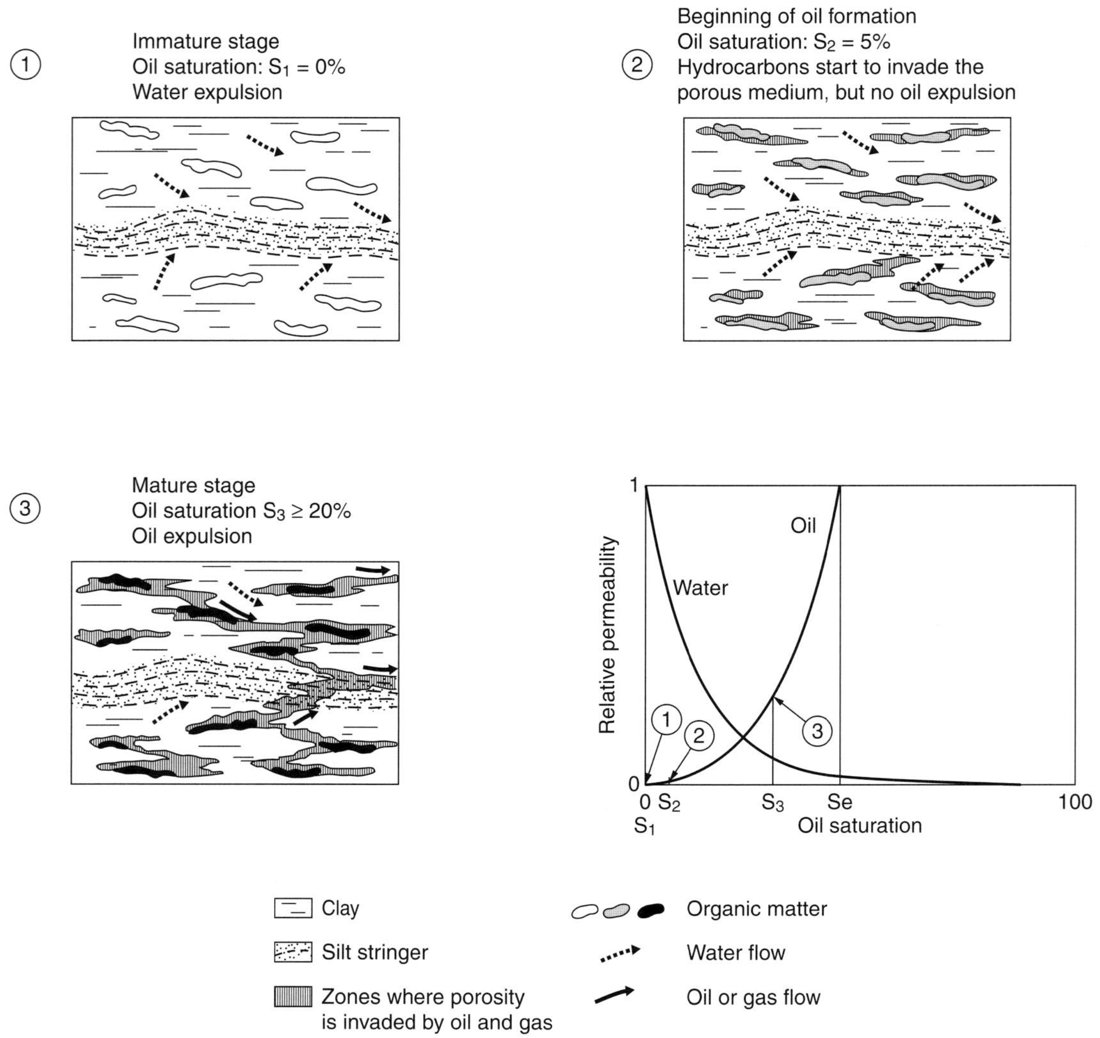

Figure 12

Sketch describing the mechanism of oil and gas expulsion from their source rock: 1) during diagenesis, no oil or gas is produced in the source rocks; the porewater is expelled by compaction; 2) at the start of catagenesis, not enough oil is formed to produce a continuous network in the source rock. The capillary pressure prevents its expulsion, but water continues to be expelled; 3) once an oil network is formed in the source rock, the oil starts to be expelled. An oil saturation threshold in the pores of the source rock must be reached, expressed by the curves of permeability to oil and water in the inset. After Ungerer (1993), in Bordenave *et al.* (1993).
With permission of Éditions Technip.

in appreciable quantities from the source rocks. Besides, the hydrocarbon distribution in the reservoirs absolutely failed to correspond to that which their solubility would have dictated, and the observation and modeling of oil formation showed that this formation took place at depths where compaction had already expelled from the source rocks most of the water that they initially contained.

It was therefore necessary to conclude that the expulsion of oil took place in a separate phase from water. But how, since considerable capillary forces existed in the source rock opposing its movement?

This dilemma was resolved by Ph. Ungerer at Institut français du pétrole. *The diagram in Figure 12 shows how he claims expulsion takes place: the source rock, which originally contains an immature kerogen, is progressively buried and exposed to increasing temperature and geostatic pressure. The pressure in the more porous and permeable rocks surrounding it is hydrostatic, which means that the pressure in the pore fluids of the source rock can, if they are expelled, reach up to 2.5 times the pressure of the fluids in the pores of the drains. As long as the kerogen is in the immature zone, there is no oil in the pores of the source rock. Since the pressure in the source rock pores is higher than the pressure in the pores of the surrounding rocks, the water they contain flows into the neighboring permeable rocks, which play the role of drains, and the volume of the source rock therefore decreases.*

Since the depth continues to increase, oil begins to form in the source rock, but not sufficiently to form a continuous phase. Hence the capillary pressure opposes its expulsion. The water continues to be expelled. In a final stage, most of the water having been expelled and the oil having finally formed a continuous phase in the source rock or at least in certain portions thereof, no capillary forces exist opposing the expulsion of the oil, and it is finally expelled from the source rock by the excess pressure of fluid over the surrounding permeable rock. Clearly the oil is expelled later after its formation as the organic matter of the source rock is lower.

These mechanisms have the advantage of being mathematically formalizable by equations of two-phase flow and the concept of relative permeability of the water and oil, as currently used in oil production modeling. An oil saturation threshold had to be reached in the source rock pores for expulsion to take place. The first mathematical model coupling oil formation and migration was then created and presented for the first time at the 11th World Petroleum Congress in London (Durand et al., *1984). Its principles provided the basis for all the models that were subsequently developed by the oil companies.*

Progress in modeling oil formation soon helped to determine not only the quantities of oil formed during maturation, but also its composition. It then became possible to model interferences between oil formation and its expulsion from the source rock: clearly the greater or lesser effectiveness of expulsion governs the residence time of the oil that is formed in the source rock. If expulsion is very easy, the system is open and most of the oil formed is found in the drains as it is formed; if expulsion is very difficult, the system approaches a closed system: the oil formed remains longer in the source rock: as the temperature and/or time increase, the oil is cracked into lighter products before being expelled (Ungerer *et al.*, 1988).

3.7 The Creation of Organic Petrology

At the beginning was palynology and coal petrography!

Palynology etymologically means science of dust: its object is to inventory the microscopic organic debris observable in the rocks and, inter alia, the spores and pollens which are excellent markers of climates and fossil environments. This is why it is widely used in archeology and sedimentology. It is part of the routine panoply of the sedimentologists who investigate the rocks crossed by petroleum boreholes.

Palynology became important for organic geochemistry following the work of Combaz (1964) at *Compagnie française des pétroles*, of Gutjahr (1966) at *Shell*, and of Correia (1967) at *Institut français du pétrole*.

The first used a microscope to observe, in organic debris isolated from sedimentary rocks after the destruction of their minerals (by hydrochloric acid for carbonates and by hydrofluoric acid for silicates), regularities in proportions of the different components: spores, pollens, wood fragments, plant cells, amorphous organic matter which is the majority component in most cases but which may have widely varying appearances: floc, granular, film, etc. (Combaz, 1980). He called palynofacies the diagram of the proportions of the different components. This palynofacies is a signature of the sedimentary formations crossed in drilling and can therefore serve, inter alia, to correlate these formations together. To some extent, it is also possible to approximate the petroleum potential of the organic matter present. For example, an organic matter very rich in higher plant debris very probably contains a type 3 kerogen, hence with low petroleum potential.

The latter two, by observing spores and pollens isolated from samples of sediments from petroleum basins, found a progressive blackening of them with depth, and identified classes of increasing coloration which they then compared with the location of the fields. They demonstrated the existence of a correlation between these classes and the steps of oil formation. This approach followed the same principle as those of D. White and M. Teichmüller.

These methods were soon used systematically and improved in the oil companies. A number of coal petrographers then took an interest in oil. We have already mentioned the work of M. Teichmüller at the *Krefeld Research Center* in Germany. From 1965, steady cooperation was established between certain coal petrographers and petroleum geochemists, leading to the emergence of a synthetic discipline, organic petrography, which includes all microscopic methods of investigation of rock organic matter, whether dispersed in the rocks or concentrated as in the case of coals. Pioneers of this organic petrography, apart from M. Teichmüller and her colleagues (Teichmüller, 1986), include B. Alpern at the *Charbonnages de France Research Center (CERCHAR)* at Creil (Alpern *et al.*, 1972), A. Combaz at *CFP*, *P*. Robert at *SNEA*, P. van Gijzel at the *University of Nijmegen* in Holland, N.H. Bostick of the *Illinois State Geological Survey* in the United States, and H. Jacob at the *BGR* in Germany.

1973 witnessed in the premises of the CNRS *in Paris, under the leadership of B. Alpern, a symposium entitled: "Petrography of organic matter of sediments, relations with paleotemperature and petroleum potential", whose title clearly explains the major concerns of the discipline: apart from the need to class the organic material of the rocks as a function of its morphology and origin, the aim was to measure parameters under the microscope connected with the effect of temperature on the organic matter of the rocks and hence the maturation, and parameters allowing an estimate of the quantities of petroleum that this organic matter could produce during its burial, which is called the petroleum potential. The proceedings of this symposium were published in 1975 (Alpern, 1975).*

The *International Committee for Coal Petrography (ICCP)* then created a "Dispersed Organic Matter (DOM)" commission, in which the organic petrographers presented their work. A rival society, the *Organic Petrology Society* was created later.

Vitrinite reflectance then became the reference parameter for petroleum geochemists to calibrate the advancement of maturation. Other parameters consequently came into routine use, such as the Thermal Alteration Index (TAI) which measures, following the work of M. Correia, the degree of blackening of organites in relation to the stage of maturation, or measurements of the intensity or color of fluorescence of these organites. Correspondence tables between these parameters and stages of oil formation were proposed.

Maceral analysis, by which the proportions of the different macerals of a kerogen are measured, was used to assess the petroleum potential, and then, for the study of organic matter by reflection microscopy, became the equivalent of the palynofacies for the study by transmission microscopy.

Organic petrography then became an indispensable complement to the physicochemical methods for the analysis of organic matter, because it allowed the observation of the variety and heterogeneity of this matter.

3.8 Towards a Broad View of Organic Geochemistry: Organic Sedimentology, Biogeochemistry and Organomineral Geochemistry

The progress of chemical analysis and the overall approach of geological phenomena permitted by the developments of the theory of plate tectonics allowed organic geochemistry in the 1960s to go beyond the relatively narrow framework of the study of fossil fuels. In a broader view, it gradually became a science of chemical interactions between the biosphere, consisting of living organisms and the products of their degradation after their death, and the geosphere. In these interactions, microorganisms play an important role. It is already in this "biogeochemical" form that Verdnasky had imagined organic geochemistry.

Considered in this way, the formation of fossil fuels is a biogeochemical process. This also applies to the formation of coral reefs and stromatolits, the formation of mollusk shells, the precipitation of carbonates in the oceans, the formation of organic soils and sediments, and the concentration of chemical elements that leads to the formation of useful minerals (Disnar and Sureau, 1990). In the latter case, organomineral geochemistry is alluded to rather than biogeochemistry. We shall illustrate the development of this organic geochemistry in the broad sense of the term by the following examples.

3.8.1 Organic Sedimentology

Most geologists evidenced very slight interest in the study of sediments containing organic matter for a long time, even after they had realized their importance in the formation of oil. Some exceptions were D. White and H. Potonié, who tried in the early 20th century to understand how peats were formed and to identify the factors responsible for their variety. Potonié in 1908 coined the term sapropel for muds rich in organic matter formed in reducing environments.

Also noteworthy is the work of Trask and Patnode (1942), who used the resources of their time to analyze a large number of sediments rich in organic matter, and later on, the work of Hunt (1963, 1972).

However, all of this failed to create a genuine organic sedimentology, in other words, the study of the logic of distribution in time and space of the sediments rich in organic matter.

An important step in this direction was made by the oceanographers, who started by measuring the organic carbon content of recent sediments and relating them to the productivity of surface waters for phytoplankton and the physiography of the sedimentary environments. In France, a pioneer was J. Debyser, a pupil of Bourcart, who was an oceanographer with an interest in depositional modes of marine muds. J. Debyser in 1959 presented his thesis on the geochemistry of marine muds. He was also one of the promotors of the creation of the organic geochemistry department of *IFP*.

After this, Datsko (1959, 1961), Pakhomova (1961), Bordovsky (1965), Calvert and Price (1971), Degens *et al.* (1971, 1973, 1974), among others, investigated the distribution of organic carbon in sediments of a number of bays and closed seas, the Black Sea, the Caspian Sea, the Bering Sea, Walvis Bay, etc., and related them to the productivity of phytoplankton, the oxygen contents of the water column, the direction of currents, the mineralogy of sediments *(Fig. 13)* etc. The Russian oceanographer Romankevitch (1977) plotted a map of the organic carbon content of the sediments sampled in all the oceanographic surveys throughout the world and related them with that of the oceanic productivity of plankton previously determined by E.T. Degens and his colleagues. The oceanographers were joined in the 1970s by organic geochemists, under the pressure of petroleum explorers who wanted to anticipate the presence and extension of organic rich sediments likely to be source rocks for oil, to secure permits in presumably favorable zones of the basins being explored. The organic geochemists had very little experience at the time with recent sediments, but enjoyed access to ancient sediments and, thanks to Rock-Eval®, possessed a rapid method of investigation. Some of them adapted the soil analysis methods to the sediments (Brown *et al.*, 1972; Huc and Durand, 1974).

Among the most prominent work of the geochemists was that of the *ORGON* and *MISEDOR* research groups (1977-1983), coordinated by R. Pelet and Y. Debyser of *Institut français du pétrole* and A. Combaz and J.L. Oudin of

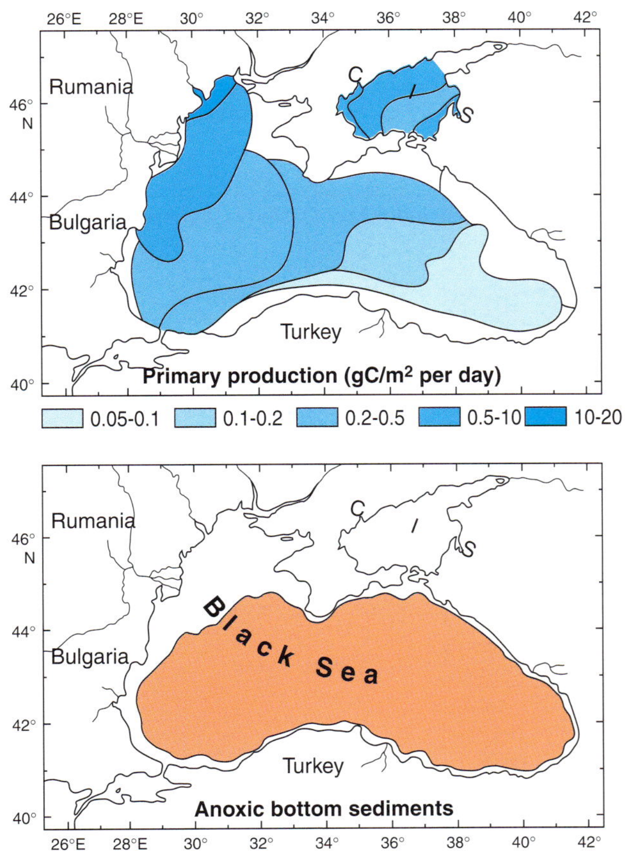

Figure 13

Primary productivity, grain size and organic carbon content of recent sediments, and location of zones where the bottom is currently anoxic in the Black Sea. Document A.Y. Huc after Finenko (1967), Shimkus and Trimonis (1974) and Müller and Stoffers (1974).

Compagnie française des pétroles, which, alongside the researchers of these two institutions, included researchers of several *CNRS* laboratories. These groups organized ocean surveys designed to analyze the composition of organic matter of recent sediments sampled by cores in sedimentary environments of which the physiography was well known. They also made a microbiological analysis of the sediments.

The DSDP (Deep Sea Drilling Project) and ODP (Ocean Drilling Project) campaigns of the American deep water drill ship *Glomar Challenger,* although mainly focused on the study of continental drift, allowed the geochemical analysis of organic rich sediments, recent and less recent, at many places of the globe. Also noteworthy was the work of B. Tissot, G. Deroo and J.P. Herbin at *IFP* and of P.C. de Graciansky at *ENS Mines de Paris* on the formation of organic matter rich sediments associated with the opening of the North Atlantic (Tissot *et al.*, 1979; de Graciansky *et al.*, 1982).

Then G. Demaison of the US company *Chevron* formally established the links between the plankton productivity of the water, the oxygen content of the sedimentation environment, the transport distance of organic debris from the surface to the sediment, and the quantity of organic matter preserved in the sediments (Demaison and Moore, 1980). Demaison highlighted the vital role played by anoxia, whether created by the lack of oxygen in the water of the sedimentation environment, or by the abundance of organic debris, in the formation of sediments rich in organic matter. In fact, in anoxic conditions, aerobic microorganisms, those with the most destructive action on organic debris, cannot develop.

It was found that only 1% on average of the carbon in the biomass annually produced at the surface of the oceans (primary productivity) escaped mineralization to carbon dioxide to be incorporated in the sediments, and that the sediments containing organic matter, future source rocks for oil and natural gas, were very fine-grained sediments, shales or marls; partly because the hydrodynamic properties of organic debris are close to those very fine minerals and because they can adsorb organic molecules, and also, and above all, because the fineness of the grains prevents the water from circulating in the sediment and hence restricts the access of oxygen to the organic matter.

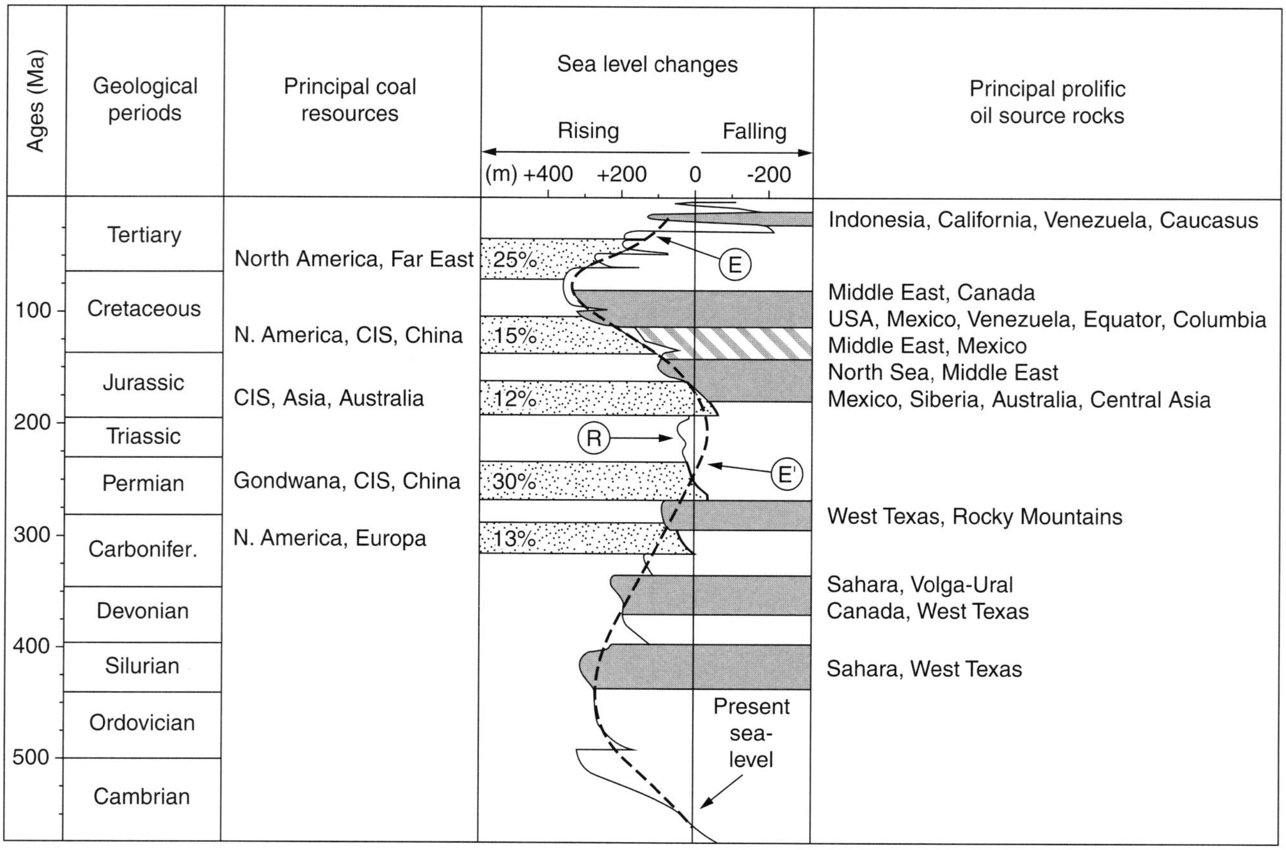

Figure 14

Distribution of the main oil source rocks and main coal series as a function of geological time, related to fluctuations in sea level. After Tissot (1979).

This gave rise to a much more accurate idea of environments favorable to the formation of sediments rich in organic matter, enabling the geologists of the oil companies to anticipate their presence in the basins from paleogeographic findings: these environments are closed seas, bays, lagoons, lakes, where the access of oxygen to the sedimentation environment is restricted, and coastal zones with upwelling. River deltas installed in areas of abundant vegetation are also very favorable environments, because the very abundant plant debris is quickly buried and therefore escapes the action of aerobic bacteria. The reference types of kerogen corresponded to three different types of sedimentation: evaporitic lacustrine environment for type 1, anoxic marine environment for type 2, deltaic or detrital coastal environment for type 3.

It was finally recognized that certain geological eras were much more favorable than others to the formation of organic matter rich sediments, which obviously have consequences on the distribution of oil and coal: in particular, the periods of the great transgressions, because they caused the formation of shallow closed seas where anoxic conditions were easily created, were favorable to the existence of source rocks for type 1 or 2 kerogen, potentially the most productive for oil; while the periods of the great regressions, during which detrital sedimentation developed, were favorable to the formation of coal (Tissot, 1979, Fig. 14*), Bois* et al. *(1982).*

3.8.2 Sedimentary Microbiology

Microorganisms were found to play a major role not in the formation of oil as believed by ZoBell, but in that of organic sediments for oil, and in the degradation of oils.

Bacterial debris can form an important share of the kerogen. This is the case for example of type 1, largely formed of membranes of bacteria which developed on debris of lacustrine algae. Aerobic microorganisms are highly active in the destruction of organic debris, which are mineralized into carbon dioxide. *Methanogenic bacteria develop in sedimentary environments devoid of oxygen, sulfates and nitrates, and produce methane* (CH_4). *This methane, whose*

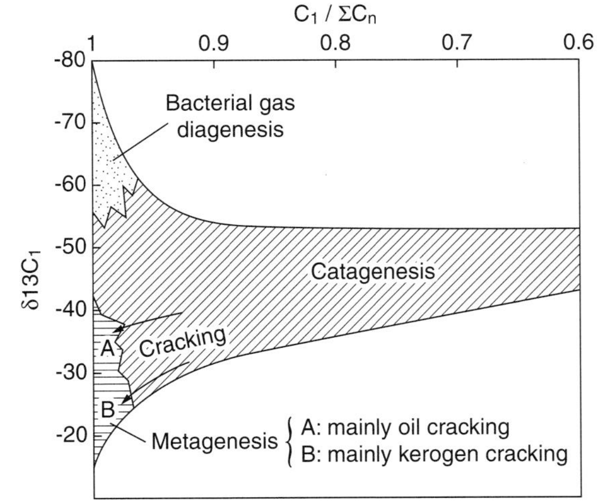

Figure 15

Situation of bacterial gas and thermal gas in a diagram where the proportions of methane in the gaseous hydrocarbons in the *x*-axis and the $\delta\ ^{13}C$ of methane carbon on the *y*-axis. The bacterial gas is characterized by a very negative $\delta\ ^{13}C$ and the absence of gaseous hydrocarbons other than methane (after Tissot and Bessereau, 1982; Tissot and Welte, 1984).

With permission of Springer-Verlag.

formation mode is similar to that of biogas produced by anaerobic fermentation of organic waste, is according to Rice and Claypool (1981), making up to 20% of the gas fields, and according to Matavelli and Novelli (1988), it even forms most of the gas found in Italy. This "bacterial" gas (Vially et al., *1992) also called "biogenic" gas, is distinguished from the "thermal" gas formed by the maturation of kerogen by the absence of other hydrocarbons than methane, and by a smaller C_{13} content of the carbon in the methane ($\delta^{13}C$ much more negative)* (Fig. 15).

Sulfate reducing bacteria, which develop in sulfate rich highly reducing environments, have the property of producing hydrogen sulfide: in detrital environments rich in iron, these can combine with the iron to form hydrotroilite which is rapidly converted to pyrite. This mineral is embarrassing for the coal industry because it liberates sulfur dioxide during its combustion. This gas represents the source of acid rain, which has acidified the lake waters in northern Europe and which has, probably wrongly, been blamed for the destruction of the forests in these regions. In calcareous environments poor in iron, hydrogen sulfide combines with organic matter to yield sulfur rich kerogens, which during their maturation produce sulfur containing oils unpopular with refiners.

Microorganisms have the capacity to degrade oils in their reservoirs (Evans et al., *1971; Deroo* et al., *1974; Connan, 1984), to the point where 20 to 30% of the oils have thus been degraded at one time or another in their history. They are also capable of degrading the oil spread during the spills as well as fuels stored in precarious conditions. These microorganisms are still insufficiently known at the present time.*

It was recently suggested that about half of the earth's biomass could be found in the form of microorganisms present in sediments. Hence a great deal still needs to be done in sedimentary microbiology.

3.8.3 Two Examples of Organomineral Geochemistry: Thermosulfate Reduction and the Formation of Uranium Deposits

Thermosulfate reduction is the reduction of the sulfates present in evaporated sediments by hydrocarbons, as soon as the temperature reaches about 100 to 150°C. Hence the process takes place at great depth in reservoirs where the reservoir rock contains sulfates. The reduction of sulfates produces carbon dioxide and hydrogen sulfide. Since the reservoir rocks of this type are generally carbonates or dolomites which contain no iron, the hydrogen sulfide cannot form pyrite and remains stable. These reservoirs hence contain large proportions of hydrogen sulfide. Some examples are the Lacq reservoir and the reservoirs of the north of the Caspian Sea. It is still difficult to describe this process other than qualitatively (Goldstein and Aizenshtat, 1994).

The formation of uranium minerals owes a great deal to organic matter (Landais *et al.*, 1990; Oh *et al.*, 1990), which acts specifically in two ways in the concentration of uranium:

- "Passively" by creating reducing conditions to facilitate the precipitation of uranium, which is known to be soluble in oxidizing medium and insoluble in reducing medium. For example, petroleum explorers are fully aware that certain organic sediments, like the Kimmeridge-clays in the North Sea and the Silurian clays in the Sahara can be identified easily by γ-ray probe because they contain significant amounts of uranium. Wireline loggers have not hesitated to generalize this observation by recommending the use of this probe to detect formations rich in organic matter. However, the presence of uranium also depends on the uranium supply by the leaching of an emerged granitic formation located nearby during sedimentation, which is not the general case. This process, however, is the origin of many uranium deposits.
- "Actively" by its adsorption and chelation properties with respect to certain metals, and particularly uranium.

3.8.4 The Dustbins of Geological History and the Evolution of Living Beings

The chemical structure of the biomarkers found in sediments and oils is, as we have seen, related to that of the functional molecules of organisms, and, by digging the dustbins of geological history, it is possible somehow to track the evolution and activity of living beings.

Thus for example, biochemists have explored and identified hopanoids in living beings after they were found in sediments (Ourisson et al., *1979), and a marker of the appearance of flower plants, 18* α (H) *– Oleanane, was identified in the tertiary sediments of the Niger delta (Ekweozor* et al., *1979). McKirdy* et al. *in 1983 identified markers of the organisms of the time in pre-Ordovician oils. This aspect was also illustrated by Seifert and Moldowan (1981) and by Reed* et al. *(1986). It deserves further development.*

A much longer list could be compiled of the study topics of this organic geochemistry in the broad sense of the term, and the lack of a greater concentration of research resources in this field at the present time is to be regretted.

The period of expansion of organic geochemistry also saw the emergence of a number of reference works: in 1978 *Petroleum Formation and Occurrence* by B. Tissot and D.H. Welte, a book that played a considerable role in the development of organic geochemistry, in 1979 *Petroleum Geochemistry and Geology* by J.M. Hunt, in 1980 *Kerogen, Insoluble Organic Matter of Sedimentary Rocks,* by B. Durand *et al.* and in 1985 *Geochemistry in Petroleum Exploration* by D. Waples.

4 1985 - MATURITY AND INTEGRATION

In 1985, all the major oil companies, including the national companies of the leading producing countries, had organic geochemistry laboratories, and the teams were often trained in the pioneer laboratories. Laboratories were also found in the public research institutions dedicated to fossil fuels just about everywhere. At the Petroleum Congress, R.J. Murris of *Shell* made a flattering assessment of the impact of organic geochemistry on exploration results. The *Afro-Asian Society of Organic Geochemistry* was created at Dehra Dun in India in 1986, and the *Latin American Society of Organic Geochemistry* was formed in 1988 at Rio de Janeiro in Brazil.

1989 saw the 14th International Meeting on Organic Geochemistry at the *Maison de la Chimie* in Paris. It gathered together 560 participants, a historic record. Half of the proceedings of this congress was devoted to petroleum geochemistry alone, and the remainder to other subjects, mainly molecular geochemistry, as well as organic sedimentolgy, structure of kerogens, geochemistry of organic sulfur, organomineral geochemistry, isotopic analysis, and analytical techniques.

Books were published such as the *Applied Organic Geochemistry* by Bordenave *et al.* (1993), *Organic Geochemistry, Principles and Applications* by Engel and Macko (1993), *Petroleum and Basin Evolution. Insights from Petroleum Geochemistry, Geology and Basin Modelling* by Welte *et al.* (1997) and works on biomarkers and their use (Connan, 1987; Moldowan *et al.*, 1992; Peters and Moldowan, 1993).

4.1 Petroleum Systems and Basin Simulators

The 1990s witnessed the triumph of the concept of the petroleum system among petroleum explorers, a concept dating back to the 1970s, but only formalized much later by G.J. Demaison (Demaison, 1984; Demaison and Huizinga, 1991).

In this concept, the basins are considered as immense physicochemical reactors, which change slowly during geological history; they contain one or more petroleum systems consisting of a source rock, reservoirs, drains allowing the circulation of oil from the source rock to the reservoir (permeable sediments such as sandstones and coarse grain limestones, faults) and permeability barriers (impermeable sediments such as shales and evaporites, fault closures). The presence of oil in a reservoir is the result of a whole sequence of more or less probable events spreading over time intervals from several tens to several hundred millions of years. This sequence includes:

- *deposition of a sediment rich in organic matter;*
- *burial of this sediment in thermal conditions allowing maturation of the organic matter that is adequate for the formation of oil and gas;*
- *migration and accumulation, thanks to the existence of a system of drains, of the oil and gas formed towards structures permitting their reception (traps);*
- *preservation of the oil and gas accumulated in these structures (good tightness of the permeability barriers, absence of biodegradation).*

The absence of a link in the chain makes the system inoperative. The "load" of these petroleum systems consists of the oil and gas produced by maturation of the kerogens present in the source rocks, and varies considerably according to the systems. The "impedance" of the petroleum systems is the resistance of the sedimentary layers to the circulation of the oil; it also varies widely from one system to another. The petroleum interest of a sedimentary basin can be assessed from the geological parameters with respect to known petroleum systems and also by the product of the probabilities of existence of each of the links in the chain.

The quantification of the functioning of petroleum systems during their geological history, and hence the location of the basin areas that are most favorable to the accumulation of oil, can be formed using "basin simulators", which offer a deterministic computer simulation of the entire sequence of events.

Such simulators were developed in particular at *IFP* by Ungerer *et al.* (1990), Burrus (1997) and Schneider *et al.* (2000) and at the *KFA* in Jülich in Germany under the direction of D. Welte. The oil majors also developed their own tools, first with the help of the above two groups.

The construction and application of these models, comparable in size to meteorological models, demands outstanding computer engineering and the availability of very powerful computers. They also demand the availability of geological and physical data that are as numerous and accurate as possible, whereof the consistency can be checked by the model. Hence basin models are a powerful incentive to construct high quality databases in the companies. They are also a powerful incentive to research in relatively unknown areas, such as the thermicity of sedimentary basins, as the thermal regime closely conditions the rate of oil formation.

These simulators are also places for the integration of a large number of disciplines. Petroleum geochemists are therefore only one of the parties involved. One of their specific roles is to supply the most accurate possible parameters of the kinetics of oil formation. To escape the uncertainties and slowness of methods for validating kinetic models by using measurements on samples, they had already developed methods to calculate kinetic parameters from laboratory experiments (Ungerer and Pelet, 1987). These methods saw considerable development and became increasingly accurate (Béhar *et al.*, 1997). They were also expected to provide a number of controls by "field" measurements such as that of the composition of the oil and organic matter in the sediments, and of the maturation parameters.

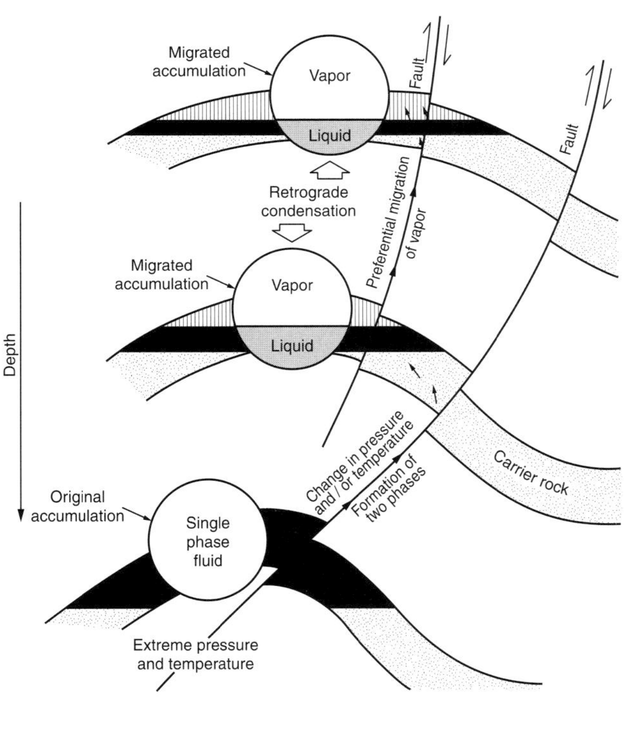

Figure 16

Sketch showing an example of multiphase behavior of hydrocarbons during their rise to the surface. The initial accumulation is single-phase because of the high temperatures and pressures at depth. The hydrocarbons escape this initial reservoir and join another structure where they form a two-phase reservoir due to decreases in temperature and pressure, etc. In Tissot and Welte (1984) after Silverman (1965).

With permission of Springer-Verlag.

While oil migration was understood in its main features, must work still needed to be done to fully transcribe the mechanisms into a basin simulator. One of the major problems is to take account of the often multiphase character of the movement of petroleum fluids in the sediments. In fact, part of the oil formed in the source rocks, primarily resins and asphaltenes, remains there in a solid or adsorbed state. The mobile portion can form one or more "oil", "gas", or "bitumen" phases in the pores of the drains and reservoirs, depending on its composition, the local thermodynamic conditions and the residence time in these thermodynamic conditions. These processes, previously investigated by Silverman (1965), *(Fig. 16),* Nogaret (1983) and Ungerer *et al.* (1983, 1988), among others, have to be "attuned" throughout the migration paths from the source rock to the reservoirs. The inclusion of this "geochemical thermodynamic" is a particularly difficult task, which is not yet fully achieved at the present time.

4.2 Sequential Stratigraphy and Organic Sedimentology

In the 1990s, interest in the organic aspects of sedimentology grew among sedimentologists, as part of a general framework of the questioning of the methods of sedimentology, initiated by the work of P. Vail of *Exxon,* which led to a renaissance of sequential stratigraphy. *This consists in predicting, at a given place, the type, succession and extension of sedimentary deposits as a function of the type and scale of the inputs, the nature of the transport mechanisms, and the space available at any time to receive the sediments (accommodation). This space is itself a function of the subsidence of the basin, the compaction rate of the sediments, and fluctuations in sea level (eustatism).* Renewed interest in organic sedimentology also emerged among oceanographers and climatologists for the reconstruction of ancient climates and geographys, in view of the relations existing between paleogeography, paleoclimates and the preservation of organic matter in the sediments. The major contributions to these researches included J. Parrish, E. Barron and W. Hay in the United States (Huc, 1995).

Several important works on organic sedimentology were published: *Petroleum Source Rocks* by Katz (1995), and special publications of the major petroleum journals: *AAPG Studies in Geology* no. 37 *Source Rocks in a Sequence Stratigraphic Framework* published by Katz and Pratt (1993), the *AAPG Studies in Geology* no. 40 *Paleogeography, Paleoclimate, and Source Rocks* published by Huc (1995) and the *SEPM Special Publication* no. 60 *Mesozoic and Cenozoic Sequence Stratigraphic of European Basins* by de Gracianski *et al.* (1998).

It became possible to clarify the paleogeography and the quantitative importance of the sediments with organic matter at different geological times (Klemme and Ulmishek, 1991). The most favorable were the Silurian, the Upper Devonian-Lower-Carboniferous sequence, the Upper-Carboniferous-Lower-Permian sequence, the Upper Jurassic, the Middle Cretaceous and the Oligocene-Miocene. At a smaller scale, relations between climatic cycles, variations in sea level and deposits of organic matter were identified, and precise correlations between levels rich in organic matter were made at the scale of the sedimentary basins (G. Bessereau et al., *1995; van Buchem* et al., *1995; B. Duval* et al., *1998; among others).*

4.3 The Decline of Petroleum Exploration

In the 1990s, discoveries of major fields became rare. Then the sudden drop in the price of oil in 1998 led the oil companies to reorganize to contend with a low price per barrel. This was the time of the mergers and acquisitions. Despite the strategic importance for the companies of the renewal of their reserves, they progressively focused on production and made swingeing cuts in exploration. Most of the organic geochemistry laboratories were closed down. In the remaining laboratories, the staff was reduced and the work almost exclusively focused on service activities for subsidiaries.

Yet new topics emerged almost immediately in petroleum exploration:

- After long years of a nearly total disinterest, particularly due to the cost of its transport, natural gas very quickly became an energy source competing with oil. In fact, the development of cogeneration techniques with gas helped to obtain very high energy efficiencies in power generation, and gas demand grew sharply. Its flexibility of use and less pollutant combustion than that of oil also favored it.
- Following the opening to international companies of a number of major producing countries, the western companies explored new regions, such as the basins of the Caspian Sea.
- Following the discoveries of large deep sea reservoirs off the Gulf of Guinea and Gulf of Mexico, specific deep offshore exploration was developed.
- Following the discovery of oil reservoirs in the deep zones of certain basins where high pressures and high temperatures prevailed, whereas gas was expected to be found, these zones aroused renewed interest in exploration.

Petroleum geochemistry adapted to this new situation by developing specific researches on these topics. At *Institut français du pétrole,* this included the work of Prinzhofer and Huc (1995) and Prinzhofer *et al.* (2000) on the geochemistry of hydrocarbon and non-hydrocarbon gases, the work of Huc *et al.* (2001) on deep offshore organic sedimentation, and the work of the modeling group on basin modeling in high temperature and high pressure zones.

It went also beyond its traditional field, exploration, to develop a production geochemistry, in which the priority topics are the identification of the different compartments in the same reservoir (England, 1990; Smalley and England, 1992; Kaufman *et al.,* 1990), the location of asphaltic levels (Huc *et al.,* 1999), and the study of biodegradation mechanisms. These studies were aimed to facilitate the recovery of the oil present in the reservoirs.

Petroleum geochemistry also developed into the field of the detection and remediation of soils polluted by hydrocarbons.

CONCLUSIONS

Organic geochemistry is a very young field compared to most other sciences. Its history has only lasted two to three generations of researchers for the time being. It is therefore possible at one glance to encompass the conditions in which it was created and developed.

It is to petroleum, particularly to petroleum exploration, that it owes its birth and its main development. It is therefore economic demand and not a search for basic knowledge that was the main drive of its creation. It is the rational observation of nature, more than theoretical developments, which played a decisive role in the understanding of the mechanisms of oil formation. The spectacular expansion of the technology of analytical methods since 1960 (separation techniques, elementary analysis, gas chromatography, all sorts of spectrometric methods, GC-MS coupling) played a vital role by allowing the separation and identification of a large number of molecules and characterization of complex products. The development of data processing helped to manage the abundance of data thus gathered, to construct models and to apply them to concrete cases.

It was observed that many researchers who had been interested in the origin of oil and the mechanisms of its formation were confused because they remained imprisoned by the knowledge and ways of thought of their discipline and their specialty: chemists against geologists (and even "chemist" geochemists against "geological" geochemists) specialists in coal, oil shales, of upstream and downstream petroleum operations.

It is the researchers of the industry or close to the industry rather than university researchers who were responsible for the shaping of organic geochemistry. This is no doubt largely explained by the fact that university teams, unless they have created links with the industry, rarely had the essential samples and geological documents. We also find that the university system, organized into small teams specialized and put in competition within the same discipline, does not foster multidisciplinarity. This is better organized in industry, where work is done by project rather than by specialty. University geologists, much more than university chemists who have made considerable achievements in molecular geochemistry, were absent to this adventure. It is also observed that at the present time, organic geochemistry is still only rarely taught in geology departments, despite the scale of the documentation available today.

Even more than those of the oil companies, the research teams of the large professional institutions were responsible for the main breakthroughs, as was also the case for coal science. This is partly explained by their capacity to mobilize a sufficient number of researchers on a project and to understand industrial concerns, and also by the time they had in comparison with the oil company researchers who had immediate and operational concerns. This probably explains the preponderance of Europe, with its numerous institutes, in the development of organic geochemistry from 1960. Among these institutes, *Institut français du pétrole* at Rueil-Malmaison *(IFP)* and *Kernforschungsanlage (KFA)* at Jülich in Germany, were particularly active and productive. And if we only mention two names among the great founders and leaders of organic geochemistry, they are certainly those of B. Tissot, who created the organic geochemistry department of *IFP,* and D. Welte, creator of the *Institut für Erdöl und Organische Geochimie* of the *KFA*.

Organic chemistry was born of oil and will perhaps disappear with it. But it still has a fine season ahead of it.

The death of oil has been announced, but not before the end of the century. It will first be necessary to maintain an exploration effort to delay the inevitable hegemony of the Middle East and this effort will be made in increasingly difficult conditions. Petroleum geochemistry in this framework will be more than ever useful for economizing very costly drilling operations. The knowhow acquired will then have to be transposed to the prospects of the Middle East, which is an underexplored region. Production geochemistry will also be to some extent the successor of exploration geochemistry.

Heavy oil, tar sands, oil shales and coals will by the end of the century make up for most of the remaining reserves of fossil fuels. Organic geochemistry can play a role in their utilization, if they still turn out to be competitive with the other forms of energy.

Organic geochemistry in the broad sense of the term: organic sedimentology, sediment microbiology, molecular geochemistry, organomineral geochemistry, etc. is bound for major developments. It is important to secure a better understanding of climatic processes, of what happens to anthropic pollutants in the water, soil and sediments, of the evolution of living beings. The long term future of organic geochemistry probably lies in the contribution to the resolution of these highly complex problems, in which a geological as well as chemical view is too little emphasized for the time being, and the drivers of this research will be the growing concern for the environment and, we hope, the taste for knowledge.

REFERENCES

Abelson, P.P. (1963) Organic Geochemistry and the Formation of Petroleum. *6 th World Petr. Cong. Proc.*, **1,** 397-407.

Albrecht, P. and Ourisson, G. (1969) Diagenèse des hydrocarbures saturés dans une série sédimentaire épaisse (Douala, Cameroun). *Geoch. Cosmochim. Act.*, **33,** 138-142.

Albrecht, P. and Ourisson, G. (1971) Biogenic substances in sediments and fossils. *Angew. Chem.*, Internat. Ed., **10**, 209- 286.

Allred, V.D. (1966) Kinetics of Oil Shale Pyrolysis. *Chem. Eng. Prog.*, **62**, 55-60.

Alpern, B. (ed.) (1975) *Pétrographie de la matière organique des sédiments, relations avec la paléotempérature et le potentiel pétrolier*, Éditions du CNRS, Paris.

Alpern, B., Durand, B., Espitalié, J. and Tissot, B. (1972) Localisation, caractérisation et classification pétrographique des substances organiques sédimentaires fossiles. In *Advances in Organic Geochemistry 1971* (H.R. von Gaertner and H.R. Wehner eds.), Pergamon Press, 1-28.

Baker, E.G. (1959) Origin and Migration of Oil. *Science*, **129,** 871-874.

Béhar, F. and Vandenbroucke M. (1987) Chemical Modelling of Kerogens. *Org. Geochem.*, **11,** 15-24.

Béhar, F., Vandenbroucke, M., Tang,Y., Marquis, F. and Espitalié, J. (1997) Thermal Cracking of Kerogen in Open and Closed Systems: Determination of Kinetics Parameters and Stoechiometric Coefficients for Oil and Gas Generation. *Org. Geochem.*, **26,** 321-339.

Berthelot, M. (1866) Sur l'origine des carbures et des combustibles minéraux. *Ann. Chim. et Phys., 4e série*, **9,** 481-483.

Bessereau, G., Guillocheau, F. and Huc, A.Y. (1995) Source Rock Occurrence in a Sequence Stratigraphic Framework: The Example of the Lias of the Paris Basin. In *AAPG Studies in Geology,* **40,** A.Y. Huc Ed., Tulsa, Oklahoma.

Bois, C., Bouché, and Pelet, R. (1982). Global Geological History and the Distribution of Hydrocarbon Reserves. *Bull. Am. Assoc. Petr. Geol.*, **66,** 1248-1270.

Bordenave, M., Combaz, A. and Giraud A. (1970) Influence de l'origine des matières organiques et de leur degré d'évolution sur les produits de pyrolyse du kérogène. In *Advances in Organic Geochemistry, 1966*, G.D. Hobson and G.C. Speers Eds., Pergamon Press, London, 389-405.

Bordenave, M.L. (ed.) (1993) *Applied Petroleum Geochemistry*, Éditions Technip, Paris.

Bordovsky, O.K. (1965) Accumulation and Transformation of Organic Substances in Marine Sediments. *Marine Geology*, **3,** Elsevier, Amsterdam, 3-114.

Braun, R.L. and Rothman, A.J. (1975) Oil Shale Pyrolysis: Kinetics and Mechanism of Oil Production. *Fuel*, **54,** 129-131.

Bray, E.E. and Evans, E.D. (1961) Distribution of n-Paraffins as a Clue to Recognition of Source Beds. *Geochim. Cosmochim. Acta*, **22,** 2-15.

Breger, I. (ed.) (1963) *Organic Geochemistry*, Pergamon Press, Oxford.

Brown, F.S., Baedeker, M.J., Nissenbaum, A. and Kaplan, I.R. (1972) Early Diagenesis in a Reducing Fjord, Saanich Iinlet, British Columbia III. Changes in Organic Constituents of Sediment. *Geochim. Cosmochim. Acta,* **36,** 1185-1203.

Buchem van, F.S., Boer de, L., McCave, I.N. and Herbin, J.P. (1995) The Organic Carbon Distribution in Mesozoïc Marine Sediments and the Influence of Orbital Climatic Cycles (England and the Western North Atlantic), In *AAPG Studies*, 40, A.Y. Huc Ed., Am. Assoc. Petr. Geol., Tulsa, Oklahoma.

Burrus, J. (1997) Contribution à l'étude du fonctionnement des systèmes pétroliers : Apport d'une modélisation bidimensionnelle. *PhD Thesis*, École des Mines de Paris, 473.

Calvert, S.E. and Price, N.B. (1971) Recent Sediments of the South-West African Shelf. In *Atlantic Continental Margins, London Inst. Geol. Sc.*, 175-185.

Chiarelli, A. (1973) Étude de nappes aquifères profondes-Contribution de l'hydrogéologie à la connaissance d'un bassin sédimentaire et à l'exploration pétrolière. *PhD Thesis*, Univ. Bordeaux, 401.

Claypool, G.E. and Reed, R. (1976) Thermal-Analysis Technique for Source Rock Evaluation: Quantitative Estimate of Organic Richness and Effects of Lithologic Variation. *Bull. Am. Assoc. Petr. Geol.*, **60,** 608-626.

Combaz, A. (1964) Les palynofaciès. *Rev. Micropal.*, 205-208.

Combaz, A. (1980) Les kérogènes vus au microscope. In *Kerogen* (B. Durand Ed.), Éditions Technip, Paris, 55-111.

Connan, J. (1984) Biodegradation of Crude Oils in Reservoir. In *Advances in Petroleum Geochemistry,* Ed. by Brooks, J. and Welte, D.H., Academic Press, London, **1,** 299-335.

Connan, J. (1987) Regards sur la géochimie pétrolière d'aujourd'hui. *Bull. Centres Rech. Exploration-Production* Elf-Aquitaine, **11,** 103-115.

Correia, M. (1967) Relations possibles entre l'état de conservation des éléments figurés de la matière organique et l'existence de gisements d'hydrocarbures. *Oil & Gas Science and Technology,* **22,** 1285-1306.

Datsko, V.G. (1959) *Organic Matter in Soviet Southern Waters,* Ird. Akad. Nauk SSSR Moscow (in Russian).

Datsko, V.G. (1961) Value of Phytoplankton Production in Balance of Organic Material in Marine Reservoir in Azov, Black and Caspian Seas. In *Primary Production of Seas and Inland Waters-Minsk* (in Russian).

Debyser, J. (1959) Contribution à l'étude géochimique des vases marines. *Thèse,* Université de Paris.

Debyser, J. and Deroo, G. (1969) Faits d'observation sur la genèse du pétrole. *Oil & Gas Science and Technology,* **24,** 21-48.

Degens, E.T., Herzen von, R. and Wong H.K. (1971) Lake Tanganyika: Water Chemistry, Sediments, Geological Structure. *Naturwissenschaften*, **58,** 224-291.

Degens, E.T., Herzen von R., Wong, H.K., Deuser, W.G. and Janash, H.W. (1973) Lake Kivu: Structure, Chemistry and Biology of an East African Rift Lake. *Geol. Rundschau,* **62,** 245-277.

Degens, E.T. and Ross, D.A. (ed) (1974) The Black Sea: Geology, Chemistry and Biology. *Am. Assoc. Petr. Geol.* Memoir 20, Tulsa, 629.

De Leeuw, J.W. (1986) High Molecular Weight Markers.In *Biological Markers in the Sedimentary Record* (R.B. Johns Ed.) Elsevier, Amsterdam, 256-258.

Demaison, G.J. and Moore, G.T. (1980) Anoxic Environments and Oil Source Bed Genesis. *Bull. Am. Assoc. Petr. Geol.*, **64,** 1179-1209.

Demaison, G.J. (1984) The Generative Basin Concept. In *Petroleum Geochemistry and Basin Evaluation,* Ed. by Demaison G.J. and Murris, R.J., *Am. Assoc. Petr. Geol.* Memoir 35, 1-14.

Demaison, G.J. and Huizinga, B.J. (1991) Genetic Classification of Petroleum Systems. *Bull. Am. Assoc. Petr. Geol.*, **75,** 1626-1643.

Deroo, G. (1976) Corrélations huiles brutes - roches-mères à l'échelle des bassins sédimentaires. *Bull. Centre Rech.* Pau, **10,** 317-335.

Deroo, G., Tissot, B., McCrossan, R.G.and Der, F. (1974) Geochemistry of Heavy Oils of Alberta. *Canadian Soc. Petr. Geol.* Memoir 3, Calgary, Canada, Stacs Data Services, 148-167.

Disnar, J.R. and Sureau, J.F. (1990) Organic Matter in Ore Genesis: Progress and Perspectives. In *Advances in Organic Geochemistry* 1989 (B. Durand and F. Béhar Eds), Pergamon Press, 577-590.

Durand, B., Fripiat, J.J. and Pelet, R. (1972) Alkylammonium Decomposition on Montmorillonite Surfaces in an Inert Atmosphere. *Clays and Clay Minerals,* **20,** 19-33.

Durand, B. and Espitalié, J. (1973) Évolution de la matière organique au cours de l'enfouissement des sédiments. *C.R. Acad. Sci.*, Paris, **276,** 2253-2256.

Durand, B. and Espitalié, J. (1975) Méthode et appareil permettant l'évaluation rapide du potentiel de production pétrolière de sédiments géologiques sur la base de faibles prélèvements. *Brevet national 74/14347* (Rock-Eval).

Durand, B. and Espitalié, J. (1976) Geochemical Studies on the Organic Matter from the Douala Basin (Cameroon) II. Evolution of Kerogen. *Geochim. et Cosmochim. Acta,* **40,** 801-808.

Durand, B., (ed.) (1980) *Kerogen, Insoluble Organic Matter from Sedimentary Rocks,* Éditions Technip, Paris.

Durand, B., Ungerer, P., Chiarelli, A. and Oudin, J.L. (1984) Modélisation de la migration de l'huile. Application à deux exemples de bassins sédimentaires. *Proceedings of the 11th World Petr. Cong.*, London, 1981, **1,** 3-15.

Durand, B. (1987) Du kérogène au pétrole et au charbon: les voies et les mécanismes des transformations des matières organiques sédimentaires au cours de l'enfouissement. *Mém. Soc. Géol.*, France, nouvelle série, **151,** 77-95.

Duval, B.C., Cramez, C.and Vail, R. (1998) Stratigraphic Cycles and Major Marine Source Rocks. In de Graciansky C., Hardenbol J., Jacquin T. and Vail, R. (eds.), *Mesozoïc and Cenozoïc Sequence Stratigraphy of European Basins, SEPM Special Publication,* **60,** 43-51.

Ekweozor, C.M., Okogun, G.I., Ekong, D.E.U. and Maxwell, J.E. (1979) Preliminary Organic Geochemical Studies of Samples from the Niger Delta, Nigeria. *Chem. Geol.*, **27,** 39-37.

Engel, M.H. and Macko, S.A. (1993) *Organic Geochemistry, Principles and Applications*, Plenum Press, 871.

England, W.A. (1990) The Organic Geochemistry of Petroleum Reservoirs. *Organic Geochemistry,* **16,** 415-425.

Engler, C. (1888) *D. Chem. Ger.,* 1816.

Engler, C. (1912) Die Bildung der Haupt-Bestandteile des Erdöls. *Petroleum,* **7.**

Erdman, J.G., Marlett, E.M. and Hanson, W.E. (1958) The Occurrence and Distribution of Low Molecular Weight Aromatic Hydrocarbons in Recent and Ancient Sediments. *Am. Chem. Soc., Petroleum Chemistry Div.,* Preprints Chicago Mtg., **3,** 4, 39.

Espitalié, J., Laporte, J.L., Madec, M., Marquis, F., Leplat, P., Paulet, J. and Boutefeu, A. (1977) Méthode rapide de caractérisation des roches-mères, de leur potentiel pétrolier et de leur degré d'évolution. *Oil & Gas Science and Technology,* **32,** 23-42.

Evans, C.R., Rogers, M.A. and Bailey, N.J.L. (1971) Evolution and Alteration of Petroleum in Western Canada. *Chem. Geol.*, **8,** 147-170.

Fausett, D.W. (1964) Mathematical Study of an Oil Shale Retort. *PhD Thesis,* Univ. of Wyoming.

Finenko, Z.Z. (1967) Primary Production in Southern Seas. In *Problems of Bio-Oceanography* (V.A. Vodianitsky, Ird Nau Kova Dumka, Kiev (in Russian).

Forsman, J. and Hunt, J.M. (1958) Insoluble Organic Matter (Kerogen) in Sedimentary Rocks of Marine Origin. In L.G. Weeks (Ed.), Habitat of Oil: A Symposium. *Am. Assoc. Petr. Geol.*, Tulsa, Oklahoma, 747-778.

Fougeroux de Bondaroy, M. (1770) *Second mémoire sur le Pétrole et sur les vapeurs inflammables connues dans quelques parties de l'Italie*, Histoire de l'Académie Royale des Sciences, **91**.

Galimov, E.M. (1980) C_{13}/C_{12} in Kerogen. In *Kerogen* (B. Durand Ed.), Editions Technip, Paris.

Gold, T. (1979) Terrestrial Sources of Carbon and Earthquake Outgasing. *J. of Petr. Geol.*, **1**, 3, 1-19.

Goldstein, T. and Aizenshtat, Z. (1994) Thermochemical Sulfate Reduction, *A Review. J. Therm. Anal.*, **42**, 241-290.

Gracianski de, P.C., Deroo, G., Herbin, J., Jacquin, T., Magniez, F., Montadert, C., Müller, C., Ponsot, C., Schaaf, A., and Sigal, J. (1982) Ocean-Wide Stagnation Episodes in the Late Cretaceous. *Geol. Rundschau*, **75**, 17-41.

Graciansky de, P.C., Hardenbol, J., Jacquin, T. and Vail, R. (eds.) (1998) Mesozoïc and Cenozoïc Sequence Statigraphy of European Basins. *SEPM Special Publication*, **60**, 43-51.

Gransch, J.A. and Eisma, E. (1970) Characterisation of the Insoluble Organic Matter of Sediments by Pyrolysis. In *Advances in Organic Geochemistry*, 1966, (G.D. Hobson and G.C. Speers Eds.) 407-426.

Gutjahr, C.C.M. (1966) Carbonisation Measurements of Pollen Grains and Spores and their Application. *Thesis*, Leiden.

Huc, A.Y. and Durand, B. (1974). Étude des acides humiques et de l'humine de sediments récents considérés comme précurseurs des kérogènes. In *Advances in Organic Geochemistry*, 1973, Actes du 6e congrès de géochimie organique, Éditions Technip, Paris, 53-72.

Huc, A.Y. (ed.) (1995). AAPG Studies in Geology No. 40. *Am. Assoc. Petr. Geol.*, Tulsa, Oklahoma.

Huc, A.Y., Carpentier, B., Guehenneux, G. and Magnier, C. (1999) Geochemistry in a Reservoir and Production Perspective. *SPE Paper 53146*.

Huc, A.Y., Vandenbroucke, M., Bertrand, Ph., Gayet, J. and Stow, D.A.V. (2001) Organic Sedimentation in Deep Offshore Settings: The Quaternary Sediments Approach. *Marine and Petroleum Geology*, **18**, 513-517.

Hubbert, M.K. (1953). Entrapment of Oil under Hydrodynamic Conditions. *Bull. Am. Assoc. Petr. Geol.*, **37**, 1954-2026.

Huck, G., and Karweil, J. (1955) Physikalische Probleme der Inkohlung. *Brennst. Chem.*, **36**, 1-11.

Hunt, T.S. (1863) Report on the Geology of Canada. In *Canadian Geol. Surv., Rep.* 983

Hunt, J.M. (1963) Geochemical Data on Organic Matter in Sediments. In V. Bese (Ed.) 3rd International Scientific Conference on Geochemistry, *Microbiology and Petroleum Chemistry Proc.*, KULTURA, Budapest, **2**, 394-412.

Hunt, J.M. (1972) Distribution of Carbon in Crust of Earth. *Bull. Am. Assoc. Petr. Geol.*, **56**, 2273-2277.

Hunt, J.M. (1979) *Petroleum Geochemistry and Geology*, Freeman, San Francisco.

Jaccard, A. (1895) *Le pétrole, l'asphalte et le bitume au point de vue géologique*, Félix Alcan, Paris.

Jeffrey, W.A. and Kaplan, I.R. (1988) Hydrocarbons and Inorganic Gases in the Gravberg-1 Well, Siljan Ring, Sweden. *Chemical Geology*, **71**, 1-3, 237-257.

Kartsev, A.A., Vassoevitch N.B., Geodekian, A.A., Neruchev, S.G., and Sokolov, V.A. (1971) The Principal Stage of Oil Formation. In *8th World Petrol., Cong. Proc., Applied Science Publishers*, London, **2**, 3-11.

Karweil, J. (1955) Die Metamorphose der Kohlen vom Standpunkt der physicalischen Chemie. *Z. Dtsch. Geol. Ges.*, **107**, 132-139.

Kaufman, R.L., Ahmed, A.S. and Elsinger, R.L. (1990) Gas Chromatography as a Development and Production Tool for Fingerprinting Oils from Individual Reservoirs: Application in the Gulf Coast of Mexico. *Gulf Coast Section of the SEPM Foundation 9th Annual Research Conference Proceedings*, 263-282.

Katz, B.J. and Pratt, L. (1993) Source Rocks in a Sequence Stratigraphic Framework. *AAPG* Studies in Geology, Tulsa, Oklahoma, 37, 237.

Katz, B.J. (1995) *Petroleum Source Rocks*, Springer-Verlag, 327.

Klemme, H.D. and Ulmishek, G.F. (1991) Effective Petroleum Source Rocks of the World: Stratigraphic Distribution and Controlling Depositional Factor. *Bull. Am. Assoc. Petr. Geol.*, **75**, 1809-1851.

Krevelen van, D.W. (1950) Graphical-Statistical Method for the Study of Structure and Reaction Processes of Coal. *Fuel*, **29**, 269.

Krevelen van, D.W. (1961, 2nd ed. 1981) *Coal: Typology, Chemistry, Physics, Constitution*, Elsevier, Amsterdam.

Krooss, B. (1986) Diffusion of C_1 to C_5 Hydrocarbons in Water-Saturated Sedimentary Rocks. *Erdöl und Kohle-Erdgas-Petrochemie*, **39**, 399-402.

Landais, P., Dubessy, J., Poty, B. and Robb, J.R. (1990) Three Examples Illustrating the Analysis of Organic Matter Associated with Uranium Ores. In *Advances in Organic Geochemistry*, 1989 (B. Durand and F. Béhar Eds.), Pergamon Press, 601-608.

Largeau, C., Derenne, S., Casadevall, E., Berkaloff, C., Corolleur, M., Lugardon, B., Raynaud, J.F., and Connan, J. (1990) In *Advances in Organic Geochemistry*, 1989 (B. Durand and F. Béhar Eds.), Pergamon Press, 889-895.

Larter, S.R., Horsfield, B. and Douglas, A.G. (1977) Pyrolysis as a Possible Mean of Determining the Petroleum Generation Potential of Sedimentary Organic Matter. In *3rd International Symposium of Analytical Pyrolysis*, Elsevier, Amsterdam, 189-202.

Leplat, P. (1967) Application of Pyrolysis Gas Chromatography to the Study of the Non Volatile Petroleum Fraction. *J. Gas Chromatography*, **5**, 128-135.

Lewan, M.D., Winters, J.C. and MacDonald, J.H. (1979) Generation of Oil-Like Pyrolysates from Organic Rich Shales. *Science*, **203**, 897-899.

Leythaeuser, D., Schaefer, R.G. and Yükler, A. (1982) Role of Diffusion in Primary Migration of Hydrocarbons. *Bull. Am. Assoc. Petr. Geol.*, **66**, 408-429.

Lind, S.C. (1938) *The Science of Petroleum*, Oxford University Press, New York, 39-41

Lopatin, N.V. (1971) *Temperature and Geological Time as Factors of Carbonification*. Akad. Nauk SSSR., *Izs. Ser. Geol.*, **3**, 95-106.

Louis, M. and Tissot, B. (1967) Influence de la température et de la pression sur la formation des hydrocarbures dans les argiles à kérogène. *7th World Petr. Cong. Proc.*, **2**, 47-60.

Mackenzie, A.S. (1984) Application of Biological Markers in Petroleum Geochemistry. In *Advances in Petroleum Organic Geochemistry* (J. Brooks and D.H. Welte Eds.), Academic Press, 115-214.

Matavelli, L. and Novelli L. (1988) Geochemistry and Habitat of Matural Gas in Italy. *Organic Geoch.*, **13**, 1-13.

McAuliffe, C.D. (1966) Solubility in Water of Paraffin, Cyclo-Paraffin, Olefin, Acetylene, Cycloolefin and Aromatic Hydrocarbons. *J. Phys. Chem.*, **70**, 1267-1275.

McAuliffe, C.D. (1980) Oil and Gas Migration: Chemical and Physical Constraints. In *AAPG Studies in Geology no. 10, Problems of Petroleum Migration* (W.H. Roberts III and J.R. Cordell Eds), 89-107.

Mc Coy, A.W. and Trager, P. (1919) Notes on Principles of Oil Accumulation. *J. Geol.*, **27,** 252-262.

McIver, R.D. (1967) Composition of Kerogen Clue to its Role in the Origin of Petroleum. *7th World Petr. Cong. Proc.*, **2,** 25-36.

McKirdy, D.M., Aldridge, A.K. and Ypma, J.M. (1983) A Geochemical Comparison of Some Crude Oils from Pre-Ordovician Carbonate Rocks. In *Advances in Organic Geochemistry,* 1981 (M. Bjoroy *et al.* Ed.), Wiley, Chichester, 99-107.

Mendeleiev, D.I. (1878) *Revue Scientif.*, XIII, 409.

Menschein, W.G. (1959) Origin of Petroleum. *Bull. Am. Assoc. Petr. Geol.*, **43,** 925-943.

Moldowan, J.M., Albrecht, P., Philp, R. (1992) *Biological Markers in Sediments and Petroleum*, Prentice Hall, 411.

Müller, J. and Stoffers, P (1974) Mineralogy and Petrology of Black Sea Sediments. In the Black Sea, Geology, Chemistry and Biology (E.T. Degens and D.A. Ross, Eds.). *Am. Assoc. Petr. Geol.,* Memoir 20, 200-248.

Nogaret, E. (1983) Solubilité des hydrocarbures dans le gaz naturel comprimé. Application à la migration du pétrole dans les bassins sédimentaires. *PhD Thesis,* École des Mines de Paris.

Oberlin, A., Boulmier, J.L. and Durand, B. (1974) Electron Microscopic Investigation of the Structure of Naturally and Artificially Metamorphosed Kerogen. *Geochimica et Cosmochimica Acta,* **38,** 647-650.

Oh, C.H., Trichet, J., Oh, J.H. and Rouzaud, J.N. (1990) Relation Between the Thermal Behaviour and the Structure of Kerogen in the Gosean Uranium Deposit, Korea. In *Advances in Organic Geochemistry,* 1989 (B. Durand and F. Béhar Eds.), Pergamon Press, 609-620.

Ourisson, G., Albrecht, P. and Rohmer, M. (1979) The Hopanoids. Paleochemistry and Biochemistry of a Group of Natural Products. *Pure Appl. Chem.*, **51,** 709-729.

Pakhomova, A.S. (1961) *Organic Matter in the Bottom Deposits of the Caspian,* Tr. Gos. Okeanogr. Inst., 6, 18, (in Russian).

Peake, E., and Hodgson, G.W. (1965) Alkanes in Natural Aqueous Systems: Accommodation of C_{20}-C33 *n*-Alkanes in Distilled Water and Occurrence in Natural Water Systems. *Bull. Am. Oil Chem. Soc.*, 56th Annual Spring Meeting, Houston, Texas.

Pelet, R., Debyser, Y., Combaz, A. *et al.* (1977, 1978, 1979, 1981, 1983) Géochimie organique des sédiments marins profonds. CEPM-CNEXO, *Comité d'études géochimiques marines,* Éditions du CNRS.

Peters, K.E. and Moldowan, J.M. (1993) *The Biomarkers Guide,* Prentice Hall, 363.

Philippi, G.T. (1965) On the Depth, Time and Mechanism of Petroleum Generation. *Geochim. Cosmochim. Acta,* **29,** 1021-1049.

Petrov, A.A. (1987) *Petroleum Hydrocarbons,* Springer Verlag.

Pratt, W.E. (1943) *Oil in the Earth*, Lawrence, Univ. of Kansas Press, 106.

Prinzhofer, A. and Huc, A.Y. (1995) Genetic and Post-Genetic Molecular and Isotopic Fractionations in Natural Gases. *Chemical Geology,* **126,** 281-290.

Prinzhofer, A., Mello, M.R. and Takaki, T. (2000) Geochemical Characterization of Natural Gas: A Physical Multivariable Approach and its Applications in Maturity and Migration Estimates. *Bull. Am. Assoc. Petr. Geol.*, **84,** 1152-1172.

Radke, M. (1987) Organic Geochemistry of Aromatic Hydrocarbons. In Advances in *Petroleum Organic Geochemistry* (J. Brooks, and D.H. Welte, Eds.), Academic Press, London, **2,** 141-217.

Reed, J.D., Illich, H.A. and Horsfield, B. (1986) Biochemical Evolutionary Significance of Ordovician oils and their Sources. In *Advances in Organic Geochemistry, 1985* (L.D. Leythaeuser and J. Rullkötter Eds.), Academic Press, London, 141-217.

Rice, D.D. and Claypool, G.E. (1981) Generation, Accumulation and Resource Potential of Biogenic Gas. Bull. *Am. Assoc. Petr. Geol.*, **65,** 5-25.

Romankevitch, Y.A. (1977) *Geochemistry of Organic Matter in the Oceans*, Dokl. Acad. Nauk SSSR (in Russian).

Robert, P. (1985) Histoire géothermique et diagenèse organique. *Bull. Centres Rech. Explor.-Prod.,* Elf-Aquitaine, Mém. 8, Pau.

Rullkötter, J. and Michaelis, W. (1990) The Structure of Kerogen and Related Materials. A Review of Recent Progress and Future Trends. In *Advances in Organic Geochemistry,* 1989 (B. Durand and F. Béhar eds.), Pergamon Press, 829-852.

Sackett, W.M., Bernard, B. and Brooks, J. (1975) Isotopic and Molecular Compositions of Seep Gases in the Gulf of Mexico. Paper Presented at the *Southern Regional Organic Geochemists' Meeting,* Nov. 1975, Port Aransas, Texas.

Schoell, M. (1983) Genetic Characterization of Natural Gases. *Bull. Am. Assoc. Petr. Geol.,* **67,** 2225-2238.

Schneider, F., Wolf, S., Faille, I. and Pot, D. (2000) A 3D Model for Hydrocarbon Potential Evaluation: Application to Congo Offshore. *Oil and Gas Science and Technology,* **1,** 3-13.

Seifert, W.K. and Moldowan, J.M. (1981) Paleoreconstruction by Biological Markers. *Geochim. Cosmochim. Acta,* **45,** 783-794.

Schochardt, M. (1943) *Grunlagen und neuere Erkenntnisse der angewandten Braunkohlenpetrographie,* Knapp, Halle.

Silverman, S.R. and Epstein (1958) Carbon Isotopic Composition of Petroleums and Other Sedimentary Organic Materials. *Bull. Am. Assoc. Petr. Geol.,* **42,** 998-1012.

Silverman, S.R. (1965) Migration and Segregation of Oil and Gas. In *Fluids in Subsurface Environments, Am. Assoc. Petr. Geol.,* Memoir 4, Tulsa, Oklahoma, 53-65.

Shimkus, K.M. and Trimonis, E.S. (1974) Modern Sedimentation in Black Sea. In the Black Sea, Geology, Chemistry and Biology (E.T. Degens and D.A. Ross, Eds). *Am. Assoc. Petr. Geol.,* Memoir 20, 249-278.

Smalley, C. and England, W.A. (1992) Assessing Reservoir Compartimentalisation During Field Appraisal: How Geochemistry Can Help. *SPE Paper 25005.*

Smith, V. (1954) Studies on the Origin of Petroleum, Occurrence of Hydrocarbons in Recent Sediments. *Bull. Am. Assoc. Petr. Geol.,* **38,** 377-404.

Sokolov, V. (1974) *Géochimie des gaz naturels,* Éditions Mir, Moscou.

Stach, E. (1935) *Lehrbuch des Kohlenpetrographie*, Borntraeger, Berlin.

Stach, E., Mackowsky, M.Th., Teichmüller, M., Taylor, G.H., Chandra, D. and Teichmüller, R. (1982) *Coal Petrology,* Borntraeger, Berlin.

Stahl, W.J. (1977) Carbon and Nitrogen Isotopes in Hydrocarbon Research and Exploration. *Chem. Geol.,* **20,** 121-140.

Stevens, N., (1956) Origin of Petroleum, A Review. *Bull. Am. Assoc. Petr. Geol.,* **40,** 51-61.

Stevens, N., Bray, E.E. and Evans, D. (1956) Hydrocarbons in Sediments of Gulf of Mexico. *Bull. Am. Assoc. Petr. Geol.,* **40,** 975-983.

Stone, R.W. and ZoBell, C.E. (1952) Bacterial Aspects of the Origin of Petroleum. *Indus. Eng. Chem.,* **44,** 2564-2567.

Stopes, M.C. (1919) *Proc. Roy. Soc.* London, B 90, 470.

Stopes, M.C. (1935) *Fuel,* 14, 4.

Teichmüller, M. (1958) Métamorphisme du charbon et prospection du pétrole. *Rev. Ind. Minér.,* numéro spécial, 99-113.

Teichmüller, M. (1986) Organic Petrology of Source Rocks, State of the Art. In *Organic Geochemistry,* **10,** 581-599.

Tissot, B. (1969) Premières données sur les mécanismes et la cinétique de la formation du pétrole dans les bassins sédimentaires. Simulation d'un schéma réactionnel sur ordinateur. *Oil & Gas Science and Technology,* **24,** 470-501.

Tissot, B. (1973) Vers l'évaluation quantitative du pétrole formé dans les bassins sédimentaires. *Revue Assoc. Fr. Tech. Pétr.,* **222,** 27-31.

Tissot, B., Durand, B., Espitalié, J. and Combaz, A (1974) Influence of the Nature and Diagenesis of Organic Matter in Formation of Petroleum. *Am. Assoc. Petr. Geol.,* **58,** 499-506.

Tissot, B., Deroo, G. and Espitalié, J. (1975) Étude comparée de l'époque de formation et d'expulsion du pétrole dans diverses provinces géologiques. Proc. 9th World Pet. Congr., *Applied Science Publ.,* **2,** 159-169.

Tissot, B. and Welte, D.H. (1978, 2nd edition 1984) *Petroleum Formation and Occurrence,* Springer Verlag.

Tissot, B. (1979) Effect on Prolific Petroleum Source Rocks and Major Coal Deposits Caused by Sea Level Changes. *Nature,* **277,** 462-465.

Tissot, B., Deroo, G. and Herbin, J. (1979) Organic Matter in Cretaceous Sediments of the North Atlantic: Contribution to Sedimentology and Paleogeography. In *Deep Drilling Results in the Atlantic Ocean: Continental Margins and Paleoenvironments,* Talwian, M., Hay W., Ryan, W.B.F. Maurice Ewing Series, 3, *Am. Geoph. Union,* 362-374.

Tissot, B. and Bessereau, G. (1982) Géochimie des gaz naturels et origine des gisements de gaz en Europe occidentale. *Oil & Gas Science and Technology,* **37,** 63-77.

Trask, D. and Wu, C.C. (1930) Does Petroleum Forms in Sediments at Time of Deposition? *Bull. Am. Assoc. Petr. Geol.,* **14,** 1451-1463.

Trask, D., Hammar, H.E. and Wu, C.C. (1932) *Origin and Environment of Source Sediments of Petroleum,* Gulf Publications, Houston.

Trask, D., and Patnode, H.W. (1942) Source Beds of Petroleum. *Am. Assoc. Petr. Geol.,* Tulsa, Oklahoma, 566.

Treibs, A. (1934) Chlorophyll und Häminderivate in bituminösen Gesteinen, Erdölen, Erdwachsen und Asphalten. *Ann. Chem.,* **510,** 42-62.

Ungerer, P., Béhar, E. and Discamps, D. (1983) Tentative Calculation of the Overall Expansion of Organic Matter During Hydrocarbon Genesis from Geochemistry Data. Implications for Primary Migration. In *Advances in Organic Geochemistry,* 1981, (M. Bjoroy Ed.), Wiley, 129-135.

Ungerer, P. and Pelet, R. (1987) Extrapolation of Kinetics of Oil and Gas Formation from Laboratory Experiments to Sedimentary Basins. *Nature,* **327,** 52-54.

Ungerer, P., Espitalié, J., Béhar, F. and Eggen, S. (1988) Modélisation mathématique des interactions entre craquage thermique et migration lors de la formation du pétrole et du gaz. *C.R. Acad. Sc. Ser. II,* Paris, 927-934.

Ungerer, P., Burrus, J., Doligez, B., Chenet, Y. and Bessis, F. (1990) Basin Evaluation by Integrated 2D Modelling of Heat Transfer, Fluid Flow, Hydrocarbon Generation and Migration. *Bull. Am. Assoc. Petr. Geol.,* **74,** 309-335.

Ungerer, P. (1993) Modelling of Petroleum Generation and Migration. In *Applied Petroleum Geochemistry* (M.L. Bordenave Ed.), Éditions Technip, Paris.

Vassoevich, N.B., Korchagina, Yu.I., Lopatin, N.V. and Chernyshev, V.V. (1969) Principal Phase of oil Formation. Moskow Univ. Vestnik 6, 3-6 (in Russian), English Translation, *Int. Geol. Rev.* 1970, 12, 11,1276-1296.

Vandenbroucke, M., Bordenave, M.L. and Durand, B. (1993) Transformation of Organic Matter with Increasing Burial of Sediments and the Formation of Petroleum in Source Rocks (1993) In *Applied Petroleum Geochemistry* (M.L. Bordenave Ed.), Éditions Technip, Paris.

Vernadsky, V.I. (1934) *Outlines of Geochemistry ONTI,* Gornogeolog. Neft. Izd.

Vially, R. (ed.) (1992) *Bacterial Gas,* Éditions Technip, Paris.

Waples, D.W. (1980) Time and Temperature in Petroleum Formation/Application of Lopatin's Method to Petroleum Exploration. *Bull. Am. Assoc. Petr. Geol.,* **64,** 916-926.

Waples, D.W. (1985) *Geochemistry in Petroleum Exploration,* Reidel Publish. Cy, Dordrechts, 232.

Welte, D.H., Hagemann, H.W., Hollerbach, A., Leythaeuser, D. and Stahl, W. (1975) Correlation Between Petroleum and Source Rock. *9th World Petr. Cong. Proc.,* **2,** 179-191.

Welte, D.H., Horsfield, B. and Baker, D.R. (1997) *Petroleum and Basin Evolution. Insights from Petroleum Geochemistry, Geology and Basin Modelling,* Springer-Verlag.

White, D. (1915) Geology: Some Relations in Origin Between Coal and Petroleum. *J. Wash. Acad. Sci.,* **6,** 189-212.

White, D. and Thiessen (1913) The Origin of Coal. *USBM Bull.,* 38.

Whitehead, W.L. (1951) *Studies of the Effect of Radio Activity in the Transformation of Marine Organic Materials into Petroleum Hydrocarbons. Fundamental Research on Occurrence and Recovery of Petroleum,* American Petroleum Institute, 192-201.

ZoBell, C.E. (1947) Microbial Transformation of Molecular Hydrogen in Marine Sediments with Particular Reference to Petroleum. *Bull. Am. Assoc. Petr. Geol.,* **31,** 1709-1751.

Final manuscript received in January 2003

Oil & Gas Science and Technology – Rev. IFP, Vol. 58 (2003), No. 2, pp. 233-241
Copyright © 2003, Éditions Technip

Petroleum Geochemistry at the Dawn of the 21st Century

A.Y. Huc[1]

1 Institut français du pétrole, 1 et 4, avenue de Bois-Préau, 92852 Rueil-Malmaison Cedex - France
e-mail: a-yves.huc@ifp.fr

Résumé — La géochimie organique pétrolière à l'aube du XXIe siècle — Après un bref historique de l'évolution de la géochimie organique pétrolière au cours des dernières décennies, cet article résume l'état actuel de la recherche dans ce domaine et présente les orientations dictées par les besoins des opérateurs. Aujourd'hui, les acteurs industriels s'intéressent à des prospects de plus en plus profonds (haute pression, haute température), à l'exploration des grands fonds marins, à la recherche de gaz et à la production des énormes réserves d'huiles lourdes de la planète. Dans ce contexte, l'enjeu est de contribuer d'une manière efficace à la prédiction quantitative des réserves en huile et gaz à l'échelle d'un bassin (volume, composition, incertitudes) et à la prédiction de la qualité des fluides contenus dans un gisement, et cela par une meilleure connaissance des événements géochimiques affectant leur préservation et leur altération.

Abstract* — *Petroleum Geochemistry at the Dawn of the 21st Century — *Following a short review of the evolution of petroleum organic geochemistry in recent decades, this paper depicts the present status of research in the field and describes the guidelines dictated by the needs of the operators. Today, the industrial players are interested in increasingly deep prospects (high pressure, high temperature), in the exploration of the deep offshore, the search for gas and in the production of the planet's enormous heavy oil reserves. In this context, the challenge is to contribute effectively to the quantitative prediction of the oil and gas reserves at the scale of a basin (volume, composition, uncertainties) and to predict the quality of the fluids present in a reservoir, through a closer knowledge of the geochemical events affecting their preservation and their alteration.*

1 BACKGROUND AND TRENDS

1.1 Geochemistry

Geochemistry is a discipline devoted to the study of the chemistry of geological objects. Accordingly, it shares its analytical and experimental approach with chemistry, and its consideration of time and space with geology. This specificity makes geochemistry an irreplaceable investigative tool for all phenomena that involve transformations or interactions of matter in the geosphere.

The geochemist must constantly adapt his analytical and experimental approach to the complex and often poorly isolated systems that he studies (chemical reactor of the size of a sedimentary basin, heterogeneous medium, generally open system, out-of-equilibrium reactions, absence of steady state conditions, number and chemical structure of reactants and products difficult if not impossible to define accurately, representativity of samples, extrapolation to geological dimensions of experiments conducted at laboratory space-time scales, etc.). He must also develop concepts that incorporate the scale and geological history of the problems investigated.

To contend with the difficulties inherent in the chemistry of geological material (contamination, trace species, structural complexities of certain substances, etc), geochemists have been forced to develop experimental protocols and very high technology specific analytical techniques that are routinely used today in the petroleum industry.

The contribution of geochemistry is considerable for the petroleum geosciences because it supplies information that serves to define quantitative constraints in domains where they are often lacking. Geochemistry has been instrumental in defining the concept of "petroleum system", which no explorer can do without today.

1.2 Petroleum Geochemistry

Petroleum geochemistry is concerned with the geological evolution of the very substances making up the sedimentary rocks: formation of kerogen (fossilized sedimentary organic matter), followed by its decomposition into hydrocarbons; fluid migration; hydrocarbon transformations within the reservoirs; transformations of the mineral suite of the reservoirs and their petrophysical consequences. This kind of geochemistry serves petroleum exploration because it develops tools to describe, understand and predict:
- the formation of oil and gas and their migration;
- the thermal history of the basins and the composition of the fluids that have passed through them;
- the distribution of porosities in the drains and reservoirs.

It also serves for the characterization and production of the hydrocarbons fields, because it provides the keys to a closer understanding of the distribution of important properties at the scale of the reservoir: composition of the fluids, and petrophysical properties (porosity, permeability, wettability, irreducible saturation, resistivity, water geochemistry, etc.).

The tools of geochemistry are measurement techniques, as well as conceptual and numerical models constructed in particular through the extrapolation of the laboratory experiments.

1.3 Petroleum Organic Geochemistry

1.3.1 The Origin of Oil, Oil/Source Rock Correlations and Kinetic Formalism

Between 1970 and 1980, organic geochemists played a decisive role in the evolution of ideas concerning the origin of fossil hydrocarbons, and the identification and formalization of the chemical processes involved in the formation of oil and gas.

The modern vision of the "petroleum system" is based on the presence of ingredients (all) necessary for the presence of hydrocarbon accumulation, manely, a source rock, a reservoir rock, a cap rock and a trap, and also on the fact that these elements must fit into a dynamic scenario in which the major role is held by the source rock, whose thermal history (primarily controlled by its burial over time) orchestrates the "petroleum" life of the basin.

It is the recognition of this geochemical process of kinetic cracking (formalized as a first order kinetics) of the organic matter (kerogen) and the products of its transformation (primary and secondary cracking), present in the source rocks, which crystallized the idea of the "petroleum system". It also led to the development of the numerical basin models. It was at that time that the concept of "types" of kerogens, reflecting the variability of the properties of the fossilized organic matter in the sediments, and its fundamental role for the petroleum potential of the source rocks, was proposed.

One of the many benefits of the conceptual breakthrough achieved in this way was the availability to the industry of the Rock-Eval pyrolyzer which elegantly incorporated the knowledge about the properties of the sedimentary organic matter and its evolution through its thermal history, but above all, helped to translate them into simple and readily usable parameters for the exploration teams, such as the identification of potential "source rock" intervals, organic matter content, type and thermal maturity.

It was also at this time that molecular geochemistry was established by developing the concept of "biomarkers". These molecules, which are found in oils, have retained sufficiently well preserved chemical structures that help to affiliate them with biological molecules present in the tissues of living organisms whose remains generated the organic matter of the source rocks. The distribution of these biomarkers can be considered as a genuine "fingerprint" for an oil. The comparison of this molecular imprint with those that can be obtained from organic extracts of source rocks existing in the same sedimentary basin then often enabled the explorer to recognize the actual source rock(s) that charge the reservoirs of a petroleum system (the "oil/source rock correlation" principle). The biomarkers, which, at the time of their fossilization, possess a spatial molecular configuration corresponding to those that are specific to the biomolecules from which they originated, also proved to be invaluable for estimating the thermal maturity of an oil. In fact, the unique initial molecular configuration is progressively replaced by a mixture of molecules due to the appearance of molecules of identical formulas but whose spatial configurations are thermodynamically favored (these molecules are either produced from the previous molecules by isomerization or added by liberation during the thermal cracking of the kerogen). The ratio of the molecules displaying a stable conformation increases through the thermal history of the source rock (formation of more stable and degradation of less stable molecules). This principle underlies the broad set of "molecular maturity parameters" available to the practitioner.

1.3.2 Understanding the Source Term, Quantifying and Predicting the Composition of the Hydrocarbons Formed

The second investigation phase (1980-1990) was aimed to better constrain the kinetic equations, which were crucially important for the basin models (1D, 2D), currently under development. The basic difficulty was to determine the quantitative kinetic parameters for organic matter (kerogens of different types and associated oils) which have an extremely complex chemical structure. The adaptation of "Rock-Eval" type "open system" pyrolyzers and the use of complementary "closed system" approaches, in anhydrous conditions or in the presence of water, helped to develop procedures to obtain these parameters, which have since been applied successfully by exploration professionals. It was also important to have a better handle on the composition of the fluids formed by cracking from the different "types" of source rocks, according to the advancement of the reaction. This made it possible, at any time in the history of the basin, to determine the composition of the fluids available for expulsion to the reservoir rocks. The results obtained, at the cost of huge experimental programs, are operational today, and have allowed the development of a new generation of operational numerical basin models (Temiscomp, Petromod, Basinmod, Sextant, etc.).

At the same time, it was necessary to understand the factors responsible for the existence of source rocks in the sedimentary stack of a basin and those controlling their regional extension, in order to reduce the risks of a permit and to supply information on the geological sections with a reasonable representation of the source term (the source rock) in the basin models. This led to major research efforts in the area of the sedimentology of source rocks.

1.3.3 Gas and Reservoir

In recent years, in response to major trends in the policy of the industry (greater interest in gas exploration, deep prospects, focus on production), the research programs were substantially refurbished:

- The formation of hydrocarbon gases, heretofore considered as by-products in the oil genesis schemes, was investigated *per se*.

  Significant breakthroughs were achieved in this field, particularly with the establishment of diagnostic methods and numerical models taking account of the variety of sources of natural gas: biogenic, produced by early or late cracking of kerogen, secondary cracking of the oil. The results of these studies, which have powerful implications for the appraisal of the gas potential of the basins, are currently undergoing validation. The displacement (migration) of the gas in the basins, which must account for aspects of physics that are negligible in the case of oil (solubility in water, diffusion) is the subject of ongoing research and the development of quantitative models. This specific behavior of gas is now included in certain four-phase (solid, water, oil, gas) basin simulators.

  One of the major risks in gas exploration is the presence of nonhydrocarbon species in variable amounts (CO_2, N_2, H_2S). In this area, where the knowledge is still rudimentary, emphasis is laid on the development of tools to diagnose the origin of these gases, and the design of the numerical models describing the formation of these gases, issued from the thermal decomposition of organic matter (CO_2, N_2, H_2S), and also from inorganic sources (mantle CO_2, thermal decomposition of carbonates, N_2 from ammoniacal nitrogen associated with clays, H_2S from low temperature bacterial sulfate reduction and from high temperature thermal sulfate reduction).
- The spatial change in the composition of the hydrocarbons present in the reservoirs is an irreplaceable source of information for understanding the supply history of a field: source of the oils, filling scenario (supply direction, mixture of several feeds, replacement of a pre-existing oil charge, etc). It also helps to constrain the state of connectivity of the reservoirs (compartmentalization) and thereby supplement the geometric information provided by geologic approaches. The aim is to identify a spatial heterogeneity of the composition (molecular imprint) of the oil which, if it exists, reflects the existence of permeability barriers that prevented the compositional homogenization since the charging time of the reservoir.

To do this, a panoply of analytical methods and data processings designed to identify the presence of an heterogeneity and thereby constrain the image of a field is available today to the industry under the generic term of "reservoir geochemistry".

- The presence of solid or semi-solid organic phases (asphalts, bitumens) is fairly widespread in petroleum reservoirs (Ghawar, the world's biggest field, contains an accumulation of bituminous material several tens of meters thick in its central portion). Methodologies designed to locate these bitumens and to identify their extension have been developed and applied successfully. The source of these bitumens is still the subject of investigations that have already demonstrated the wide variety of geological situations causing their appearance. The thermal cracking of oils in place is one of these processes. It is accompanied by the formation of a solid carbonaceous residue (pyrobitumen), which alters the properties of the reservoir (reduced porosity and permeability, modification of grain surface properties, migration of fine solid particles). The growing interest of explorers in deep prospects evidently increases the "pyrobitumen risk", making a better knowledge of these products and their formation more necessary than ever.
- A special effort has been mounted in recent years to found a genuine "geothermodynamics", aimed to integrate the geochemical approach and the thermodynamic approach for the study of petroleum fluids. Promising results have been obtained, as for example, the formalization of phase changes occurring during the migration of the hydrocarbons, which takes account of the composition of the fluids as calculated by the compositional kinetic models discussed earlier (this formalism is incorporated today in certain basin models like Temiscom). Research projects underway adapt (meshes, time steps) the thermodynamic calculations of the engineering type to the geological evolution of the reservoir fluids. The aim of this new approach is to provide explorers and producers with diagnostic and predictive tools concerning the properties of the hydrocarbons trapped.
- Our knowledge of source rocks has grown considerably thanks to recent studies taking account of the sedimentology of these deposits in terms of sequential analysis and cyclostratigraphy. It is now becoming possible to define, indeed to predict, the arrangement, volume and internal heterogeneity of these particular sedimentary intervals, as well as the type of associated organic matter. This is a major advance in the assessment of the basins: *a priori* forecasting of the petroleum potential, and spatial relations between source rocks and drains at the scale of the petroleum system.
- The Rock-Eval 6, marketed in 1996, represents an undisputed advance over its predecessors. Beyond its entirely reviewed technology (high temperature furnace, sample loading robot, infrared detectors, etc), the new functionalities it features help to obtain original supplementary parameters, as for example, the quantification of the carbonates.

1.4 The New Situation

1.4.1 Quantifying Migration and Constraining the Geological History of a Reservoir

The needs of the industry which, while maintaining strong demand on the reservoir and gas, fits into a context of enlargement of the permit area as well as interest in new types of prospects (high pressure/high temperature, deep offshore, production of the enormous heavy oil reserves, etc), demand a scientific and technical investment that is capable of addressing the revival of activity in exploration/production while creating new challenges for petroleum geochemistry. In this connection, the state of knowledge fairly clearly determines the projects that must absolutely be undertaken:

- exploration: quantitative prediction of oil and gas reserves at the scale of a basin (volume, composition, uncertainties);
- reservoir: geological events affecting the qualities of a reservoir and the preservation and alteration of the composition of the fluids it contains.

- The interest directed to deeply buried objectives increases the risk associated with the composition of the fluid anticipated in a prospect. The accuracy of the compositional numerical models must therefore be enhanced and the stability of the hydrocarbons in high pressure/high temperature conditions better constrained. It is therefore necessary to add, to the now conventional approaches of the laboratory simulations (pyrolysis), which are sometimes difficult to extrapolate to natural conditions for the composition of the fluids, a theoretical approach relying on molecular numerical modeling applied to representative chemical species, whereof the behavior could serve as a reference for "lumping" operations.
- The quantitative prediction of reserves in a petroleum system is closely linked today to our ability to formalize the hydrocarbon retention and migration mechanisms in the basins. In fact, although explicitly accounted for in present approaches, the real mobility of all the petroleum products and the characteristics of the displacements are poorly constrained in terms of physics. The physical displacement thresholds (for example, factors limiting expulsion from the source rock), the flow of petroleum products during the different steps of migration, the losses and modifications of composition caused, represent essentially empirical information today. The advent of the 3D basin models, one of their *raisons d'être* being ultimate volumetric quantification, must be accompanied by progress in this field. This makes it necessary to:
  - quantify (with help from the thermodynamic approach) the role of adsorption/desorption mechanisms along the

migration chain: kerogen-mineral surface-petroleum products for expulsion from the source rock, adsorption or insolubilization by thermodynamic disequilibrium along the sedimentary drains for secondary migration;
- make better use of information connected with natural tracers present in the oils and gases (trace metallic elements, rare gases, etc.);
- integrate at the basin scale, the information potentially contained in the hydrocarbon flows reaching the surface (surface shows, mud volcanoes, etc.).

- The preservation of the fluids in a trap over geological time scales is an important factor for analyzing exploration risks. A reservoir is basically a transitory object. Its existence at a given moment depends on the residence time in the trap, the quality of the trap, and the dispersion or alteration mechanisms at play. The determination of the residence time is based on the possibility of dating the charging(s) by the hydrocarbons. This chronology is determined today:
  - from the results of numerical basin models (definition of the "critical moment" at which the hydrocarbons begin to reach the trap) which, to be usable, imply a good knowledge of the geological history of the source rock (burial sequences, heat flux history). The existence of erosional phases, a long history or poorly constrained thermal regimes, induce inaccuracy that sometimes makes the model unusable;
  - and whenever possible, from absolute dating of the start of filling, presumed to arrest the mineral diagenesis (for example, dating of the latest diagenetic shales by the K/Ar method).

  It may be observed that no constraint is supplied by the fluid itself, and this is particularly problematic today insofar as the fluid present may not be the one emplaced during the last trap filling phase, which is the one dated by the K/Ar method.

  The use (hitherto unexplored) of certain radiogenic elements, for example, iodine 129 and xenon 129, which tend to display affinities with mobile hydrocarbons, and which could have descended from radioactive isotopes present in the source rocks (U, Th), could serve as a basis for an original methodology for dating the reservoir charging. Other "natural clocks" are also conceivable, all associated with radioactive decays yielding "daughter" isotopes like ^{4}He, ^{40}Ar, ^{187}O. One could understand the importance of developing a method for direct dating of the emplacement of hydrocarbon accumulations and its integration within the dynamic framework of the petroleum system: fitting of hydrocarbon formation models, information on the time-table of migration episodes, grouping of fields by supply periods, relationship to the structural history (activation of migration paths, temporal correlation with families of traps).

- The changes in composition of the fluids occurring in the traps during geological timescales have strong implications for the commercial value of the hydrocarbons of a reservoir as well as the conditions of their production. The knowledge of the mechanisms involved and their prediction at the regional scale thus appear to be necessary for taking industrial decisions. In this area of fluid alteration, which becomes crucial in a context of deep exploration, an approach that combines thermochemistry and experiments, supplemented by field data, should permit a decisive breakthrough. Many factors affect the composition of the fluids, and some of them are very poorly known. The thermal reduction of sulfates, for example, is a mechanism identified in numerous fields (deep Lacq, North Caspian, Abu Dhabi, Alberta, Gulf Coast) and which, in the presence of anhydrite, causes the oxidation of the hydrocarbons and the formation of H_2S. The economic consequences are considerable due to the alteration of the reserves and the presence of sour gas in the fields. Our knowledge about this process is embryonic and is mainly based on naturalistic observations. The reactions involved are not fully established, realistic experimental attempts have hitherto failed, and prediction remains empirical. Note that in this connection, the interactions occurring in the subsurface between hydrocarbons and sulfur species, ignored for many years, are becoming a research subject with many applications: prediction of the oil sulfur content, prediction of the presence of H_2S in the reservoirs.

  As opposed to deep objectives, shallow fields are subject to alterations by the action of bacteria, and this for example, is one of the processes which led to the Alberta and Orinoco heavy crudes and tar sands. This widely described biodegradation has never been the subject of a quantitative treatment due to the difficulty of modeling the biological factors, especially when they concern yet incompletely identified organisms, living in unusual environments. This represents a whole field of investigation barely touched by preliminary studies, which have nonetheless demonstrated the vast potential of coupling geochemistry with microbiology. Considerable advances achieved recently in microbiology (molecular genetics, study of "extremophiles") are opening attractive prospects for the study of the biodegradation of oils in reservoir conditions. The prediction of the spatial variation of composition and properties of the heavy oils in a reservoir would particularly help to optimize the architecture of the horizontal drains used for the cold production of these fields. In this respect, the understanding of the origin of gases (products of the biodegradation itself?) often dissolved in heavy oils is of prime importance because these gases impacted the natural foaming capacity of the oils and therefore the conditions of production.

- The need for a more accurate characterization of the oils produced during the history of a field, in order to anticipate water inflows, to identify the possible participation of a neighboring compartment, to better quantify the allocations during a multiple completion and sometimes to locate damage on a casing, leads the geochemist to place his molecular imprint techniques at the disposal of the producer. The geochemistry of organic deposits (paraffins, asphaltenes, diamandoids), actively contributes to the research effort aimed to explain and predict (at the exploration and production scale) the risks associated with the appearance of these mechanisms. "Reservoir geochemistry" and "Production geochemistry" are largely operational today and mobilize many geochemists in the oil companies.

This brief review of petroleum organic geochemistry and its evolution over the years, driven by advances in knowledge, technology and industrial orientations, illustrates the important technical role that it can and must play in taking decisions and reducing the risks during exploration, reservoir characterization and production. The considerable expertise of organic geochemistry, concerning the knowledge of hydrocarbons, their reactivity and their role in the biogeological cycle, must, even more than in the past, be placed at the service of public demand in the field of environmental conservation. Helping the operators to better evaluate the impact of petroleum activities and, alongside them, participating in installing the right prevention and remediation methods, should more than ever be included in the missions assigned to geochemists.

2 A TIMELY CONCERN INVOLVING GEOCHEMISTRY: THE DEEP OFFSHORE

Most of the oil majors are involved today in the exploration and production of fields in the deep ocean depths. The high cost of the operations reduces the number of boreholes drilled, and demands better predictability of the prospects with a minimum quantity of information. Many risks are at least partly associated with geochemistry (viable petroleum system, fluid composition, biodegradation, use of the information offered by surface shows, organic deposits during production, etc.). Reducing these risks hence demands rapid progress in areas still inadequately known, such as the habitat of source rocks in deep offshore conditions, the biodegradation of oils or the use of surface shows.

2.1 Biodegradation

The biodegradation of oils is currently one of the major risks in deep offshore operations. This mechanism of oil alteration by the action of bacteria is reflected by a decrease in the API density of the oils and often by an increase in their *in situ* viscosity. The merchant value of the oil and the insufficient yields can accordingly compromise the economic value of a discovery. Our knowledge of this mechanism is largely inadequate today in view of the industrial challenges.

What we know:

By definition, biodegradation demands the presence of active microorganisms, at least during a part of the geological history of a reservoir. Hence temperature is a crucial factor, and the experience gained makes it possible today to propose guides to the limits beyond which biodegradation is a minor risk (75°C) or none at all (95°C). The consideration of the temperature of a field, or even better, the modeling of the regional thermal history, is hence an indispensable prerequisite to assessing the "biodegradation" risk in exploration.

What we do not know:

Conventionally speaking, biodegradation was considered as the result of the activity of aerobic bacteria in the presence of oxygen supplied by meteoric waters. The latest studies tend to show that the alteration of the oil is probably the result of anaerobic organisms. It is extremely important to verify this new concept, because the circulation of meteoric waters, the only ones that could contain oxygen, in contact with the oil accumulation, would become a non necessary prerequisite. This does not necessarily mean that the circulation of an aquifer is not instrumental to renew the nutrients and the electron acceptors, while preventing the self-poisoning of the system by eliminating the toxic products resulting from microbial activity. The microbial mechanisms involved in this anaerobic degradation (sulfate-reduction, fermentation, methanogenesis), the electron acceptors necessary (sulfates, metals, etc.), and the limiting factors, other than temperature (nutrients, trace elements, salinity?) remain theoretical. Basically, the kinetics of the mechanism is unknown. A better prediction of biodegradation, relying on a deterministic model, therefore depends on the answers provided to these questions.

In this respect, basin models should ultimately be enriched with biodegradation modules. Already, the functionalities of these models, explicitly taking account of the hydrodynamism, the thermal history, the formation of the hydrocarbons and their composition, help to calculate a number of boundary conditions of the prospects and to assess the age of filling of the fields, parameters whose knowledge contributes to a better understanding of the risks associated with biodegradation, or to the effective testing of the assumptions made.

2.2 Source Rocks

The presence, distribution and quality of the source rocks condition the viability and effectiveness of a petroleum system. Beyond the deep sea extension of source rocks already accounted by the models recognized in the systems explored in shallower depths (*i.e.* lacustrine source rocks of Neocomian and Barremian age of the Western African and Brazilian margins, for example), the problem of the occurrence of source rocks deposited under a large depth of water must be faced, with even greater insistence with increasing distance form the continental shelf. The accumulation of organic matter in a deep setting is still little investigated, but seems possible. For instance sediments rich in marine organic matter have been deposited since the Late Miocene offshore Namibia, with a thickness of up to tens of meter, and to water depths exceeding 3000 m. A type III source rock model associated with Miocene low sea level deposits has been proposed for the Mahakam delta system (Indonesia) in order to explain the discoveries made in the deep Strait of Macassar.

2.3 Surface Shows

The seepage of hydrocarbons on the surface (surface shows) has always been regarded with close attention by explorers because they offer invaluable indications at the regional scale concerning the existence of an active petroleum system. The geochemistry of these surface shows has never truly been developed and its practical application (surface geochemistry) mainly consists in systematically mapping surface manifestations reflecting the arrival of hydrocarbons, using different direct techniques (for example, macro-show records, analysis of soil or sediment gases) and indirect methods (*e.g.* microbiology). This prospecting technique, often challenged when addressing traces, recently enjoyed a revival of interest aroused by the deep sea exploration of relatively shallow prospects, of which the low grade cap rocks permits substantial dismigration. Extensive reconnaissance surveys of the shows, by coring, have thus been carried out in recent years. In such situations, however, it is clear that not only the mapping but also the composition of the fluids dissipated at the surface potentially offer invaluable information to anticipate the quality of the oils in the reservoir. Technological breakthroughs in the field of sampling, and conceptual breakthroughs in the interpretation of the composition of the shows, are necessary to make the most of this information.

2.4. Organic Deposits

The circulation of oceanic waters at the global scale involves downwelling of the polar waters towards the deep ocean floor and their rise in the tropical and equatorial upwelling zones. The consequence of this hydrodynamic system is the low temperature (~4°C) of the ocean floor waters. The production facilities installed in great depths are hence exposed to the formation of temperature sensitive organic deposits like paraffins or waxes. Hence the prediction of the "paraffin" risk is a major economic challenge, considering that remediation operations cost up to twenty times more below 2000 m. The genetic origin of the oils containing these paraffinic molecules, often combined with type III source rocks (continental organic matter) and type I rocks (lacustrine organic matter) is therefore important to take into account in an early stage of field assessment. Besides, some oils may contain very high molecular weight paraffins (up to 100 carbon atoms) which already raise problems by obstructing the production facilities in conventional operating conditions. The origin of these molecules, which mainly appear to be inherited from lacustrine source rocks (type I), is far from established. Recent analytical advances (high temperature chromatography) which have demonstrated the existence of these hydrocarbons, and the beginning of their identification, have shown that their composition, unlike the lower molecular weight paraffins primarily formed of *n*-alkanes (linear hydrocarbons) was complex and often included branched (and cyclic?) molecules, sometimes predominant. The thermodynamic properties of these compounds are still unknown, making the existing paraffin (*n*-paraffins) precipitation models inappropriate. A research effort addressing the characterization and origin of these molecules, as well their thermodynamic properties, hence appears to be indispensable.

Over and above the technology employed, the "deep offshore" challenge has only been faced thanks to the concepts and geological models developed in recent decades, to which geochemistry has made a powerful contribution. However, as stated earlier, the specificities of the petroleum systems concerned, and the conditions of existence of the deposits and their production, demand additions to this body of knowledge and the rapid availability of prediction guides and tools for the operators, unfortunately still lacking.

ACKNOWLEDGEMENTS

This document considerably benefited from the critical reading and the constructive comments of Mireille Vandenbroucke and Alain Prinzhofer.

REFERENCES

General Readings

Bordenave, M.L. (1993) *Applied Petroleum Geochemistry*, Éditions Technip, Paris.

Durand, B. (1980) *Kerogen: Insoluble Organic Matter from Sedimentary Rocks*, Éditions Technip, Paris.

Engel, M.H. and Macko, S.A. (1993) *Organic Geochemistry, Principles and Applications*, Plenum Press, New York and London.

Hunt, J.M. (1995) *Petroleum Geochemistry and Geology*, W.H. Freeman and Company, New York, 2nd edition.

Tissot, B.P. and Welte, D.H. (1984) *Petroleum Formation and Occurrence*, 2nd edition, Springer-Verlag.

Source Rocks

Demaison, G. and Moore, G.T. (1980) Anoxic Environments and Oil Source Bed Genesis. *AAPG Bull.*, **64**, 1179-1209.

Duval, B.C., Cramez, C.and Vail, P.R. (1998) Stratigraphic Cycles and Major Marine Source Rocks. In *Mesozoic and Cenozoic Sequence Stratigraphy Of European Basins*, De Graciansky, P.C., Hardenbol, J., Jacquin, T. and Vail, P.R. eds., *SEPM Special Publication*, **60**, 43-51.

Hedges, J.I. and Oades, J.M. (1997) Comparative Organic Geochemistries of Soils and Marine Sediments. *Organic Geochemistry*, **27**, 319-361.

Huc, A.Y. (1988) Sedimentology of Organic Matter. *Dahlem Workshop Report on «Humic Substances and their Role in the Environment»*, F.H. Frimmel and R.F. Christman eds., John Wiley and Sons Ltd, Chichester, 1988, 215-243.

Huc, A.Y. (1995) Paleogeography, Paleoclimate, and Source Rocks. *AAPG Studies in Geology*, **40**, Tulsa, USA.

Huc, A.Y., Vandenbroucke, M.. Bertrand, P., Gayet, J.and Stow, D.A.V. (2001) Organic Sedimentation in Deep Offshore Settings: The Quaternary Sediments Approach. *Marine and Petroleum Geology*, **18**, 513-517.

Katz, B.J. and Pratt, L. (1993) : Source Rocks in a Sequence Stratigraphic Framework. *AAPG Studies in Geology,* **37**.

Katz, B.J. (1995) *Petroleum Source Rocks*, Springer-Verlag.

Tyson, R.V. (1995) *Sedimentary Organic Matter,* Chapman and Hall, London.

Kinetics and Modeling of Hydrocarbon Composition

Behar, F., Vandenbroucke, M., Tang, Y., Marquis, F.and Espitalié, J. (1997) Thermal Cracking of Kerogen in Open and Closed Systems: Determination of Kinetic Parameters and Stoichiometric Coefficients for Oil and Gas Generation. *Org. Geochem.*, **26**, 5/6, 321-339.

Braun, R.L. and Burnham, A.K. (1990) Mathematical Model of Oil Generation, Degradation, and Expulsion. *Energy and Fuels,* **4**, 132-146.

Espitale, J., Ungerer, P., Irwin, H. and Marquis, F. (1988) Primary Cracking of Kerogens. Experimenting and Modelling C_1, C_2-C_5, C_6-C_{14} and C_{14+} Classes of Hydrocarbons Formed. *Org. Geochem.*, **13**, 893-899.

Lewan, M.D. (1997) Experiments on the Role of Water in Petroleum Formation. *Geochimica and Cosmochimica Acta*, **61,** 3691-3723.

McNeil, R.I. and BeMent, W.O. (1996) Thermal Stability of Hydrocarbons: Laboratory Criteria and Field Examples. *Energy and Fuels*, **10**, 60-67.

Pepper, A.S. and Corvit, P.J. (1995) Simple Kinetic Models of Petroleum Formation. Part I: Oil and Gas Generation from Kerogen. *Marine and Petroleum Geology,* **12**, 3, 291-319.

Pepper, A.S. and Dodd, T.A. (1995) Simple Kinetic Models of Petroleum Formation. Part I: Oil and Gas Cracking. *Marine and Petroleum Geology,* **12**, 3, 321-340.

Sweeney, J.J., Burnham, A.K. and Braun, R.L. (1987) A Model of Hydrocarbon Generation from Type I Kerogen: Application to Uinta Basin, Utah. *AAPG Bull.*, **71**, 8, 967-985.

Tissot, B.P. and Espitalié, J. (1975) L'évolution de la matière organique des sédiments : application d'une simulation mathématique. *Revue de l'Institut français dupPétrole*, **30**, 743-777.

Ungerer, P. and Pelet, R. (1987) Extrapolation of Kinetics of Oil and Gas Formation from Laboratory Experiments to Sedimentary Basins. *Nature,* **327,** 52-54.

Ungerer, P. (1990) Kinetic Modelling of Oil Formation and Expulsion. *Org. Geochem.*, **16**, 1-3, 1-25.

Migration of Hydrocarbons

Doligez, B. (1987) Migration of Hydrocarbons in Sedimentary Basins. *Colloques et séminaires,* 45, Éditions Technip, Paris.

England, W.A. and Fleet, A.J. (1991) Petroleum Migration. *Geological Society Special Publication,* 59, Geological Society, London.

Migration Tracers

Ballentine, C.J. and O'Nions, R.K.(1994 The Use of Natural He, Ne and Ar Isotopes to Study Hydrocarbon-Related Fluid Provenance, Migration and Mass Balance in Sedimentary Basins. *Geofluids*, **78**, 347-361.

Battani, A., Sarda, P. and Prinzhofer, A. (2000) Basin Scale Natural Gas Source, Migration and Trapping Traced by Noble Gases and Major Elements: The Pakistan Indus Basin. *Earth Planet. Sci. Lett.*, **181**, 1-2, 229-249.

Bosch, A. and Mazor, E.(1988) Natural Gas Association with Water and Oil as Depicted by Atmospheric Noble Gases: Case Study From the Southeastern Mediterranean Coastal Plain. *Earth Planet. Sci. Lett.,* **87**, 338-346.

Zartman, R.E, Wasserburg, G.J. and Reynolds, J.H. (1961) Helium and Argon in Some Natural Gases. *Journal of Geophysical Research*, **66,** 277-306.

Basin Modeling

Burrus, J. (1997) Contribution à l'étude du fonctionnement des systèmes pétroliers : apport d'une modélisation bidimentionnelle. *PhD Thesis*, École des Mines de Paris.

Rudkiewicz, J.L, Penteado, H.L.B, Vear, A., Vandenbroucke, M., Brigaud, F., Wendebourg, J. and Düppenbecker, S. (2000) Integrated Basin Modeling Helps to Decipher Petroleum Systems. In *AAPG Memoir,* 73, *"Petroleum Systems of the South Atlantic Margins"*, 27-40.

Schneider, F., Wolf, S., Faille, I. and Pot, D. (2000) A 3D Model for Hydrocarbon Potential Evaluation: Application to Congo Offshore. *Oil & Gas Science and Technology - Rev. IFP,* **55**, 3-13.

Ungerer, P., Burrus, J., Doligez, B., Chenet, P.Y. and Bessis, F. (1990) Basin Evaluation by Integrated 2D Modelling of Heat Transfer, Fluid Flow, Hydrocarbon Generation and Migration. *AAPG Bull.*, **74,** 309-335.

Welte, D.H., Horsfield and B., Baker, D.R. (1997) *Petroleum and Basin Evolution. Insights from Petroleum Geochemistry, Geology and Basin Modeling*, Springer-Verlag, Berlin, Heildeberg.

Molecular Markers

Connan, J. (1987) Regard sur la géochimie organique pétrolière d'aujourd'hui. *Bulletin des centres de recherches Exploration-Production Elf-Aquitaine*, **11,** 2, 181-219.

Moldowan, J.M., Albrecht, P. and Philp, R.P. (1992) *Biological Markers in Sediments and Petroleum*, Prentice Hall, New Jersey, USA.

Peters, K.E. and Moldowan, J.M. (1993) *The Biomarkers Guide*, Prentice Hall, New Jersey, USA.

Geochemistry of Gases

Clayton, C. (1991) Carbon Isotope Fractionation During Natural Gas Generation from Kerogen. *MarineaAnd Petroleum Geology*, **8,** 232-240.

Cramer, B., Poelchau, H.S., Gerling, P., Lopatin, N.V. and Littke, R. (1999) Methane Released from Groundwater: The Source of Natural Gas Accumulations in Northern West Siberia. *Marine and Petrol. Geol.*, **16**, 225-244.

Prinzhofer, A. and Huc, A.Y. (1995) Genetic and Post-Genetic Molecular and Isotopic Fractionations in Natural Gases. *Chemical Geology*, **126**, 281-290.

Prinzhofer, A. and Pernaton, E. (1997) Isotopically Light Methane in Natural Gases: Bacterial Imprint or Segregative Migration? *Chemical Geology*, **142**, 193-200.

Prinzhofer, A., Mello, M.R. and Takaki, T. (2000) Geochemical Characterization of Natural Gas: A Physical Multivariable Approach and its Applications in Maturity and Migration Estimates. *AAPG Bull.*, **84**, 8, 1152-1172.

Ricchiuto, T. and Schoell, M. (1988) Origin of Natural Gases in the Apulian Basin in South Italy: A Case History of Mixing of Gases of Deep and Shallow Origin. In *Advances in Organic Geochemistry*, Mattavelli, L. and Novelli, L. eds., 311-318.

Schoell, M. (1983) Genetic Characterization of Natural Gases. *AAPG Bull.*, **67**, 12, 2225-2238.

Alteration of Hydrocarbons in Reservoirs

Anisimov, L.A. (1978) Conditions of Abiogenic Reduction of Sulfates in Oil and Gas Bearing Basins. *Geokhimiya*, **1**, 1692-1702.

Behar, F., Ungerer, P., Kressman, S., Rudkiewicz, J.L. (1991) Thermal Evolution of Crude Oils in Sedimentary Basins: Experimental Simulation in a Confined System and Kinetic Modeling. *Revue de l'Institut français du pétrole*, **46,** 2, 151-181.

Bernard, F., Connan, J. and Magot, M. (1992) Indigenous Microorganisms in Connate Water of Many Oil Fields: A New Tool in Exploration and Production Techniques. *SPE, Paper 24811.*

Connan, J. (1984) *Biodegradation of Crude Oils in Reservoirs, Advances in Petroleum Geochemistry,* eds. J. Brooks and Welte, **1**, 299-335.

Cross, M. (1999) Rates and Mechanisms of Thermochemical Sulphate Reduction. *PhD Thesis,* University of Manchester.

Deroo, G., Tissot, B., Mc Grossan, R.G. and Der, F. (1974) Geochemistry of Heavy Oils of Alberta, Oil Sand, Fuel of the Future. *Can. Soc. Pet. Geol. Mem.*, **3**, 148-189.

Goldstein, T.P. and Aizenshtat, Z. (1994) Thermochemical Sulfate Reduction, A Review. *J. Therm. Anal.*, **42,** 241-290.

Hanin, S., Adam, P., Kowalewski, I., Huc, A.Y., Carpentier, B. and Albrecht, P. (2002) Bridgehead Alkylated 2-Thiaadamantanes: A Novel Markers for Sulfurisation Processes Occurring under High Thermal Stress in Deep Petroleum Reservoirs. *J. Chem. Soc. Chem. Commun., 1750-1751.*

Huc, A.Y., Nederlof, P., Debarre, R., Carpentier, B., Boussafir, S. and Laggoun-Desfarge, F. (2000) Pyrobitumen Occurrence and Formation in a Cambro-Ordovician Sandstone Reservoir, Fahud Salt Basin, North Oman. *Chemical Geology*, **168**, 99-112.

Lafargue, E. and Barker, C. (1988) Effect of Water Washing on Crude Oil Composition. *AAPG Bull.*, **72**, 263-276.

Machel, H.G. (2001) Bacterial and Thermochemical Sulfate Reduction in Diagenesis Settings-Old and New Insights. *Sediment. Geol.*, **140**, 143-175.

Orr, W.L. (1977) Geologic and Geochemical Controls on the Distribution of Hydrogen Sulfide in Natural Gas. In *Advances In Organic Geochemistry*, R. Campo and J. Goni eds., Enadimsa, 571-597.

Schenk, H.J. and Horsfield, B. (1995) Simulating the Conversion of Oil into Gas in Reservoir: The Influence of Frequency Factors on Kinetics Predictions. *17th International Meeting on Organic Geochemistry,* 4-8 September, San Sebastian, Spain, AIGOA, 1102-1113.

Winters, J.C. and Williams, J.A. (1969) Microbiological Alteration of Crude Oil in the Reservoir. Symposium on Petroleum Transformation in Geological Environments. *Div. Pet. Chem. Am. Chem. Soc., Paper Petro.*, **86**, E22-E31.

Woorden, R.H., Smalley, P.C. and Oxtoby, N.H. (1995) Gas Souring by Thermochemical Sulfate Reduction at 140°C. *AAPG Bull.*, **79**, 6, 854-863.

Reservoir Geochemistry

Cubit, J.M. and England, W.A. (1995) The Geochemistry of Reservoirs. *Geological Society Special Publication,* 86, Alden Press, Oxford, UK.

England, W.A. (1990) The Organic Geochemistry of Petroleum Reservoirs. *Organic Geochemistry*, **16,** 1-3, 415-425.

Fowler, M.G. and Peters, K.E. (2002) Application of Petroleum Geochemistry to Exploration and Reservoir Management. *Organic Geochemistry,* **33,** 1, 5-36.

Huc, A.Y., Carpentier, B., Guehenneux, G. and Magnier, C. (1999) Geochemistry in a Reservoir and Production Perspective. *SPE, Paper* 53146.

Kaufman, R.L., Ahmed, A.S. and Elsinger, R.L. (1990) Gas Chromatography as a Development and Production Tool for Fingerprinting Oils from Individual Reservoirs: Applications in the Gulf Coast Of Mexico. *Gulf Coast Section of The Society of Economic Paleontologists and Mineralogists Foundation, 9th Annual Research Conference Proceedings*, 263-282.

Lomando, A.J. (1992) The Influence of Solid Reservoir Bitumen on Reservoir Quality. *AAPG Bull.*, **76,** 8, 1137-1152.

Smalley, P.C. and England, W.A. (1992) Assessing Reservoir Compartimentalisation During Field Appraisal: How Geochemistry Can Help. *SPE, Paper 25005.*

Rock-Eval

Behar, F., Beaumont, V. and Penteado, H. (2001) Rock-Eval 6 Technology: Performances and Developments. *Revue de l'Institut français du Pétrole,* **56,** 111-134.

Espitalié, J., Laporte, J.L., Madec, M., Marquis, F., Leplat P. and Paulet, J. (1977) Méthode rapide de caractérisation des roches mères, de leur potentiel pétrolier et de leur degré d'évolution. *Revue de l'Institut français du pétrole*, **32**, 23-45.

Espitalié, J., Marquis, F. and Barsony, I. (1984) *Geochemical Logging; in Analytical Pyrolysis-Techniques and Applications,* K.J. Voorhees ed., Boston, Butterworth, 276-304.

Lafargue, E., Marquis, F. and Pillot, D. (1998) Rock-Eval 6 Applications in Hydrocarbon Exploration, Production, and Soil Contamination Studies. *Revue de l'Institut français du pétrole,* **53**, 4, 421-437.

Peters, K. (1986) Guidelines for Evaluating Petroleum Source Rock Using Programmed Pyrolysis. *AAPG Bull.*, **70**, 318-329.

Final manuscript received in January 2003

Oil & Gas Science and Technology – Rev. IFP, Vol. 58 (2003), No. 2, pp. 243-269
Copyright © 2003, Éditions Technip

Kerogen: from Types to Models of Chemical Structure

M. Vandenbroucke[1]

1 *Institut français du pétrole, division Géologie-Géochimie, 1 et 4, avenue de Bois-Préau, 92852 Rueil-Malmaison Cedex - France*
e-mail: mireille.vandenbroucke@ifp.fr

Résumé — Les kérogènes : de la notion de types aux modèles de structure chimique — Le but de cet article est de présenter, à travers une revue des développements conceptuels et analytiques en géochimie dans la période 1970-1990, les principales étapes qui ont conduit, à travers la préparation et l'analyse des kérogènes d'une part, et l'étude de systèmes pétroliers de référence d'autre part, aux notions de types, chemins d'évolution et modèles statistiques de structure chimique des kérogènes.

Le kérogène est défini comme la matière organique des sédiments qui produit le pétrole, opposant ainsi le pétrole, soluble dans les solvants organiques courants, à son complément insoluble. Ce kérogène étant une matière organique complexe intimement mélangée aux minéraux des sédiments, la première tâche a été de mettre au point un protocole fiable pour son isolement, permettant ensuite son étude par des méthodes physicochimiques variées. En parallèle, le développement des permis d'exploration pétrolière, donnant accès aux échantillons de forage, permettait la comparaison géochimique de divers systèmes pétroliers. Les comparaisons furent faites non seulement entre les compositions chimiques des huiles, roches mères et kérogènes, mais aussi sur l'époque de production du pétrole. La notion de cinétique de craquage du kérogène en pétrole résultant de ces études de systèmes pétroliers, associée à l'observation d'une compensation des facteurs temps et température, conduisit à utiliser la pyrolyse en laboratoire pour évaluer le potentiel pétrolier restant à produire par le kérogène, et donc, à la construction de l'appareil Rock-Eval. Au milieu des années 1970, les principaux paramètres géochimiques des kérogènes de certaines séries pétrolières de référence étaient disponibles pour la définition des notions de type et de chemin d'évolution des kérogènes au cours de l'enfouissement géologique des sédiments.

Un pas important dans la connaissance de la composition chimique du kérogène fut franchi dans les années 1980 avec l'utilisation de nouvelles techniques de pyrolyse analytique et préparative, et leur couplage avec différents détecteurs. Les produits de pyrolyse, considérés comme briques élémentaires issues du craquage thermique du kérogène, pouvaient ainsi être analysés et quantifiés au niveau moléculaire, sans les problèmes de représentativité rencontrés avec l'analyse des extraits naturels et dus aux pertes de composés volatils et de produits ayant migré hors des roches mères. Le développement parallèle de la résonance magnétique nucléaire du ^{13}C en phase solide permit l'analyse et la quantification des diverses formes du carbone et de son environnement chimique dans le kérogène, ce qui n'était pas possible auparavant sous forme quantitative par la spectroscopie infrarouge ou ultraviolette. L'analyse quantitative des briques moléculaires et des groupes fonctionnels qui les assemblaient dans le kérogène permit ainsi d'élaborer des modèles moléculaires statistiques de kérogènes dans le but de visualiser l'ensemble de leur composition atomique et moléculaire, ainsi que les changements de composition liés aux types et à la maturité géologique. Il ne sera jamais possible de représenter une structure réelle de kérogène, tout simplement parce que le kérogène est un mélange de macromolécules variées, et non de polymères, et donc leur analyse individuelle est impossible. Cependant, une structure hypothétique moyenne, prenant en compte un grand nombre d'informations issues d'analyses variées, permet de visualiser de façon synthétique les principales ressemblances et différences entre les matières organiques sédimentaires.

Abstract* — *Kerogen: From Types to Models of Chemical Structure — *The aim of the present paper is to review the conceptual and analytical developments in the period 1970-1990 which have led, through kerogen isolation and analysis on one hand, and case studies of petroleum systems on the other hand, to the concept of kerogen types, evolution paths, and statistical chemical models.*
Kerogen is defined as the sedimentary organic matter generating petroleum, an insoluble product as opposed to its counterpart soluble in usual organic solvents, such as petroleum. As kerogen is a complex organic material intimately mixed with minerals in sediments, the first task was to set up a robust procedure for its isolation, enabling then its study by various physicochemical analyses. The parallel development of oil exploration, resulting in geological sample availability, made the geochemical comparison of various petroleum systems possible. Comparisons concerned not only oils, source rock extracts and kerogen compositions, but also the timing of petroleum generation. The notion of kinetic cracking of kerogen into petroleum stemming from these case studies, associated with the observation of time and temperature compensation, resulted in the use of pyrolysis to evaluate the oil potential still to be generated by the kerogen, and in the construction of the Rock-Eval. In the mid 70s, all main parameters on kerogens from reference series of source rocks were available to define the notions of types and evolution paths of kerogens upon geological maturation.
A further important step for improving the knowledge of kerogen composition was achieved in the 80s using new techniques of analytical and preparative pyrolysis and their coupling with different detectors. The pyrolysis products, small building blocks issued from the kerogen thermal cracking, could thus be analyzed and quantified at the molecular level, without the problems of representativity associated with natural extract analyses, such as loss of volatile fractions or product migration out of source rocks. Contemporaneous developments in solid state ^{13}C *NMR allowed quantification of the various forms of carbon and their molecular environment in the kerogen, whereas quantification was not possible with IR or UV spectroscopy. The quantification of molecular building blocks and their bonding functional groups in kerogens allowed conceptual averaged molecular models of kerogens to be proposed in order to visualize their atomic and molecular composition, and the changes occurring in this composition according to types and maturity. Although it will never be possible to represent a true kerogen structure, simply because it is a mixture of various nonpolymeric macromolecules, an hypothetical average structure of kerogen, representing a large amount of information from various analyses, can provide a synthetic view of the main resemblances and differences among sedimentary organic matters.*

INTRODUCTION

Kerogen is defined as the insoluble sedimentary organic matter capable of generating petroleum. The fact that this sedimentary organic matter is usually intimately mixed with minerals during its deposition was of course a source of difficulties because its physical isolation, or at least concentration, was required before analysis, a problem that did not arise for coal studies. The onset of research work in coal science and organic geochemistry also aimed at different industrial objectives, downstream projects with gasification and liquefaction for coal, upstream projects with petroleum exploration for kerogen. For these reasons chemical studies on coal began much earlier than those on kerogens, with the publication of a major textbook, *Coal*, by van Krevelen in 1961, whereas at the same period the main efforts on kerogen were devoted to development and comparison of isolation methods (Forsman, 1963). Without minimizing the impact of these early papers and geochemistry books on the progressive understanding of benefits that petroleum exploration would draw from organic geochemistry studies, the publication of a major textbook on the origin and occurrence of petroleum, by B. Tissot and D. Welte, occurred in 1978, and the corresponding textbook on kerogen, edited by B. Durand, was published in 1980.

The aim of the present paper is thus to review the main steps of conceptual and analytical developments in organic geochemistry during the period 1970-1990 which have led, through kerogen isolation and analysis on one hand, and case studies of petroleum systems on the other hand, to the concept of kerogen types, evolution paths, and chemical structural models. Although kerogen is also a major sink in the global carbon cycle, the processes of kerogen formation in present and past geological times, in other words the mechanisms of transformation and selective preservation of the organic matter finally incorporated into sediments, will not be discussed in this paper. Indeed these topics are still a matter of debate and active research. Despite numerous advances in knowledge of kerogen precursors, many important points are still far from being elucidated.

1 DEFINITION AND ISOLATION OF KEROGEN

1.1 Usual Definition and Related Ambiguities

1.1.1 The Broadening of Successive Kerogen Definitions

The word kerogen was coined by Crum-Brown (1912) and used to describe the organic matter of a Scottish oil shale that produced a waxy oil upon distillation (keros = wax). This definition was thus restricted to low maturity organic-rich rocks of economic importance. Further on, recognizing that organic matter in sedimentary rocks, even at concentrations ranging around 1%, can generate oil either by artificial pyrolysis or by burial for long periods of time, White (1916), then Trager (1924), both cited by Durand (1980), extended this early economic definition to all organic matter in rocks capable of oil generation.

It should be noted that this conceptual definition implies that kerogen and petroleum have complementary characteristics, as recognized by Forsman and Hunt (1958). Accordingly, petroleum being soluble in usual organic solvents, they defined in turn kerogen as the sedimentary organic matter insoluble in these solvents. However, even this now largely accepted operational definition leads to a whole range of different procedures aiming at physical or chemical concentration or isolation of kerogen, as discussed below.

The definition of Forsman and Hunt was later extended by Durand (1980) to all insoluble sedimentary organic materials, including not only materials dispersed in sedimentary rocks but also almost "pure" organic deposits such as humic and algal coals, as well as the insoluble organic matter from recent sediments obtained after removal of the humic fraction that is soluble in aqueous basic solutions. The rationale is that all these substances represent various types of sedimentary organic matter and/or different stages of its transformation process after geological burial.

1.1.2 Ambiguities Resulting from the Definition of Kerogen

It is important to keep in mind that the name kerogen, in opposition with usual chemistry nomenclature, does not represent a substance with a given chemical composition. Indeed kerogen is a generic name, in the same sense as lipids or proteins. Several organic precursors and their mixtures may lead to kerogen incorporation in sediments. Moreover, with geological burial of sediments at temperatures ranging from a few tens to less than 200°C, during time periods from tens to thousands of million years, the chemical composition of kerogen dispersed in sedimentary layers (called source rocks) progressively changes, a part of it being transformed into petroleum and gas. The operational definition of kerogen on the basis of insolubility accounts neither for these compositional variations due to source and evolution, nor for a possible mixture of the substance able to generate petroleum with other insoluble organic compounds like coke for example, resulting from further petroleum transformation into source rocks.

Another major drawback of the pragmatic definition of kerogen by its insolubility in organic solvents, in contrast with oils or bitumens solubility, is that the composition and chemical features of kerogen will be closely dependent on the organic solvent and extraction procedure used for this separation. As a consequence, analytical data on kerogens can be compared only if the same solvent is used. Unfortunately, there are no established standard protocols regarding extraction procedures, either in terms of the solvent polarity or for sample grinding, temperature, duration of extraction and stirring efficiency. The result is that extraction can never be considered as "complete" since any change in these parameters may result in extraction of further compounds. Polar solvents used in the past, such as chloroform, mixtures of benzene and methanol, or pyridine frequently used by coal scientists, have been almost completely abandoned now due to their high toxicity. Petroleum geochemists use generally slightly polar solvents with rather low boiling points, and most often dichloromethane, sometimes mixed with methanol. The reason for choosing this type of solvent is that bitumen resulting from kerogen transformation in source rocks contains mainly hydrocarbons, representative of major components of crude oils, most of which can be efficiently extracted with such solvents and further analysed by a variety of analytical methods.

Beside solvent extraction needed for separation of soluble and insoluble organic matter in rocks, analytical techniques used for kerogen characterization often require its isolation from minerals, or at least its concentration. This can be done by more or less efficient physical or chemical methods. A review of these different procedures can be found in Forsman (1963), Robinson (1969), Durand and Nicaise (1980), and for the most recent ones in a review paper on kerogen (Vandenbroucke and Largeau, in press). In consequence, only the usual chemical procedure set up at *IFP* will be described here.

1.2 *IFP* Chemical Procedure for Kerogen Isolation

1.2.1 Basic Principles of the Kerogen Preparation

This procedure was designed by Durand and Nicaise (1980) on the basis of previous work by Forsman and Hunt (1958) and Saxby (1970), and its flowchart is presented on Figure 1.

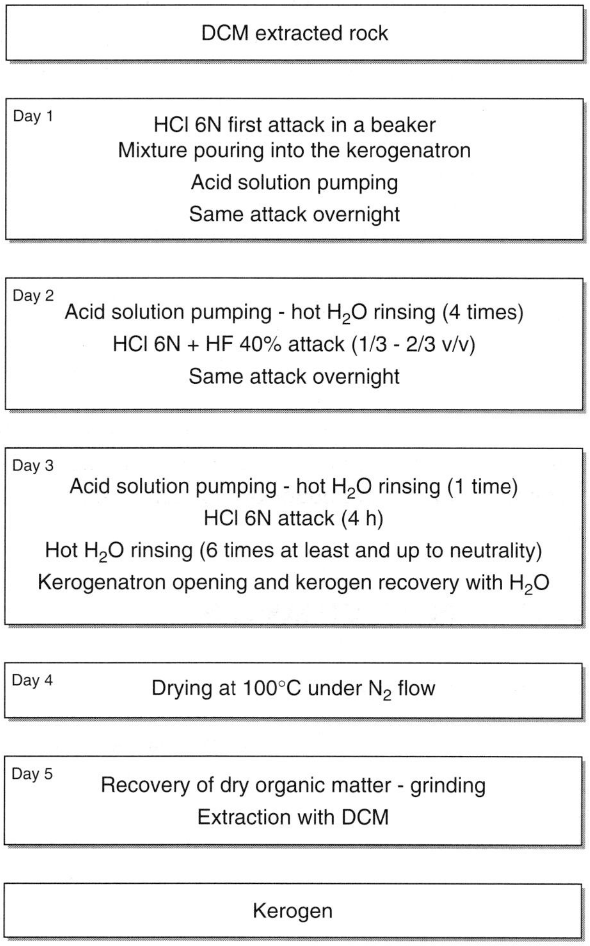

Figure 1

IFP procedure for kerogen isolation.

It is based on the destruction of major minerals from the finely ground sedimentary rock by nonoxidant acid attacks at temperatures between 60 and 70°C under an inert atmosphere, in order to prevent any oxidation of kerogen during preparation. Carbonates, sulfides, sulfates and hydroxides are first destroyed by two successive attacks with 6N hydrochloric acid, then clay minerals, quartz and silicates are destroyed also by two successive attacks using a mixture of 40% hydrofluoric and 6N hydrochloric acids (resp. 1/3 and 2/3 v/v). Each first acid attack lasts 4 h, and the second acid attack is performed overnight. In defining the analytical procedure, some minerals were observed to remain as residues with kerogen after these acid attacks. Two classes of resistant minerals were recognized by the authors cited above: firstly, newly formed fluorides resulting from the hydrofluoric acid attack such as ralstonite $Na_xMg_xAl_{2-x}(FOH)_6H_2O$ and other complex fluorides (Robinson, 1969; Durand and Nicaise, 1980); secondly, minerals resisting these acids, mainly pyrite but also minor heavy oxides such as zircon, rutile and anatase. The newly formed fluorides are complex and very difficult to redissolve once precipitated. Since they interfere with further analysis of kerogen, it is important to prevent their formation. According to Durand and Nicaise (1980), the best way to do so is first to perform several thorough rinsings with hot distilled water between acid attacks without filtering the residue to dryness, in order to eliminate a maximum amount of cations. A second and very important step of the preparation, forgotten by many users of this procedure, is to perform a new 6N hydrochloric attack during 4 h after the HF/HCl destruction of silicates, in order to eliminate the remaining ions able to form complex fluorides upon drying. After rinsing to neutrality, the kerogen concentrate can then be filtered to dryness, recovered and dried at 100°C under an inert gas flow. Once dry, it can be weighed in order to control recovery, ground and stored under inert gas for further steps as described below.

Except for very immature sediments which can loose a few percent of their organic carbon by HCl hydrolysis during the demineralization procedure, these chemical methods do not alter significantly the structure of kerogens, and do not generate newly formed solvent soluble organic material, as tested by Larsen *et al.* (1989) on a series of coals of varying maturity starting from immature lignites. However as already indicated above, solvent extraction of whole source rocks before acid attacks does not remove the totality of solvent soluble compounds. Indeed trapped extract is released during acid attacks by both destruction of minerals enclosing some extract in micropores and unfolding of kerogen molecules, due to breaking of hydrogen bonds inside the kerogen structure. It is thus necessary to perform a second extraction once the organic matter has been concentrated. Alternatively, the acid attacks may be performed directly on the whole sedimentary rock and a single extraction done afterwards, but it is not a good procedure if the source rock is rich in extractable compounds, these latter preventing a good contact of minerals by the aqueous acid solutions used for attacks.

Once isolated by this procedure, the kerogen concentrate still contains residual minerals that could not be eliminated by the nonoxidant acids selected for attacks, as noted above. The main remaining mineral is pyrite, frequently enclosed inside the organic network of kerogen, as a normal by-product of bacterial kerogen formation and incorporation in some sediments. However, many of the classical analyses on kerogens can be done even with pyrite contents up to 40 wt%. Several methods were tested for pyrite elimination from kerogen concentrates in view of specific analyses, but efficient procedures have been set up only recently. They will not be described in detail here; the reader can refer for example to the papers of Acholla and Orr (1993) for chemical destruction and of Stankievicz *et al.* (1994) for density separation.

1.2.2 Description of Materials and Procedures

The successive acid attacks and rinsing procedure is performed in a specially designed cell *(Fig. 2)* made of a fluorinated translucent polymer (Voltalef) allowing to check the kerogen *vs.* liquid level during filtrations. The cell is placed into a heating block permanently adjusting a preset temperature of cell liquids. Its upper part comprises two threaded holes, one open for introduction of initial rock and liquid solutions and escape of inert gas excess, the second connected to a distribution ramp with controlled nitrogen flow and ending by a pipe bubbling into the liquid to insure both mixing and an inert atmosphere. The bottom of the cell is threaded for opening and maintains a grid supporting millipore Teflon filters between Teflon gaskets. The solutions can be pumped through the filters by a flexible tube regulated by a cock and connected to a peristaltic pump. All connecting parts in contact with acid solutions are of course made of HF and HCl resistant materials.

The weighed finely ground rock is first placed into a glass beaker, and a preliminary 6N HCl attack is performed at room temperature outside the preparation cell, in order to control a possible abundant foam formation if the initial carbonate content is high. In this latter case, foam is decreased by addition of a few drops of acetone. The ground rock suspension is then introduced via a funnel into the cell, the initial acid solution is drained and fresh 6N HCl is added. Heating and nitrogen bubbling are then applied and the succession of attacks and rinsing is performed as described previously. At the end of the whole preparation procedure, the cell is removed form the oven, its bottom is unscrewed and the walls are rinced with deionized water over a small crystallizer to recover the kerogen. The same is done with the Teflon gaskets and the filters with the kerogen. When all kerogen is recovered, the crystallizer is placed into a heating dessicator at 100°C under a nitrogen flow to complete drying. The dry kerogen is then crushed in an agate mortar. As isolated kerogen is very prone to oxidation, it is very important to store it under nitrogen atmosphere into a tightly sealed glass container in the dark, and to repeat this storage procedure each time an aliquot of kerogen is sampled for analyses.

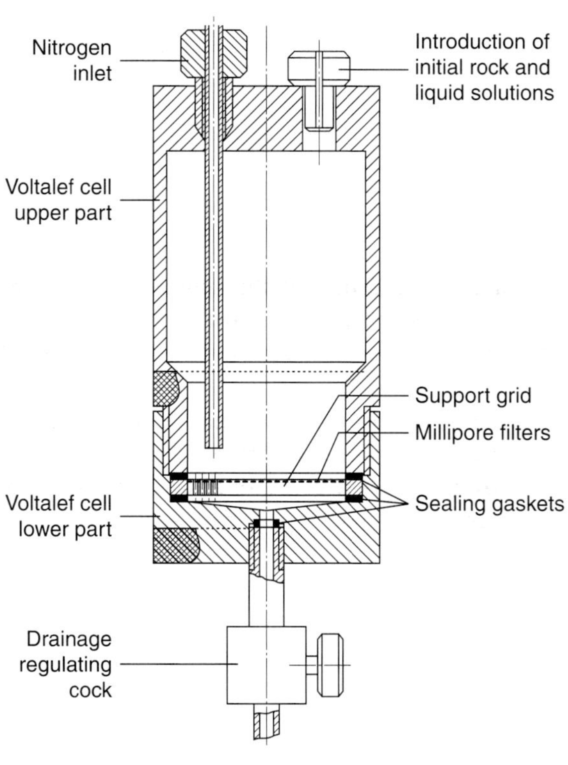

Figure 2

Schematic cross section of a cell designed at *IFP* for kerogen preparation ("kerogenatron").

2 USUAL ANALYSES ON KEROGEN

2.1 Elemental Analysis

The major elements in kerogen concentrates are the following: C, H, N, O, S, and possibly Fe from pyrite, with generally C and N amounting respectively for the highest and the lowest weight %. Elemental analysis is considered reliable if the sum of these elements accounts for more than 90 wt%, but Durand and Monin (1980) cite an average balance of 95% on 427 kerogen analyses. Given this incomplete mass balance and an average oxygen value of 11% on 324 kerogen analyses, it is very important that oxygen be measured directly and not calculated by difference. This constraint was already pointed out by coal scientists in the 50s, and a review of methods for direct determination of oxygen, and their comparison with oxygen obtained by difference on a series of reference coals, can be found in Ehmann *et al.* (1986).

The *IFP* geochemistry laboratory being not equipped with the various microanalysis apparatus needed for a correct determination of these elements in kerogens, it was decided from the beginning of our studies to carry out these analyses through a specialized laboratory, the same for now more than thirty years. C, H and N are determined on one aliquot of the kerogen by thermal conductivity detection of gases produced from combustion at 1000°C. Oxygen is measured on another aliquot by pyrolysing the kerogen under nitrogen flow, transforming oxygen-containing gases to CO, then oxidizing CO to CO_2, followed by quantification of CO_2 by coulometry. Total S including pyritic and organic S is obtained by oxidation of a third aliquot of kerogen followed by quantitative coulometry of SO_2. The determination of Fe is necessary to calculate the amount of pyrite FeS_2 in the sample; the organic sulfur may then be calculated by difference of pyritic sulfur with the measured total sulfur. This analysis is done on a fourth aliquot by mineralization of the sample by a mixture of nitric and sulfuric acids, then quantification of Fe by atomic absorption. Other minerals resisting acid attacks, such as barite, rutile and anatase, may be estimated by the difference to 100% of the sum of previously mentioned elements. Ashes after combustion may also be weighed in order to approximately crosscheck the obtained data. However if pyrite is present, this crosscheck should account for the oxidation of pyrite during combustion.

These analyses are used for classifying the kerogens in a van Krevelen diagram that will be detailed below. This diagram plots the main elements, C, H and O as atomic ratios of H/C *vs.* O/C. Starting from the wt% analyses, these atomic ratios are obtained by multiplying the weight ratio of H to C by 12, and the weight ratio of O/C by 0.75 (*i.e.* 12/16). If other atomic ratios are needed, the same type of calculation is done, multiplying the weight ratio by the reverse atomic weight ratio of the elements considered; for example the atomic N/C ratio is their weight ratio multiplied by 12/14.

2.2 Pyrolysis

2.2.1 Basic Principles of Pyrolysis Applied to Kerogen Characterization

Pyrolysis is a widely used degradation technique that allows breaking a complex substance into fragments, by heating it under an inert gas atmosphere. The small compounds thus obtained are building blocks of the complex substance, but they can often be analyzed more easily, eventually up to a molecular level, and quantified. Applying this technique to hydrocarbon generation from kerogen thermal cracking also means that geological conditions with long time intervals at low temperature can be replaced by laboratory conditions with short experiment duration at high temperature, of course in a defined domain where cracking reactions are similar. Accordingly, pyrolysis has become immediately after the development of the kerogen isolation procedure a major technique for geochemical studies on kerogen, both for characterization of its molecular building blocks and for kinetic studies considered later in this paper.

Several types of pyrolysis systems can be used for performing kerogen cracking. Open systems, where effluents are quickly swept away from the hot zone by an inert gas flow or by vacuum extraction, are the easiest for monitoring the formation of molecular building blocks from kerogen, since secondary reactions between generated products should not occur. The operating conditions of such systems are however very far from geological conditions as no pressure can be applied to the kerogen, and pyrolysis occurs simultaneously with thermal extraction, usually with a programmed temperature ramp from 300 to 650°C. The high final temperature precludes operating on a whole rock, because of catalytic artefacts by dehydrated minerals (Espitalié *et al.*, 1984). Qualitative analytical pyrolysis can be performed by coupling the pyrolysis chamber to a GC column, maintained at low temperature during the pyrolysis step to trap effluents on the column head, then transporting them selectively by the usual temperature program. However quantification is possible only for products carried through the GC column to the detector. It can be seen, by comparing GC quantification with the weight of pyrolysate fractions recovered directly in a cold trap, that recovery depends both on the molecular weight and the chemical structure of products, as shown in Table 1. Total effluent production is correctly quantified only if products are collected at the output of the pyrolysis oven, either as chemical fractions if the pyrolyzate is first recovered into a cold trap, or as carbon weight if analyzed continuously with a flame ionization detector. Two of these systems conceived at IFP are successively detailed below.

TABLE 1

Comparison of hydrocarbon amounts measured by pyrolysis-gas chromatography (Py-GC) and preparative pyrolysis in C_6-C_{13} and C_{14+} carbon ranges, for a maturity series of type II kerogens. Consistency of results for C_6-C_{13} hydrocarbons and their difference for C_{14+} hydrocarbons shows that in the latter, a part of hydrocarbons (mainly aromatics from analyses) does not pass through the chromatographic column. Higher quantities measured by Py-GC for the most mature kerogen are due to loss of very light hydrocarbons upon preparative pyrolysis work up (argon evaporation)

Type II kerogen series parameters				C_6-C_{13} pyrolysate		C_{14+} pyrolysate**	
at. H/C	at. O/C	HI (mg/g C)	T_{max} (°C)	Py-GC (wt %)*	Prepar. Py (wt%)*	Py-GC (wt%)*	Prepar. Py (wt%)*
1.24	0.114	661	423	5.68	4.58	9.76	12.04
1.29	0.101	680	426	4.32	4.35	7.38	13.35
1.28	0.084	730	435	5.84	5.48	10.86	16.93
1.07	0.056	590	442	5.40	5.38	7.64	15.33
0.86	0.057	260	446	3.37	3.49	3.86	7.90
0.6	0.070	36	456	1.02	0.63	0.82	0.50

* weight % of the pyrolysed kerogen.
** sum of saturates, unsaturates and aromatics only.

2.2.2 Preparative Pyrolysis System for Quantification of Kerogen Structural Components

The preparative pyrolysis device *(Fig. 3)* is a cylindrical minifurnace with gold coated inner wall, as described in previous publications (Béhar and Pelet, 1985; Vandenbroucke *et al.*, 1988). A weighed kerogen amount loaded on a gold rod is introduced into the pyrolysis chamber under argon flow. After preliminary heating at 300°C for 3 min to flush any remaining solvent or water, the temperature is raised to 600°C using a temperature program similar to that of classical Rock-Eval pyrolysis. Effluents are swept away and condensed, together with the carrier gas, in a glass trap cooled by liquid nitrogen. After pyrolysis completion, the trap is disconnected and open to let argon evaporate, then the pyrolysate is recovered by addition of solvent at room temperature in the trap, resulting in loss of C_1-C_5 products and nonhydrocarbon gases, that can be analyzed separately using a specific pyrolysis device connected to a vacuum line (Lorant and Béhar, 2002). Two preparative pyrolysis experiments are performed for each kerogen sample, one for pyrolysate recovery with *n*-pentane and analysis of the C_{6+} fraction (Lafargue and Béhar, 1989), the other for pyrolysate recovery with dichloromethane (DCM) and analysis of the C_{15+} fraction (Béhar *et al.*, 1989).

Fractionation and Molecular Analysis of the Pentane Solution by GC Quantification

The pentane solution is fractionated by microcolumn liquid chromatography on silica into total C_{6+} saturates (pentane elution) and C_6-C_{14} aromatics (elution by a mixture pentane/dichloromethane 90/10 v/v), then quantified by GC/FID (flame ionisation detector). Other products dissolved into pentane are not eluted using these solvent mixtures. However the quantification of the C_6-C_{14} fraction may be hampered by a partial evaporation of C_6-C_9 compounds during preparation. Two standards are used for calibrating the FID. The detector response is linear in the range 0.06 to 1.29 µg/µl and is controlled regularly. This response coefficient is used for the quantification of total amounts of C_{6+} saturates plus unsaturates and C_6-C_{14} aromatics, by difference of GC areas of fractions and blanks.

Molecular analysis of the C_{6+} *n*-alkanes and alkenes is performed by GC-FID. The C_6-C_{14} aromatics are identified by GC-MS (mass spectroscopy) analysis and their relative amount is quantified both on the total ionic current (TIC) trace and the FID trace with the same response factor as for saturates. These two quantifications give fairly similar results in this carbon range because the aromatic fraction contains mainly mono- and di- alkylaromatics plus alkylthiophenes, with a FID response close to that of saturates; however benzothiophenes have a significantly lower response factor and should be corrected.

Fractionation and Molecular Analysis of the C_{15+} Compounds

The DCM solution is evaporated and the recovered pyrolysate weighed. It is then partly redissolved into pentane and separated by microcolumn liquid chromatography on

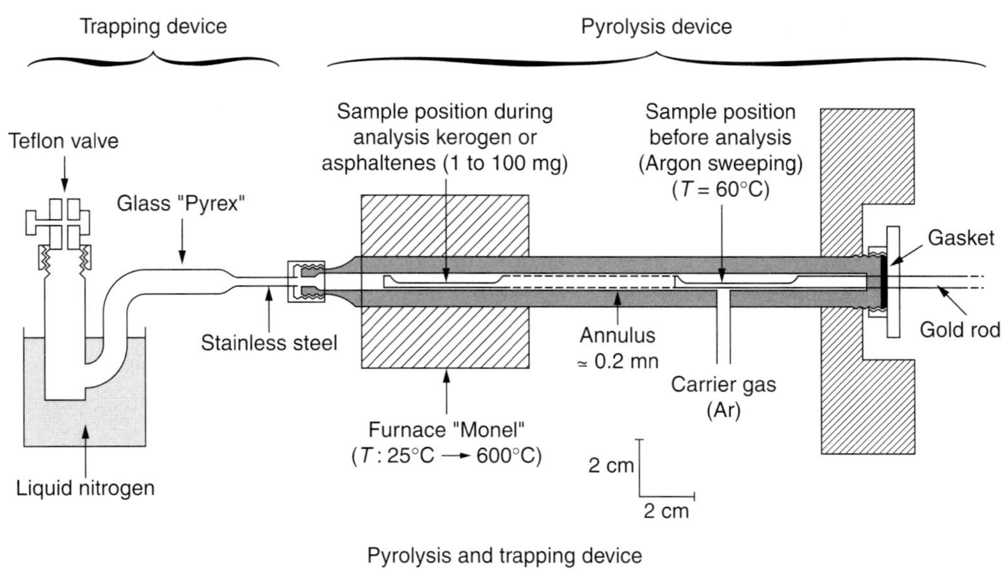

Figure 3

Preparative pyrolysis device designed at *IFP*. This device, also named "S_2 analyser", has a temperature programmable oven, allowing pyrolysis of a sample under Rock-Eval conditions, and quantitative recovery and analysis of pyrolysis effluents in the S_2 peak.

silica into saturates plus unsaturates (pentane elution), aromatics (pentane/dichloromethane elution, 65/35 v/v) and N, S and O containing compounds (NSO) (acetone elution plus pentane insolubles). After solvent evaporation, the three fractions are weighed. Mass spectrometric group type analysis of the C_{15+} saturates and aromatics may be performed on these fractions after their mild catalytic hydrogenation (5% rhodium on alumina, 120°C, pH_2 40 bar). NSOs are submitted to elemental analysis.

2.2.3 The First Rock-Eval Pyrolysis Device

It is not necessary to present once more the Rock-Eval technique and its worldwide impact on the introduction of organic geochemistry in petroleum companies. This section will thus only focus on the principal studies at the origin of the conception of this revolutionary commercial apparatus, capable of characterizing source rock type, evolution stage and present and remaining hydrocarbon potential, by an analysis lasting less than half an hour. It must be recalled that before Rock-Eval analysis, geochemical studies were performed on a limited number of oils and rock extracts which had to be prepared, fractionated and analyzed by various physicochemical techniques to obtain molecular parameters allowing classification and correlation studies. The same was true for kerogen preparation and analysis, thus the whole geochemical analysis on a sample lasted several days and was quite expensive.

Early attempts to characterize source rock properties such as origin and stage of evolution of organic matter by pyrolysis were those of Bordenave *et al.* (1970) and Giraud (1970). Isothermal pyrolyses were performed on the crushed rock at 280°C during 10 to 30 min and the effluent trapped in liquid air, then was flushed by helium into a chromatographic column, with analysis of the C_1-C_{11} fraction. They observed that continental source rocks produced mainly aromatic hydrocarbons whereas marine source rocks produced more paraffins than aromatics. Kerogen evolution stage was determined by pyrolysis of the whole source rock at 500°C, oxidation of the effluent in CO_2, with quantification by gas chromatography. The evolution stage was the ratio of this carbon pyrolysable to CO_2 to the total carbon; it was recognized to depend on the origin of the source rock, but much more on its geological evolution stage. Another pyrolysis technique where the oven containing the source rock was directly connected to a flame ionization detector (Leplat and Noël, 1972) was shown to produce pyrograms, the shape of which was characteristic of marine kerogens *vs.* coals. Their maturity was later estimated by connecting the oven to a short chromatographic column used as a cold trap during pyrolysis at 500°C, followed by GC analysis of generated CO_2 and C_2H_4 using a thermal conductivity detector (Leplat and Noël, 1974).

Many studies were also performed at *IFP* on pyrolysis effluent analysis and its comparison to natural equivalents. A precursor of the preparative pyrolysis system described above was constructed, with programmed heating of kerogen up to 500°C under helium flow, and subsequent trapping of effluents in a chloroform bath at minus 5°C. The pyrolysates of many reference source rocks, recovered after chloroform evaporation, could then be analyzed by the same methods as used for extracts, and the evolution of kerogen residues compared to that in natural series (Vandenbroucke *et al.*, 1977; Huc, 1978; Monin *et al.*, 1980). Thermogravimetry and pyrolysis effluent analysis by mass spectrometry was also a major research topic at this time. Many analyses were performed on generation of H_2O, CO_2, SO_2, H_2S, CH_4 and hydrocarbons by kerogens, associated with their source rock or isolated from minerals (Souron *et al.*, 1974; Souron *et al.*, 1977; Durand-Souron *et al.*, 1982). The studied samples were heated under vacuum in the source of a mass spectrometer, programming the temperature up to 500°C at 5°C per minute. Comparison of gases produced in presence or absence of minerals showed that CO_2 issued from the organic matter should not be analyzed above 390°C because some mineral carbonates, and particularly siderite, could decompose beyond this limit.

This series of studies on kerogen pyrolysis products enabled the conception and production of a first commercial apparatus, the Rock-Eval 1 (Espitalié *et al.*, 1977a; Espitalié *et al.*, 1977b). This first generation Rock-Eval apparatus was an open pyrolysis system, with inert gas flow extracting the effluents generated during programmed heating of the weighed whole rock placed in a porous crucible. Heating to 300°C resulted in a thermovaporization of hydrocarbons already present as such in the rock (S_1 peak), whereas hydrocarbons produced in the 300-550°C temperature range resulted from kerogen cracking (S_2 peak). This latter peak becoming smaller and produced later with increasing natural evolution, the pyrolysis temperature for which the peak was at its maximum was used as a maturity parameter. As a result of mass spectrometric studies, CO_2 from kerogen cracking was analyzed only for temperatures under 390°C. Products carried out of the heated rock by the gas were directed through a splitter with flow regulating valves, either to a flame ionization detector for on-line quantitative analysis of hydrocarbons, or to a trap with molecular sieves connected to a Porapack column and a thermal conductivity detector. At the end of the pyrolysis, the trap was heated to release CO_2 which was then quantified (S_3 peak). The main parts and operational flowsheet of this apparatus are shown on Figure 4.

Rock-Eval pyrolysis determines kerogen properties inside source rocks, without need to isolate it from minerals. Although the initial objective of Rock-Eval pyrolysis was the screening of source rocks in exploration wells, the method has been used for many other studies, including its almost

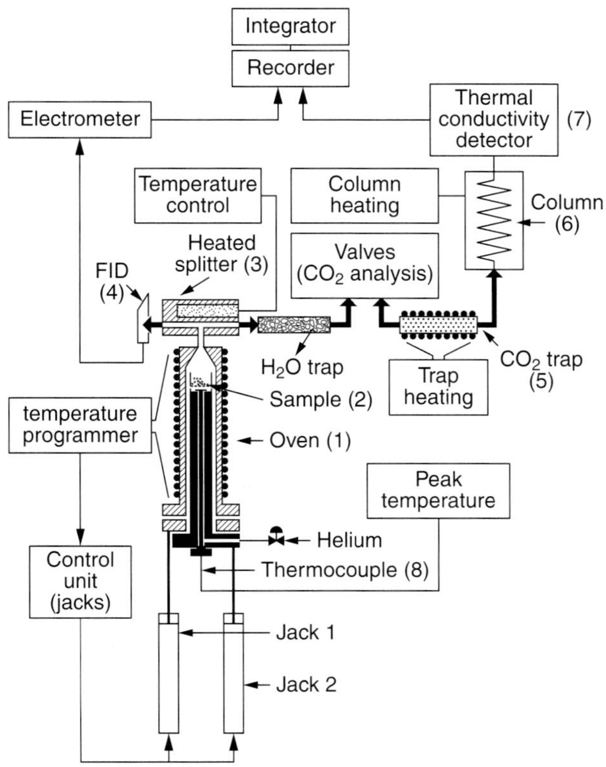

Figure 4

Schematic cross section of the first commercial Rock-Eval (from Espitalié *et al.*, 1977).

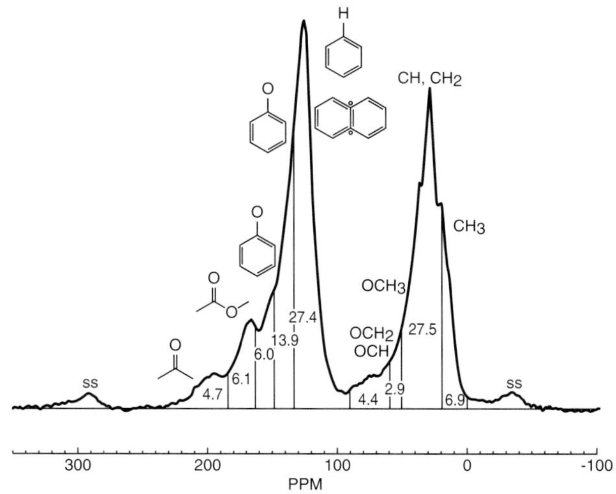

Figure 5

Example of a lignite spectrum obtained with solid state ^{13}C RMN (aromaticity factor 0.58), and identification and quantification of peaks in wt% of total carbon. ss (spinning sidebands) peaks do not belong to the coal sample.

systematic application to kerogen analysis in *IFP*, resulting in useful complementary analytical data. Several versions of the Rock-Eval apparatus have been constructed since the time of its first industrial production. The latest version, Rock-Eval 6, has several new functionalities associated with major technical modifications (Lafargue *et al.*, 1998; Béhar *et al.*, 2001), allowing much more robust and informative parameters to be obtained.

2.3 Solid State ^{13}C NMR

Adaptation of ^{13}C NMR to solid samples at the beginning of the 80s has shown a great potential for quantitative analysis of kerogen functionalities. Indeed it is the only method that can give, under controlled experimental conditions, quantitative estimates of the different carbon bond types in kerogens, as opposed to infrared spectroscopy where only qualitative comparisons are possible. Without entering into instrumental details, removal of dipolar interactions in the solid by magic angle spinning (MAS), and cross polarisation (CP) between carbons and protons using appropriate contact times (around 1 ms), enable a fairly quantitative response with good resolution of all carbon types in kerogen to be obtained (Maciel *et al.*, 1979; Hatcher, 1980). Given the low amount of ^{13}C in total carbon, a large number of spectra need to be accumulated to enhance the signal to noise ratio. It is then possible to integrate separately peak areas corresponding to given carbon environments and express them as wt% of the total carbon.

An example of a ^{13}C CP/MAS spectrum of a coal sample and the related functional quantification is shown on Figure 5. Major oxygen containing functional groups can be analyzed with this technique, beside aliphatic and aromatic structures. However heteroelements inside aromatic rings cannot generally be seen in coals and kerogens. Peaks in spectra of kerogens and coals are fairly broad due to the large number of different carbon environments for a given bond type. The uncertainty of the quantification is estimated to be around 10%.

3 CASE STUDIES ON REFERENCE PETROLEUM SYSTEMS

3.1 History of Geochemical Studies Performed at *IFP* During the Period 1965-1975

A large number of geochemical studies on petroleum systems, in collaboration with oil companies or state geological institutions, was performed at *IFP* during this period. These studies, providing a large number of natural samples of oils and source rocks from reference basins for geochemical analyses, were at the basis of major geochemical observations and concepts. In the course of these studies, detailed below, it was possible:

- to set up and apply an analytical procedure for both extraction and analysis of source rocks, and distillation and oil analysis, enabling oil-oil and oil source rock correlation (Oudin, 1970; Durand and Espitalié, 1970; Durand *et al.*, 1970);
- to validate on geological examples the kinetic models of kerogen transformation into petroleum;
- to prepare and analyse a large number of kerogens, leading to their classification according to organic source and thermal evolution.

In parallel, collaborations on more fundamental research subjects, often in the form of financial support of doctoral studies, were initiated with several universities, particularly Louvain and Strasbourg, and research institutes, such as the *Cerchar*.

3.2 Reference Petroleum Systems Analyzed at *IFP* During the Period 1965-1975

3.2.1 The Toarcian Marls (Paper Shales) from the Paris Basin (France)

Many geochemical analyses of the kerogen and extractible organic matter of this reference source rock series from the lower Toarcian paper shales were performed in *IFP* during this time (Louis, 1966; Louis and Tissot, 1967; Oudin and Califet-Debyser, 1969; Durand and Espitalié, 1971; Tissot *et al.*, 1971; Durand *et al.*, 1972; Espitalié *et al.*, 1973), and there are still ongoing analyses in *IFP* and elsewhere. Many wells were drilled during petroleum exploration in this basin, resulting in a large number of cored samples at different burial depths, from outcrops in the east of France to a maximum depth around 2500 m. Samples and related informations were kindly provided by *Erap, Mobilrex, SNPA* and *Petrorep*.

In the Paris Basin, this series with an age of 180 Ma and a thickness ranging from 10 to 20 m, rich in organic matter, was deposited in a homogeneous environment under 100 m water, and further basin subsidence resulted in transformation of kerogen into petroleum, allowing availability of a particularly homogeneous maturity suite. This series was also one of the first examples on which a kinetic simulation of the petroleum generation using the cracking scheme of Tissot (1969) was performed (Deroo *et al.*, 1969), and also one of the first where kerogens were prepared, analyzed and placed into a van Krevelen diagram to show the kerogen evolution path (Durand *et al.*, 1972), as described in a later section (Part 3.4).

3.2.2 The Devonian and Silurian Hot Shales from the Illizi Basin (Algeria)

This petroleum system contains very old and rich source rocks, mainly of lower Silurian and upper Devonian age, with local source rocks in the Ordovician and upper Silurian formations (Tissot *et al.*, 1974). Oils and source rock samples were provided by *Elf-RE*. All oils are very light, with a predominance of alkanes containing less than 15 carbons and a very low NSO content. Distinguishing criteria between oils could be found only on the basis of structural group analysis of aromatic fractions, separating oils coming from a local upper Silurian source rock from undistinguishable oils sourced by lower Silurian and upper Devonian source rocks.

The structural history of this ancient basin is complex and comprises two main subsidence periods separated by a major uplift and erosion during the Hercynian orogenesis. Modelling of oil and gas formation in this petroleum system was performed using a modified cracking scheme accounting for gas (Tissot *et al.*, 1974), on the basis of the burial history of source rocks and the timing of trap formation. The model was able to explain the absence of oil in the SW part of the basin, the presence of gas in NE (Alrar, In Amenas), condensates in SE (Zarzaïtine) and oil and gas in NW (Tin Fouyé Tabankort), this latter gas coming from a major contribution of the lower Silurian source rock.

3.2.3 The Devonian Shales from the Alberta Basin (Western Canada)

The geochemical study of this petroleum system, a joint project with the Geological Survey of Canada (Deroo *et al.*, 1977; Deroo and Powell, 1978), was performed on a large number of conventional oils (106), heavy oil extracts (63) and sediment extracts (189). Two areas of the Alberta syncline were studied: a deep central zone, and the so called "heavy oil belt" including the Athabasca tar sands.

This petroleum system study brought new insights into source rock variability, geochemical tools for oil/source rock correlation, migration distances and influence of biodegradation on geochemical parameters. Three source rock intervals were present in this basin:

- the Mannville shales (lower Cretaceous), associated with coals, were the first example of organic matter sourced by land plants, and many kerogens were prepared and analysed;
- the marine limestones of middle and upper Devonian;
- the marine shales of upper Cretaceous.

Differentiation of oil families was only possible through a systematic analysis of aromatic hydrocarbons, both by gas chromatography with a sulfur specific detector and by group type analysis with low voltage mass spectrometry. Long migration distances, possibly higher than one hundred kilometers, were deduced from correlations of oils and adequately mature source rocks, on the basis of geology and maximum burial depth. These unusually long distances were possible through a major regional unconformity acting as a route for fluids. Finally the study of water salinity associated to more or less heavy oils showed that biodegradation, and possibly water washing, occurred by infiltration, through reservoir outcrops, of fresh water bringing bacteria. The biodegradation level could be correlated to observed changes in molecular and compositional geochemical parameters, such as the disappearance of *n*-, then *iso*-alkanes, then some aromatics, the vanadium and nickel concentration, or the increase of the asphaltene content.

3.2.4 The Logbaba Silty Clays from the Douala Basin (Cameroon)

The petroleum system of the Douala basin is located SE of the present Niger river delta, and began to form in conjunction with the opening of the South Atlantic ocean. Upper Cretaceous detrital sediments, alterning sands and silty clays and carrying continental organic matter, were deposited and quickly buried in a shallow marine environment. The mineral and organic homogeneity of these sediments, associated with a significant total thickness allowing a large range of maturities to be studied, made this series particularly interesting. It was the subject of doctoral studies on the mineral (Dunoyer de Segonzac, 1969) and organic (Albrecht, 1969) contents, followed by a collaborative study including kerogens between *IFP, Strasbourg University* and *Elf-RE* and its subsidiary *Elf-Serepca* (Albrecht *et al.*, 1976; Durand and Espitalié, 1976; Vandenbroucke *et al.*, 1976).

The comparison of source and transformation characteristics between the terrestrial organic matter from this petroleum system, and the marine organic matter from Toarcian shales in both the Paris Basin and their higher maturity German equivalents, was a key point in undestanding the chemical differences in petroleum potential and oil composition between these two reference series. The preparation and analysis of a large number of kerogens in both series was also at the basis of the classification of kerogen types and evolution paths in a van Krevelen diagram of H/C *vs.* O/C atomic ratios, as detailed in Part 3.4.

3.2.5 The Green River Shales from the Uinta Basin (USA)

The Green River shales is an organic-rich formation extending over several states of USA. It was deposited in two Paleocene/Eocene alkaline palaeolakes, lake Gosiute in Wyoming, and lake Uinta in Utah and Colorado. Climatic conditions and alkaline mineral supply favored high productivity and preservation of specific microorganisms, possibly blue-green algae and bacteria resisting such agressive environments. However the exact nature of the organic supply is unknown, because no organic fossils could be identified. The geochemical study of this formation at *IFP*, in collaboration with *Shell USA*, focused on the Utah part of the Uinta basin, the main question being to clarify the respective influence of source differences and maturity on the origin and composition of crude oils (Tissot *et al.*, 1978).

Unexpected analytical difficulties caused by the highly paraffinic nature of oils and extracts had to be solved first, because C_{25+} hydrocarbons reprecipitated during compound separation with usual solvents. It appeared then that slight differences associated with changes in depositional environments could be detected at a similar maturity level, based on the composition of specific biomarkers in alkane fractions. Thanks to the large maturity range of the studied samples, it was possible to define on the related kerogen series a new type and evolution path, confirming not only the highly aliphatic nature and the hydrocarbon potential of this organic matter, but also its specific thermal behavior, characterized by a late and very narrow oil window when compared to other source rocks.

3.2.6 The Clays and Coals from the Mahakam Delta (Indonesia)

The study of this petroleum system, in collaboration with *Total CFP, Total Indonesia, Inpex* and *Pertamina,* was of particular interest due to the nature and homogeneity of its Miocene source rock associating coal beds and dispersed organic matter deposited in a deltaic setting under tropical climate, by the high quality and large maturity range of samples, and by the geological setting of oil and gas fields in a series of parallel structural trends perpendicular to the sediment supply by the Mahakam river. Numerous geochemical studies were performed on oils, source rocks, coals and kerogens, in *IFP* and *CFP* labs (Combaz and De Matharel, 1978; Durand and Oudin, 1980; Vandenbroucke *et al.*, 1983; Huc *et al.*, 1986), as doctoral studies (Boudou, 1981), and on corresponding recent sediments by the Misedor group (Pelet, 1987).

Many new insights or confirmations concerning the kerogen analysis and petroleum potential were obtained by geochemical studies on this petroleum system. Among these were:

- confirmation that terrestrial plant sourced kerogens and coals have similar composition and petroleum potential, whereas this was not possible with the Logbaba series where coals were not found;
- characterization of such post Cretaceous "perhydrous" coal series as source rocks for oil, and not only for gas as previously thought;
- observation of very easy migration of oil and gas in deltaic systems due to the close mixing of sand and shale sediments.

These observations thus raised the interest of petroleum companies for Tertiary deltaic systems under tropical climates as new zones for oil and gas exploration, with many commercial successes obtained since that time.

3.3 The Development of Kinetic Models of Kerogen Cracking into Oil and Gas

Only the main principles of early kinetic models of kerogen transformation into petroleum and the related publications using the reference petroleum system studies presented above will be described briefly here. Other papers in this journal issue, including the translation of the precursor paper of Tissot (1969), will show the applications of such models for oil exploration.

It is indeed impossible, given the variety of the chemical structures and bonds in kerogen, to describe primary cracking by elementary free-radical reactions. The main steps of kerogen cracking in early models were thus similar to those described for thermal conversion of coal, namely depolymerization, tar cracking and residue cokefaction. Depolymerization of kerogen was thus represented by a reaction producing intermediate heavy compounds extracted at this time by a mixture of methanol, acetone and benzene (MAB), together with a condensed residue and gases, both hydrocarbons and nonhydrocarbons. The MAB compounds were in turn transformed into petroleum compounds plus again condensed residue and gases. Already at this time kerogen, being recognized as a chemically complex molecule, was considered to generate MAB compounds by a number of parallel and/or successive pseudo-reactions thought to reflect the various bond energies in a given kerogen. Kerogen, MAB compounds and petroleum (chloroform extract) being respectively A, B and C, the reaction scheme is:

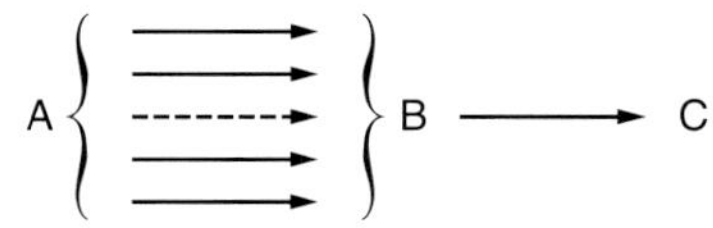

Quantitative data were calibrated in first models on the evolution with depth of extract amounts in reference petroleum systems. Assuming that all reactions are first order and that rate constants follow the Arrhenius law (see translated paper for details of equations), kinetic parameters of this scheme were calibrated on 24 samples from the Paris Basin for which the burial history, and thus the temperature applied to the system was reconstructed (Tissot and Pelet, 1971). This calibration allowed recalculation of the theoretical petroleum generation of the Dogger marls of the Paris Basin (similar to the Toarcian shales) and comparison with the presence or absence of petroleum accumulations in this same formation. The result is shown on Figure 6.

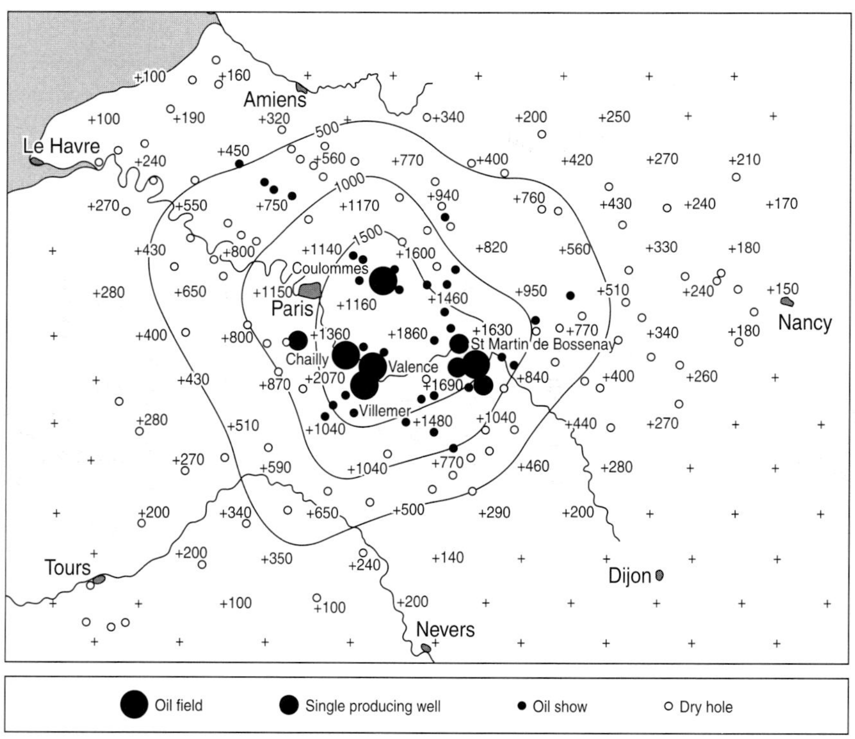

Figure 6

Modeling of oil generation (figures in gram of petroleum per ton of source rock) in the Dogger source rocks of the Paris Basin and position of oil fields, producing wells and dry holes resulting from drilling. The modeled quantities fit closely the observed accumulation zones (from Tissot and Welte, 1978, p. 510; copyright Springer-Verlag).

A new model accounting for changes in both analytical procedures (MAB extraction was abandoned due to solvent toxicity), and more important, in the classification of kerogen types as described in the next section, was proposed by Tissot and Espitalié (1975). The reaction scheme, with A, B, C being respectively kerogen, oil and gas, was:

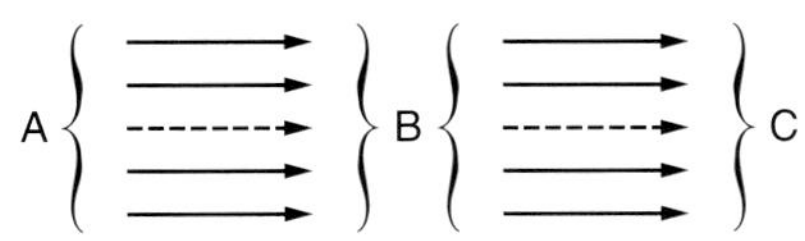

Calibration of kinetic parameters was performed not only using extract amounts produced at various evolution stages in geological conditions, but also by combining this information with thermogravimetric experiments, *i.e.* kerogen weight loss on pyrolysis under nitrogen. Six pseudo-reactions were used, with fixed increasing activation energies between 10 and 80 kcal/mol. Frequency factors of the Arrhenius law and kerogen fractions associated with each pseudo-reaction were calculated by numerical optimisation. The organic matter type was shown to influence the final weight loss of kerogen, as well as its contribution to the various pseudo-reactions. The comparison of oil and gas generation curves obtained with this kinetic model in standard geological conditions for various kerogen types is shown on Figure 7.

Further developments in kinetic models no longer use the data obtained on natural series, and are accordingly out of the scope of this review. It suffices to say that their development increased continuously in close relation with numerical modeling of petroleum systems, and that they are at the source of numerous petroleum discoveries.

3.4 The Classification of Kerogens from these Case Studies: Notion of Type and Evolution Path

3.4.1 Building up the Concepts

The classification of kerogens from sedimentary source rocks into three main types, and their evolution during maturation in geological conditions following "carbonization paths" were first proposed by Durand *et al.* (1972), on the basis of elemental analyses of kerogens from the Paris Basin and related literature data on various kerogens from different ages and locations (Mc Iver, 1967). These analyses were represented in an atomic H/C *vs.* O/C diagram similar to the one used for coals by van Krevelen in 1950 (van Krevelen, 1961), and were compared with the coalification paths of coal macerals *(Fig. 8)*. Elemental analyses of maturity series of kerogens, obtained progressively from geochemical analyses of petroleum systems described above (except for the Mahakam coals and shales and the Green River shales) were included in this diagram, allowing "evolution paths" to be defined (Durand and Espitalié, 1973; Tissot *et al.* 1974). The Green River shales (Tissot *et al.*, 1978) were then added to these reference series of kerogens *(Fig. 9)* by Tissot and Welte (1978). The three types of organic matter defined in

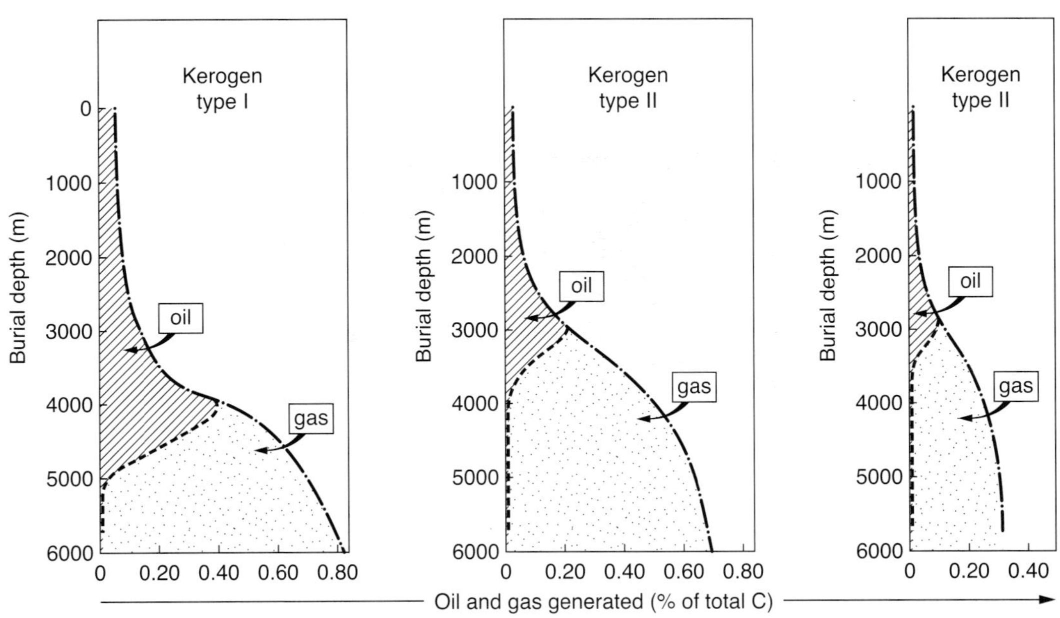

Figure 7

Comparison of oil and gas production from the three kerogen types resulting from the cracking model of Tissot and Espitalié (1975), with a constant burial of 60 m per Ma and a geothermal gradient of 35°C per km (after Tissot and Espitalié, 1975).

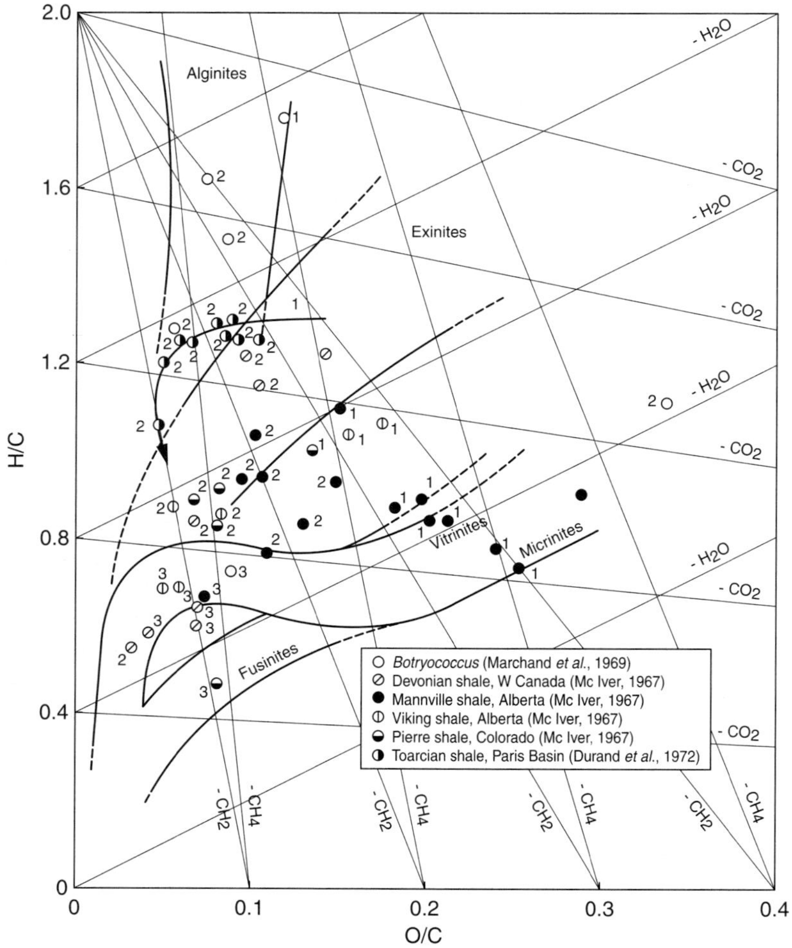

Figure 8

Comparison of elemental analyses from kerogens of the Paris Basin (Durand *et al.*, 1972) and various kerogen series (McIver, 1967) in the diagram of atomic ratios H/C and O/C for coal macerals (van Krevelen, 1961). The numbers 1, 2, 3 correspond respectively to low, medium and high maturity. The "carbonization path" is indicated for the Paris Basin kerogens (after Durand *et al.*, 1972).

this "van Krevelen diagram" were associated by Tissot *et al.* (1974) with the specific depositional environments of these petroleum systems. Other source rock samples, although available only in a limited range of maturity, were generally shown to fit the position related to the predicted depositional environment in this diagram, except for lacustrine settings as discussed below. Elemental analyses of kerogens, reflecting their chemical structure, can thus be used for characterizing both the type of the organic matter in petroleum source rocks and its evolution with increasing burial in geological conditions.

3.4.2 Kerogen Types: Chemical Characteristics and Depositional Environment

The chemical characteristics of kerogens of reference petroleum systems corresponding to the initial definition of types I to III and their associated depositional environment are, in order of decreasing petroleum potential:

- The Green River shales from the Uinta basin of Utah, as the reference type I, although some levels in this source rock formation contain classical coals. Except in these latter levels, the organic matter of immature samples was observed to be very aliphatic, with atomic H/C frequently higher than 1.5. The oxygen content is often low, although the atomic O/C ratio of immature samples may vary from 0.03 to 0.1. The analysis of oxygen functional groups in immature kerogen (Fester and Robinson, 1966) shows a major participation of unreactive oxygen assumed to be located in aliphatic ether bonds. In accordance with the high initial H/C ratio, the extracts, oils or pyrolysis products are very rich in long chain *n*-alkanes, extending over 40 carbon atoms, without marked odd or even

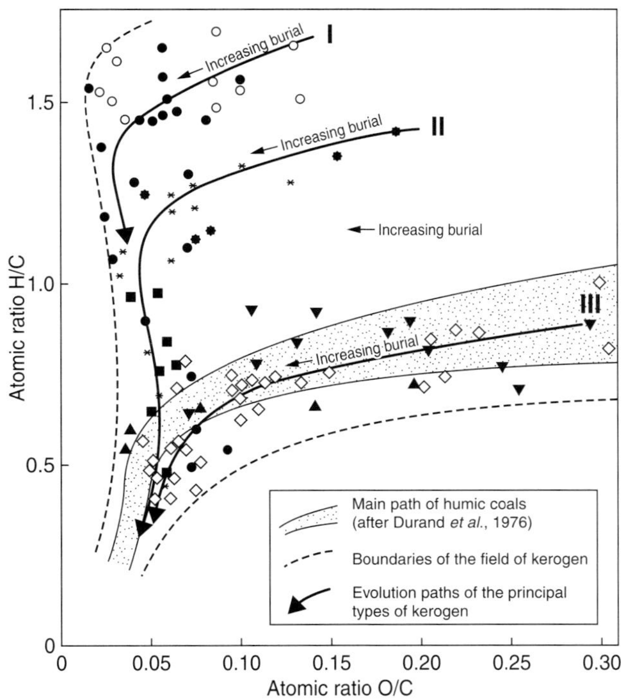

Type	Age and/or formation	Basin, country	
I	Green River shales (Paleocene-Eocene)	Uinta, Utah, USA	●
	Algal kerogens (*Botryococcus*, etc.), various oil shales		○
II	Lower Toarcian shales	Paris, France, W. Germany	*
	Silurian shales	Sahara, Algeria and Libya	■
	Various oil shales		✱
III	Upper Cretaceous	Douala, Cameroon	◇
	Lower Mannville shales	Alberta, Canada	▲
	Lower Mannville shales (Mc Iver, 1967)	Alberta, Canada	▼

Figure 9

Definition of types and evolution paths of kerogens based on petroleum reference series as indicated (from Tissot and Welte, 1978, p. 143; copyright Springer-Verlag).

predominance. The abundance of iso- and anteiso-alkanes and other biomarkers specific of microorganisms such as carotane led Tissot *et al.* (1978) to conclude that the initial organic matter input (probably *Cyanophycea*) had been strongly reworked by bacteria that could have sourced the major part of the finally sedimented organic matter. It is worth noting that the lacustrine environment associated with Green River shales is not freshwater, and that it may contain very unusual alkaline minerals. Other type I kerogens such as Autun boghead (France), torbanite (Scotland) or recent materials such as Coorongite (Australia), principally derived from *Botryococcus* remains, which also sedimented in a lacustrine setting, although in this case with freshwater, were shown to be located in the same region of the van Krevelen diagram as this Green River series. The kerogen of the organic-rich marine tasmanite from Tasmania was also classified as Type I by Tissot and Welte (1978). As very different sources, only characterized by their common highly aliphatic nature, are classified as type I, it is not really a type because of the different biological precursors, sedimentary environments, preservation conditions implicated in the formation of these kerogens. It should also be noted that source rocks containing these kerogens, although they have a high petroleum potential, are not widespread, and the geographical and geological extension of such organic deposits is generally very restricted. According to Klemme and Ulmishek (1991), they represent only 2.7% of the world's original reserves in petroleum and gas.

- Type II is based on the lower Toarcian shales of the petroleum system from the Paris Basin (France) and the equivalent Posidonienschiefer (Lias ε) formation in Germany. The atomic H/C ratio of immature kerogens is about 1.3 and the corresponding O/C ratio around 0.15. As shown by the group type analysis of kerogen pyrolysates and natural extracts (Tissot and Vandenbroucke, 1983), this organic matter contains much more aliphatic cyclic moieties than in type I, and a large amount of aromatics, the structure of which often comprises several aliphatic cycles that convert to aromatic cycles with progressive maturation. The *n*-alkane distribution in extracts or pyrolysates is generally restricted to the carbon range under 25 atoms, even at low maturity. Sulfur is always associated with this type of organic matter, either as pyrite and free sulfur, or in organic structures of kerogen such as aliphatic sulfur and thiophenic rings. The depositional environment associated with these Toarcian shales, as well as to their homologues in Germany, was a moderately deep marine environment with planktonic input as the primary source. Reworking by various bacterial communities occurred to a varying extent, depending on the location of the oxic/anoxic boundary. This Toarcian organic matter type was compared to other source rocks from marine deposits in different regions of the world (for example Devonian of Western Canada, Silurian and Devonian from North Africa). The comparison showed very homogeneous chemical characteristics both for kerogens and for extracts, oils and pyrolysates. Differences in the sulfur content of the organic matter were attributed to the different mineralogical compositions of these source rocks, the richest in carbonates being the richest in organic sulfur, because in that case iron coming from clays is not available for pyrite formation.
- Type III was initially based on the upper Cretaceous sedimentary rocks of the Douala basin (Logbaba, Cameroon), sourced by continental deltaic input and deposited in shallow marine environment (Albrecht *et al.*, 1976). This type of organic matter is frequently found in deltaic settings and derives from higher plant debris, often highly reworked due to the oxidant transportation

conditions associated with detritic sedimentation. In the case of the Logbaba silty clays, no coal beds were found at any depth. Only the most resistant chemical constituents of the terrestrial organic matter reached the sedimentation region, and once there, they were quickly buried due to the large sediment input and high subsidence rate caused by the opening of South Atlantic Ocean. Under these conditions, strong bacterial degradation probably could not occur as no sulfur was incorporated in the organic matter or as pyrite, even though sulfates were present in the seawater. As seen by both chemical and optical analyses, the organic fraction is for a large part ligneous debris with a predominantly aromatic structure, together with moieties derived from the protective constituents of higher plants with a predominantly aliphatic structure rich in long chain alkanes on the other hand. This latter part of the terrestrial organic matter is probably always a minor fraction of total kerogen. However, this double contribution explains the apparent discrepancy observed between the mostly aromatic kerogen composition as seen by elemental analysis with an atomic H/C ratio lower than 0.8 (Durand and Espitalié, 1976), and the highly paraffinic extracts and oils generated by this type III kerogen in comparison with products generated form type II (Vandenbroucke *et al.*, 1976). Other examples of type III organic matter were found also in the lower Mannville shales of Alberta, Canada. It was only with the extensive studies carried out on the organic matter maturity series of the Mahakam delta (Eocene/Miocene) in Indonesia (Combaz and de Matharel, 1978), that it was shown that pure coals and type III shales occurring together in these sedimentary rocks are derived from the same precursors. The detailed geochemical study of kerogens from shales and coals showed that there is no noticeable difference between massive coal and shale evolution at any maturity stage, although humins from very immature coals and shales are not equivalent due to differences in their sedimentological conditions (Huc *et al.*, 1986). Hence no catalytic effect of clays on kerogen evolution could be discerned in the above maturity series, although this effect was often hypothesized in the past on the basis of laboratory experiments (Sieskind and Ourisson, 1972). Successful petroleum exploration in Tertiary deltaic settings around the world in the past twenty years, resulting in intensive geochemical studies on good quality cores and cuttings, showed that dispersed type III organic matter, whether associated with humic coals or not, is, like type II, an homogeneous and chemically well defined organic matter type, in contrast with type I deposits.

3.4.3 The Main Steps of Kerogen Transformation on Burial and its Evolution Path in the van Krevelen Diagram

Early diagenesis, from the death of living organisms up to deposition of mineral and organic debris as sediment and its

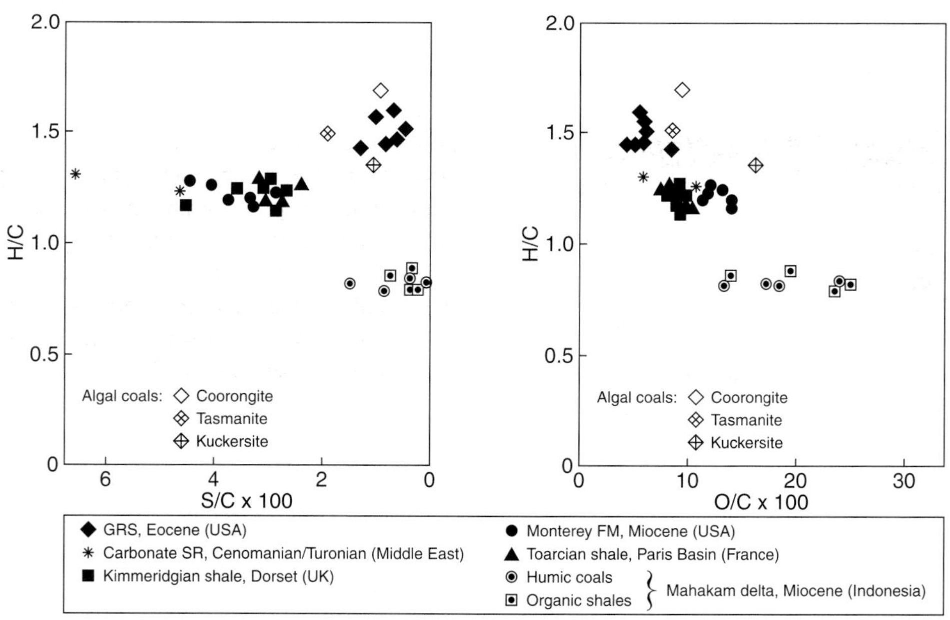

Figure 10

Elemental composition of immature kerogens in a double diagram of atomic ratios H/C *versus* O/C and S/C, showing the substantial incorporation of sulfur in marine organic matter, in contrast with S/C composition of lacustrine or continental organic matters (after Huc, 1990).

burial to a few meters depth, is a step where biological processes have a major influence on the amount and composition of finally preserved organic matter. This step, during which the main loss of nitrogen occurs due to consumption of proteins, is not accounted for in the classical van Krevelen diagram. It is also the step for possible incorporation of sulfur in the organic matter, if sulfur bacterial colonies are present. The systematic incorporation of sulfur in marine kerogens is clearly seen *(Fig. 10)* in a diagram of elemental compositions of immature kerogens, derived from the van Krevelen representation and including organic sulfur (Huc, 1990). The presence of sulfates in seawater provides the oxidant source necessary for organic matter biodegradation by sulfate-reducing bacteria. These bacteria release polysulfides, H_2S and native sulfur capable of recombining both with iron from clays to ultimately form pyrite, and with the organic matter to produce organosulfur moieties (Aizenshtat *et al.*, 1983). Preservation of organic matter during early diagenesis is strongly dependant on the oxygen content of water, a very low oxygen level (anoxia) strongly reducing the nature and efficiency of bacterial populations able to feed on organic debris during settling and early stages of sedimentation. Water circulation inside deposited sediments is also a factor controlling the bacterial destruction of organic matter. If water transported bacteria can freely circulate into the sediment porosity, which is the case in sands for example, the major part of the sedimented organic debris will be consumed. It is for this reason that high porosity reservoir rocks do not contain kerogen. The opposite situation is found for clays, not only because they may protect the organic matter by physicochemical adsorption, but also because their structure settles as "house of cards", increasing water pressure in the porosity and preventing water input during further compaction on burial. That explains why source rocks containing kerogen are for the most part fine grained sediments.

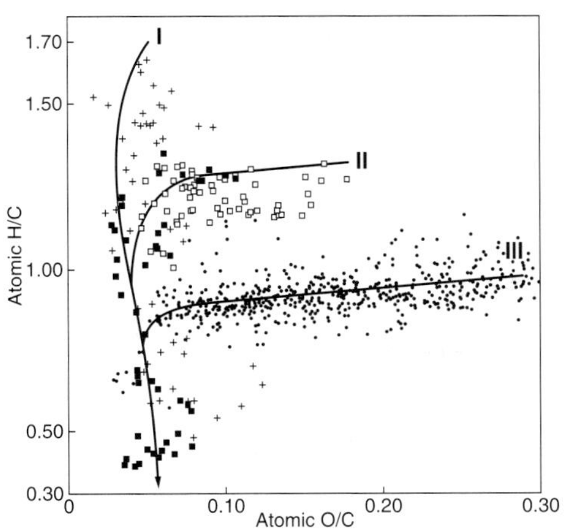

Series	Mahakam coals	Paris Basin	German Lias ε	Green River shales
Age	Miocene	Toarcian		Eocene/Paleocene
Type	III	II		I
Number of samples	807	65	46	53

Figure 11

Comparison of elemental analyses from kerogens of the three main reference series, including many samples from Lias ε from Germany (equivalent of the Toarcian of the Paris basin) with a very high maturity, and a coal series from the Mahakam delta covering a large maturity range (figure from Vandenbroucke *et al.*, 1993, copyright Éditions Technip).

Diagenesis *stricto sensu* is a step where kerogen looses large amounts of oxygen, mainly as CO_2 and H_2O (see *Fig. 8*). This step is all the more evident if the initial kerogen is rich in oxygen, and is thus mainly observed for type III organic matter, but it is also significant for type II as seen on Figure 11. In this figure a large number of elemental analyses on one reference series for each kerogen type has been reported, including the Mahakam series as type III, not yet available when Figure 9 from Tissot and Welte (1978) was published. The Mahakam coal series was used here, because kerogen preparation on very immature samples was observed to changes elemental analysis due to hydrolysis of humic compounds. It can be noticed on this series that the evolution path starts at higher H/C and the variation between the two successive steps is more marked than previously reported in the trend from the Mannville and Logbaba series.

Catagenesis is the main step of hydrogen (and carbon) loss from the kerogen, and is consequently the stage of petroleum formation. Considering a general formula of generated hydrocarbons as $(CH_2)n$, two H atoms are lost for one C atom, thus the atomic H/C of the residual kerogen decreases in the van Krevelen diagram, reflecting an increasing aromatization. It appears immediately that the petroleum potential deduced from the H/C variation range decreases strongly in the order from type I to type III. However commercial oil pool formation is not directly correlated to H/C in source rocks, because it depends also on thermodynamic and geological parameters related to the nature of generated products and their ability to migrate to existing traps. As shown by natural extract and oil analyses from petroleum case studies and thermal simulations, compounds with lighter and lighter molecular weights are formed during this stage, leading finally to gases. An insoluble aromatic residue similar to coke is generated in late steps of catagenesis as the result of secondary cracking of petroleum products issued from kerogen. This residue, which has almost no petroleum potential, can be observed alone as "pyrobitumen" in high temperature reservoirs, but is also formed in source rocks and thus mixed with the residual kerogen.

The transition between catagenesis and metagenesis zones is located for all kerogen types at an atomic H/C ratio

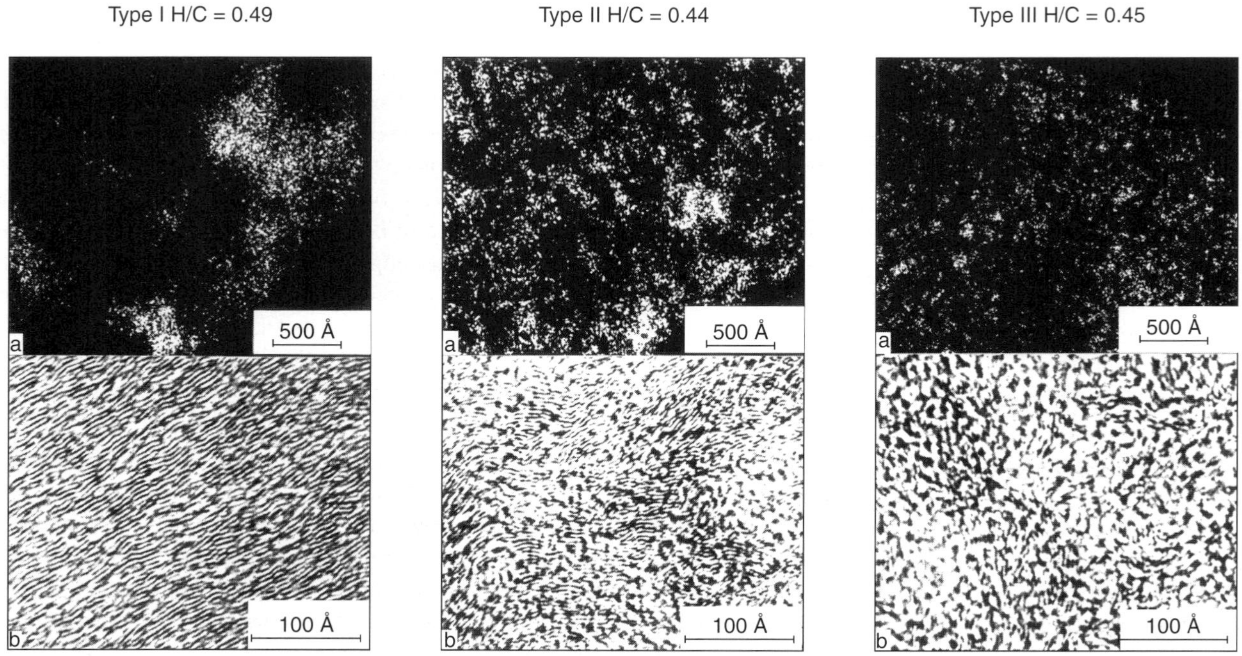

Figure 12

Comparison of the orientation of aromatic domains in the three kerogen types at high maturity, as observed with electron microscopy. (a) 00.2 dark field picture of aromatic domains (electrons diffracted by aromatic stacks). The diameter of aromatic stacks correspond to less than 10 condensed rings, piled up by 2 or 3 layers whatever the initial kerogen type. (b) 00.2 lattice fringes from aromatic domains seen edge on, showing that long distance orientation of aromatic domains depends on the initial kerogen type, with respectively for types I, II et III, average distances of 1000 Å, 250 Å and 100 Å (after Oberlin *et al.*, 1980).

around 0.6. Below this transition zone, all kerogen types defined by their elemental analysis merge into a single group. Metagenesis is the last stage of kerogen reorganization into an almost pure carbon structure. During this step, the aromatic lattice rearranges by expelling its "defects" in a manner similar to crystal formation. However, this rearrangement does not imply a substantial condensation of aromatic cycles but more a parallelization of existing aromatic structural units comprising less than 10 condensed rings into stacks associated by pi bonds, as seen by diffractometric techniques *(Fig. 12)* using transmission electron microscopy (Oberlin *et al.*, 1980). The size of domains is still, contrary to elemental analysis, a characteristic of kerogen types because it depends on the mobility of kerogen aromatic domains during catagenesis, and thus on the amount of products expelled as oil. The free radical content in kerogen, which increases continuously during the catagenesis step, reaches its maximum and then decreases, as shown on type III kerogens from the Douala basin *(Fig. 13)* using electron paramagnetic resonance (Marchand and Conard, 1980). The first eliminated product is late methane, also called dry gas, resulting from release of methyl groups on aromatics.

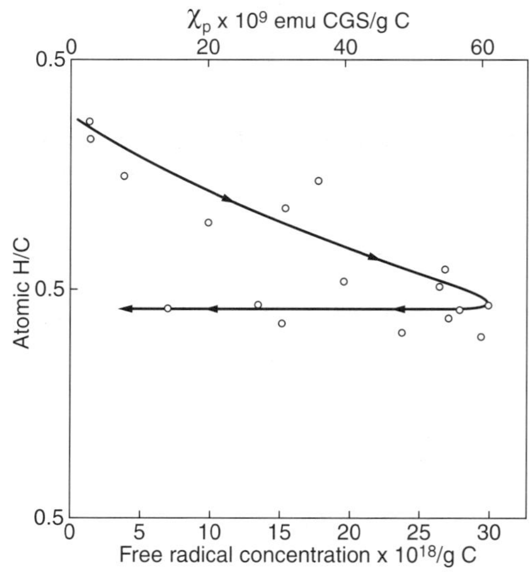

Figure 13

Evolution with increasing maturity of the concentration of free radicals in kerogens of the Logbaba series, as measured by electron paramagnetic resonance (after Marchand and Conard, 1980).

Pyrolysis of mature kerogens of all types was shown to generate the same amount of late methane, around 50 mg per g of kerogen (Béhar *et al.*, 1997). This means that the amount of methyl groups on aromatics is similar in residual kerogens whatever their initial type, and it is thus only the amount of residual kerogen left after catagenesis, higher in a type III than in a type I series, that makes terrestrial kerogens and coals good source rocks for dry gas. Nonhydrocarbon gases are also formed during metagenesis by elimination of heteroelements from resistant bonds (O from ethers) or from aromatic structures (S, N, O). Gases such as CO_2, H_2O, H_2S and N_2 were observed in both deep gas reservoirs and pyrolysis experiments (Lorant and Béhar, 2002).

3.5 The Evolution of Research in the *IFP Geochemistry Department* from 1979 to 1990

Year 1979 corresponds to the date on which B. Tissot left the *IFP Geochemistry Department* to take the responsibility of the *IFP Scientific Direction*. At this time the *Geochemistry Department* research focused on two main topics. One of them was to improve the kinetic schemes of oil and kerogen cracking, in order to include them into integrated petroleum system models. These latter aimed at reconstructing the sediment deposition, burial history and heat flow during geological time, and at calculating, from the cracking model, the amount of oil generated, the timing of its expulsion out of the source rock and its further migration and accumulation into traps. A second topic was to better understand the mechanisms of kerogen formation, in close relation with their depositional environment. It was thus necessary to follow the deposition and chemical transformation of kerogen during all steps of its geological maturation and not only during oil formation, in order to decipher the key factors of source rock formation. Accordingly, recent sediments were studied using soil science methodology, including among others isolation and analysis of humic substances, and particularly functional analysis, to understand their relationship to kerogen.

Analytical techniques had offered many new developments at this time, particularly gas chromatography for which the manufacturing of silica capillary columns with bonded phases allowed higher final temperatures and a much better peak resolution. New detectors, specific to molecular families of compounds including atoms of sulfur or nitrogen, were also developed or improved. Moreover, coupling of gas chromatography with mass spectrometry became at this time a widespread laboratory apparatus, due to the relatively low price of quadrupolar spectrometers, enabling molecular analysis of separate hydrocarbon families in complex mixtures such as petroleum compounds. Significant advances were also obtained in pyrolysis techniques for the molecular identification and quantification of building blocks of kerogen. This was achieved either by direct coupling of the pyrolysis system to the injection port of a gas chromatograph, or by preparation of pyrolysates, separated and studied in the manner described in Section 2.2.2 for extracts. Finally, major advances in organic analysis of solids resulted from developments in solid state ^{13}C NMR, enabling the quantitative analysis of carbon bond types in kerogen, as reported in Section 2.3. The resulting quantification of molecular building blocks and their bonding functional groups in kerogens thus made it possible to propose conceptual molecular models of kerogens as described below.

4 BUILDING KEROGEN CONCEPTUAL MODELS

4.1 Why Conceptual Models?

It appears clearly that given the complex macromolecular structure of kerogen, comprising both resistant biomacromolecules and recombined biodegradation products, any detailed chemical model represents only an averaged structure. To fix ideas, a detailed chemical structure which can be drawn on a single page, with a molecular mass around 25 000, makes an incredibly small amount of substance since the corresponding weight (by division by Avogadro's number) represents 4.10^{-20} g of kerogen. Any analytical technique used for kerogen characterization works on amounts in the microgram to milligram range, thus any analytical result is already an average on a large number of kerogen macromolecules. Thus a "real" representation of the chemical structure of kerogen has no meaning and only conceptual models representing at best the largest possible set of physicochemical analytical data, but including also many assumptions, can be obtained.

Although it is very difficult to describe such complex macromolecular mixtures by detailed chemical structures, many structural or chemical models of sedimentary organic matter were published in the past. Their aim was to visualize a number of its properties or analyses in a synthetic form to interpret or understand others. Thus a model aimed at representing physical or spectral properties can be a structural model whereas a model for describing chemical interactions or cracking reactions will rather be a molecular model. This rationale is very similar to that of petroleum system modeling, and in the same way the weighting and detail of input and output data will depend on the use for which these models are constructed. The first molecular models of sedimentary organic matter were built by coal scientists as early as 1942 by Fuchs and Sandhoff; then many other coal models based on an increasing number of analytical results from different techniques were built (Given, 1960; Spiro, 1981; Spiro and Kosky, 1982; Shinn, 1984; Solomon, 1981). Models for other types of sedimentary organic matter were constructed first for oil shales such as the Green River shale (Burlingame *et al.*, 1968; Yen., 1976;

Siskin *et al.*, 1995), then for the three main types of kerogen at various maturity stages (Béhar and Vandenbroucke, 1986, 1987). Finally, the development of computerized molecular modelling enabled to build 3D molecular models satisfying not only bond lengths and directions (Faulon *et al.*, 1990) but also minimum energy configurations (Kowalewski *et al.*, 1996).

Building computerized chemical models of kerogen structures may be helpful as a guide to understanding the main chemical characteristics of sedimentary organic matter, and to predicting its evolution during geological burial. Although it will never be possible to represent a true kerogen structure, because it is a mixture of various nonpolymeric macromolecules that cannot be analyzed separately, such models of average kerogen structure, by accounting for a great amount of information from various analyses, can provide a synthetic view of the main resemblances and differences among sedimentary organic matters. They may be used for comparison of organic matter types and evolution during burial, modeling of cracking and compositional yields in source rocks, evaluation of retention or expulsion of petroleum and gas as a function of the thermal maturity and related thermodynamic properties of the mobile fraction.

4.2 The Analyses used for Modeling and their Classification

4.2.1 Atomic Analyses

We will define as atomic analyses those providing quantitative information on the number of atoms, their chemical environment or their spatial distribution. For example elemental analysis, solid state ^{13}C nuclear magnetic resonance (NMR) and more generally all spectroscopic analyses, density or lattice fringe imaging with electron microscopy, are classified as atomic analyses. They are performed on the kerogen itself, characterizing a subset of its macromolecular constituents, and thus provide average structural or chemical parameters.

4.2.2 Molecular Analyses

Molecular analyses qualify and quantify small building blocks of a complex structure. They are performed on degradation products of macromolecular kerogen constituents. Either chemical degradation, by oxidation for example, or thermal degradation, both by laboratory pyrolysis (pyrolysates) and by thermal cracking in geological conditions (extracts and oils), provide some of these building blocks. The rationale for analysing degradation products and in the ideal case cracking products, the structure of which is very close to its initial counterpart in kerogen, is that their structure is small enough to be characterized by a number of analytical techniques that cannot be used for solids. Degradation products must be small molecules in order to be identified as precisely as possible, ideally at the molecular scale. However they must retain as much as possible the molecular information present in kerogen macromolecules, and thus they should not be too small. In addition, the information they carry, both qualitative and quantitative, depends directly on the yield of products obtained with the degradation procedure, and the representativity of building blocks in regard to the source macromolecule is clearly not the same if the degradation yield is 5 or 80%.

4.3 Structural Models of Kerogen Types and Evolution

Once quantitative atomic and molecular analyses are obtained on a given kerogen, the quantitative combination of molecular building blocks should ideally fit atomic data. The reality is not so simple because some information is still lacking and many hypotheses have to be made, not only on the types of bonding sites, but also on missing or incomplete data. The reverse situation can also occur, with two different techniques normally giving the same information in fact providing different analytical results. In such situations, one data set is chosen or the data is adapted on the basis of general knowledge on sedimentary organic compounds from the literature.

The main differences in structural components of kerogens according to their type at the onset of the oil window were first described in Vandenbroucke (1980) based on the composition of extracted bitumens. A structure for type II kerogen constituents was reported by Oberlin *et al.* (1980) in another chapter of the same book, using the complementary data of these authors on kerogen structural blocks, but as presented could not represent a kerogen because the building blocks had not been connected by bonds. However it was a starting point for trying to build a complete series of structures representing the three kerogen types at the main stages of their geological evolution. These structures were built by trial and error, using measured atomic and molecular parameters at the beginning of diagenesis, beginning of catagenesis (whole petroleum potential) and end of catagenesis (whole gas potential). The starting molecular weight was chosen on the basis of two constraints:

- insolubility in usual organic solvents, the limit being fixed by the measured molecular weight of dissociated asphaltene units, lower than 5000 dalton;
- the necessity to draw a kerogene structure large enough to account for statistically measured parameters, but small enough to be represented on paper and allow a reasonable number of atoms to be counted (not more than 3000).

The three kerogen type structures were thus represented with a similar number of carbon atoms around 1500 at the diagenesis stage, corresponding to molecular weights ranging from 20 000 to 25 000 dalton. Note that, as previously

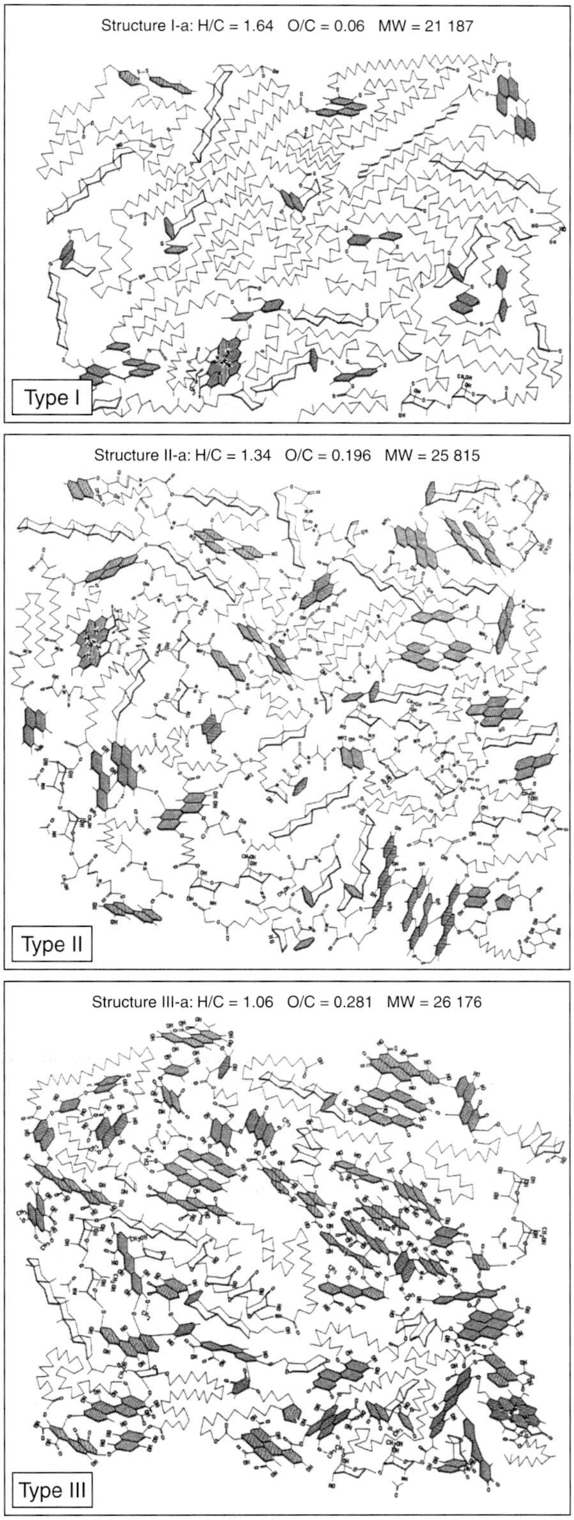

Figure 14

Structural models of kerogens: comparison of the chemical composition of kerogen types I, II and III at the beginning of diagenesis stage (after Béhar and Vandenbroucke, 1986).

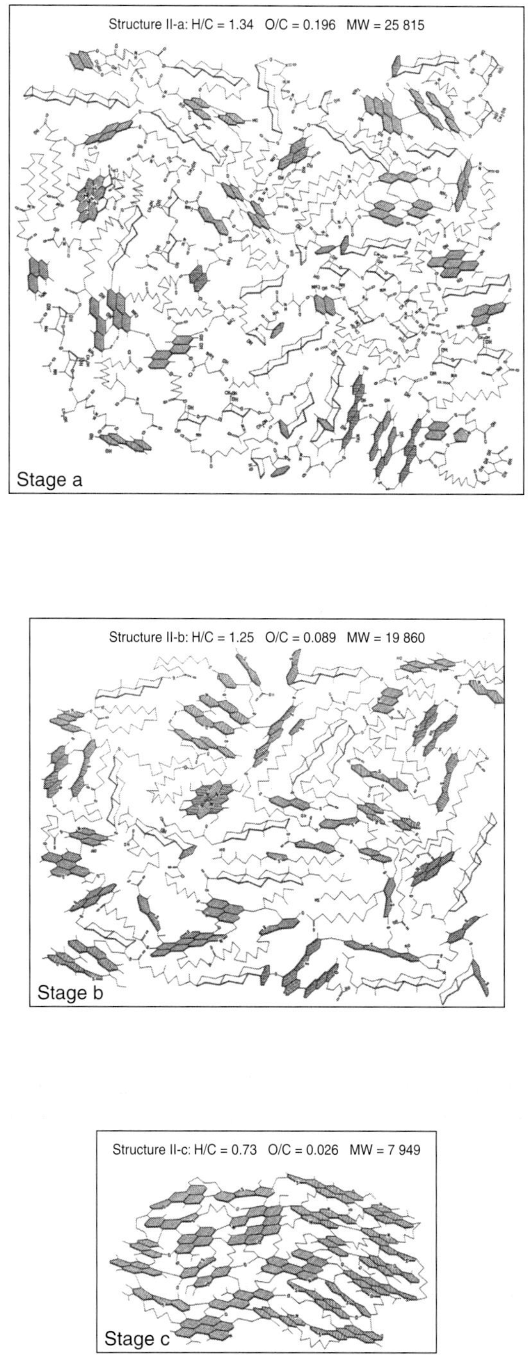

Figure 15

Structural models of kerogens: evolution of the chemical composition of type II kerogen at the evolution stages.

a: beginning of diagenesis stage;

b: beginning of catagenesis stage (beginning of oil generation);

c: end of catagenesis stage (beginning of gas generation) (after Béhar and Vandenbroucke, 1986).

remarked, this corresponds to a very small weight of kerogen (around 4.10^{-17} mg) which cannot be analysed by any chemical technique. The size of the structure on maturation was constrained by the measured weight loss of each kerogen type during thermogravimetric analysis, *i.e.* 79% for type I, 62% for type II and 34% for type III (Durand-Souron, 1980). The maximum number of condensed rings in aromatics and the orientation of aromatic clusters at the beginning of metagenesis were derived from dark-field imaging electron microscopy (Oberlin *et al.*, 1980). Randomly selected molecular building blocks from molecular analyses and functional groups from chemical functional analysis were connected by covalent bonds in order to fit elemental analysis and solid state ^{13}C NMR results, removing or adding building blocks and functional groups to improve fitting of calculated and measured atomic parameters at each further step of the construction. As already noted, the idea was not to represent an exact chemical structure of kerogens since this cannot be analysed, but rather to show in a synthetic way their main chemical and physical characteristics related to type and maturity, in order to allow visual comparison, even for non chemists. Although trapping of saturates (Béhar and Vandenbroucke, 1988) and other bitumen constituents into the kerogen network had been demonstrated, these kerogen structures deliberately omitted trapped compounds since kerogen *stricto sensu* should be totally free of soluble compounds. The resulting structures were published in two papers, one depicting the three kerogen types at various maturity stages and their associated asphaltenes (Béhar and Vandenbroucke, 1986), the other dealing exclusively with kerogen structures (Behar and Vandenbroucke, 1987). A 3D-like representation of individual cyclic building blocks was preferred to the traditional chemical representation in order to better visualize the differences in structure according to types and maturity. An example of the structures constructed for the three kerogen types at the beginning of diagenesis *stricto sensu* is shown on Figure 14. Structural transformations of type II kerogen at three increasing maturity stages, onset of diagenesis (stage a), onset of catagenesis (stage b) and end of catagenesis (stage c) is shown on Figure 15. The main structural parameters of all constructed models, fitting the analytical data of corresponding kerogen samples, are given in Table 2.

Given the tedious procedure of trial and error in assembling and counting molecular groups and bonding sites, then correcting them according to atomic data, a systematic construction and comparison of several structures fitting analyses of a given kerogen was not possible. Moreover, although the structural constituents (molecular building blocks, functional groups and bond types) can be predicted from analyses, the number of possible combinations of these elements is extremely large, and only a few models can be constructed. Computer-aided modeling was considered to help in this problem, and also in providing a more realistic view of structures by building 3D kerogen models. A computational tool for such macromolecules being not commercially available at this time, it was elaborated during a doctoral study (Faulon, 1991). Starting from the atomic and molecular analytical data previously described and from corresponding 3D structures of building blocks and bonds constructed in a library, the nature and total amount of the latter were first calculated to fit analyses. Further computation was simplified using intermediate building blocks and bonds where equivalent carbon configurations replaced heteroelements. Then, 3D carbon skeletons of kerogens were constructed by successively connecting these building blocks and bonds randomly selected from the library, assuming that any building block is connected by one or two covalent bonds. Finally, kerogen models were obtained by substituting carbons by heteroelements according to their corresponding atomic and functional analyses. A possible structure thus obtained for a type III kerogen at the end of catagenesis is shown on Figure 16. The microporosity observed in this structure is not an input constraint; it results exclusively from the construction with building blocks respecting bond lengths and directions.

TABLE 2

Main structural characteristics of kerogen models and associated natural samples corresponding to the following evolution stages:
a: beginning of diagenesis; b: beginning of catagenesis (beginning of oil generation); c: end of catagenesis (beginning of gas generation)

Kerogen	at. H/C	at. O/C	MW (dalton)	Aliphatic C (% of tot. C)	Naphthenic C (% of tot. C)	Aromatic C (fa) (% of tot. C)	Av. condensed ring number	Main funct. group	O in main f.g. (% of tot. O)
Type I-a	1.64	0.06	21 187	74	12	14	1.9	Ether	51
Type I-c	0.83	0.013	4 024	36	ε	64	3.8	Ether	50
Type II-a	1.34	0.196	25 815	51	19	30	2.8	Ester	31
Type II-b	1.25	0.089	19 860	45	14	41	3.1	Ester	48
Type II-c	0.73	0.026	7 949	28	ε	72	4.5	Ether	60
Type III-a	1.06	0.281	26 176	38	13	49	3	Phenol	25
Type III-b	0.98	0.138	18 939	34	6	60	3.4	Phenol	32
Type III-c	0.67	0.059	13 226	19	ε	81	5.7	Ether	34

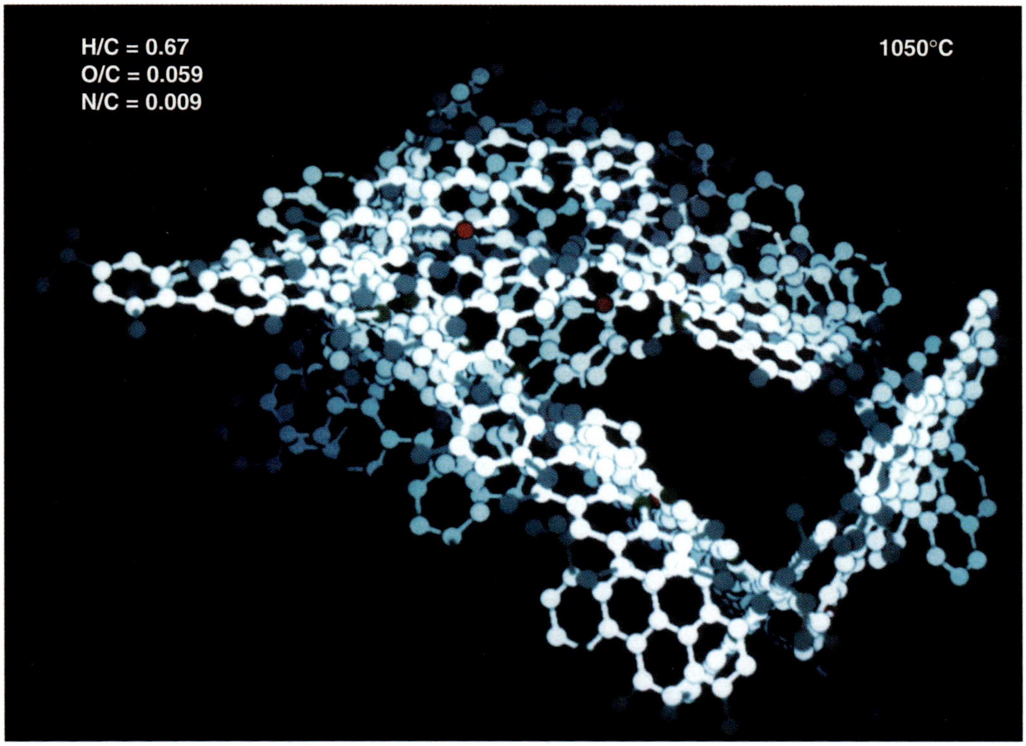

Figure 16

3D molecular structure of type III kerogen types at the evolution stage c: end of catagenesis. The energy is not minimized, but this is expected to make little difference due to the high aromaticity of this structure.

CONCLUSIONS

The increasing number of geochemical studies on petroleum systems raised in some cases questions as to the attribution of a defined type to specific source rocks, and these ambiguities were reflected by the frequent association of two types (I/II or II/III) to account for some of the measured geochemical parameters. The proposition of such associations is clearly not in line with the initial aim of characterizing sedimentary organic matter both by its original organic material and its depositional environment, as expressed by Tissot and Welte (1978, p. 142). The classification of kerogens based on reference types defined by their position in a van Krevelen diagram has indeed been of great help for oil exploration in predicting the petroleum generative capacity of source rocks and in describing the chemical characteristics of the generated oils. The notions of kerogen types and evolution paths, even though they may appear as an oversimplification of unique environmental situations, enabled a great deal of geochemical data to be condensed and explained in a simple manner to non geochemists. By achieving this, geochemistry secured a major position among the tools for oil and gas exploration. It should also be recalled that kerogen classification and association to specific depositional environments was at the source of major petroleum discoveries, and particularly those sourced by terrestrial organic matter.

In view of the extensive knowledge acquired on kerogen formation during the nearly twenty years following publication of kerogen models, thanks to studies of new samples and, more importantly, the new developments of analytical tools, one can today question their representativity. For example, the type I model presented here is likely valid only for kerogens corresponding to alkaline depositional environments similar to those occurring for the Green River Shale sedimentation. For type I organic matter resulting from eutrophication conditions, the microbial population and hence its chemical composition should be closely related to changes in the environment chemistry. Even among the Green River samples, a fairly large variability of the chemical analysis from immature kerogens was indeed observed (Tissot *et al.*, 1978), both laterally and vertically. This is also true when comparing type I kerogens from the Green River shale to *Botryococcus*-derived type I kerogens, the structure of which has been the subject of numerous studies (Béhar *et al.*, 1995, and references therein). In the case of type I paraffinic moieties, a large fraction of long chain constituents such as those observed by high temperature gas chromatography, although nonextractable with usual solvents, is probably only trapped by physical entanglement or steric hindrance and not linked by covalent bonds to the kerogen structure. The type II model could have a broader representativity, up to a certain limit of sulfur content. On the contrary, the molecular representation of coals and type III kerogens by averaged chemical structures cannot be realistic. In fact it was shown early on, by light microscopy observation of coals (Stopes, 1935), that their main constituents are physically completely separated at the microscopic level, thus implying separation at the macromolecular level. Given the common precursors of coals and type III kerogens, a similar problem should occur for the latter (Vandenbroucke, 1980). Accordingly, the organic fraction is mostly made up of ligneous debris with a predominantly aromatic structure, and a variable minor proportion of constituents derived from the protective tissues of higher plants, with a predominantly aliphatic structure rich in long chain alkanes. This double contribution from organic remains explains the apparent discrepancy observed between the mostly aromatic kerogen composition, as seen by elemental analysis with an atomic H/C ratio lower than 1, and the highly paraffinic extracts and oils generated by this type III kerogen, in comparison with type II generated products. Separation between aromatic lignous components and aliphatic debris from leaf and fruit membranes may also explain in part the very early generation and migration of heavy saturates in oils sourced by some type III kerogens, as observed in several case studies (Khorasani and Michelsen, 1991; Clayton *et al.*, 1991).

Further advances in kerogen knowledge can still be expected in the coming years from new developments in analytical techniques. However many unknowns will remain for a long time regarding the atomic structural arrangement in kerogen moieties, in part because microscopic techniques at the atomic level need to be developed and adapted to such a study, and also because the onset of kerogen formation is mainly governed by biological processes, particularly biodegradation of living organic matter, the mechanisms of which still need many years of research to be understood.

ACKNOWLEDGEMENTS

I gratefully acknowledge the pertinent comments of Françoise Béhar, Alain-Yves Huc, John Lynch and Bernard Colletta, that greatly improved earlier versions of the manuscript.

Major advances in the concepts and analytical tools described in this paper were achieved within the *IFP Geochemistry Department*. I participated from the beginning of my professional career in many of the research projects conducted by such eminent scientists as Bernard Tissot, Bernard Durand, Régis Pelet and many other colleagues from this department, and thus it is both a duty and a pleasure for me to dedicate this review to them.

REFERENCES

Acholla, F.V., Orr, W.L. (1993) Pyrite removal from kerogen without altering organic matter: the chromous chloride method. *Energy and Fuels*, **7**, 406-410.

Aizenshtat Z., Stoler, A., Cohen, Y, Nielsen, H. (1983) The geochemical sulphur enrichment of recent organic matter by polysulfides in the Solar-lake. In: Bjoroy, M. *et al.* (eds.) *Advances in Organic Geochemistry 1981*, Wiley, Chichester, 279-288.

Albrecht, P. (1969) Constituants organiques de roches sédimentaires. *PhD Thesis*, Strasbourg University.

Albrecht, P., Vandenbroucke, M., Mandengué, M. (1976) Geochemical studies on the organic matter from the Douala Basin (Cameroon) - I. Evolution of the extractable organic matter and the formation of petroleum. *Geochimica et Cosmochimica Acta*, **40**, 791-799.

Béhar, F., Pelet, R. (1985) Pyrolysis-gas chromatography applied to organic geochemistry. Structural similarities between kerogens and asphaltenes from related rock extracts and oils. *Journal of Analytical and Applied Pyrolysis*, **8**, 173-187.

Béhar, F., Vandenbroucke, M. (1986) Représentation chimique de la structure des kérogènes et des asphaltènes en fonction de leur origine et de leur degré d'évolution. *Oil & Gas Science and Technology - Rev. IFP*, **41**, 173-188.

Béhar, F., Vandenbroucke, M. (1987) Chemical modelling of kerogens. *Organic Geochemistry*, **11**, 15-24.

Béhar, F., Vandenbroucke, M. (1988) Characterization and quantification of saturates trapped inside kerogen: implications for pyrolysate composition. In: Mattavelli, L., Novelli, L. (Eds.), *Advances in Organic Geochemistry 1987, Organic Geochemistry*, **13**, 927-938.

Behar, F., Leblond, C., Saint-Paul, C. (1989) Analyse quantitative des effluents de pyrolyse en milieu ouvert et fermé. *Oil & Gas Science and Technology - Rev. IFP*, **44**, 387-411.

Béhar, F., Derenne, S., Largeau, C. (1995) Closed pyrolyses of the isoprenoid algaenan of *Botryococcus braunii*, L race: geochemical implications for derived kerogens. *Geochimica et Cosmochimica Acta*, **59**, 2983-2997.

Behar, F., Vandenbroucke, M., Tang, Y., Marquis, F., Espitalié, J. (1997) Thermal cracking of kerogen in open and closed systems: determination of kinetic parameters and stoichiometric coefficients for oil and gas generation. *Organic Geochemistry*, **26**, 321-339.

Béhar, F., Beaumont, V., De B. Penteado, H.L. (2001) Rock-Eval 6 technology: Performances and developments. *Oil & Gas Science and Technology - Rev. IFP*, **56**, 111-134.

Bordenave, M., Combaz, A., Giraud, A. (1970) Influence de l'origine des matières organiques et de leur degré d'évolution sur les produits de pyrolyse du kérogène. In: Hobson, G.D. and Speers, G.C. (eds.) *Advances in Organic Geochemistry 1966*, Pergamon Press, Oxford, 389-405.

Boudou, J.P. (1981) Diagenèse organique de sédiments deltaïques (delta de la Mahakam, Indonésie). *PhD Thesis*, Orleans University.

Burlingame, A.L., Haug, P.A., Schnoes, H.K., Simoneit, B.R. (1969) Fatty acids derived from the Green River Formation oil shale by extractions and oxidations - A review. In: Schenck, P.A., Havenaar, I. (eds.), *Advances in Organic Geochemistry 1968*, Pergamon Press, Oxford, 85-129.

Clayton, J.L., Rice, D.D., Michael, G.E. (1991) Oil-generating coals of the San Juan Basin, New Mexico and Colorado, USA *Organic Geochemistry*, **17**, 735-742.

Combaz, A., De Matharel, M. (1978) Organic sedimentation and genesis of petroleum in Mahakam delta, Borneo. *AAPG Bulletin*, **62**, 1684-1695.

Crum-Brown, A. (1912) Oil shales of the Lothians. *Memoirs Geological Survey*, 143.

Deroo, G., Durand, B., Espitalié, J., Pelet, R., Tissot, B. (1969) Possibilité d'application des modèles mathématiques de formation du pétrole à la prospection dans les bassins sédimentaires. In: Schenck, P.A., Havenaar, I. (Eds.) *Advances in Organic Geochemistry 1968*, Pergamon Press, Oxford, 345-353.

Deroo, G., Powell, T.G., Tissot, B., McCrossan, R.G.(1977) The origin and migration of petroleum in the Western Canadian sedimentary basin, Alberta - A geochemical and thermal maturation study. *Geological Survey of Canada*, Bulletin **262**.

Deroo, G., Powell, T.G. (1978) The oil sands deposits of Alberta: their origin and geochemical history. In: Strausz, O.P., Lown, E.M. (Eds.) *Oil Sand and Oil Shale Chemistry*, Verlag Chemie International, New York, 11-32.

Dunoyer de Segonzac, G. (1969) Les minéraux argileux dans la diagenèse. Passage au métamorphisme. *PhD Thesis*, Strasbourg University.

Durand, B., Espitalié, J. (1970) Analyse géochimique de la matière organique extraite des roches sédimentaires. II. Extraction et analyse quantitative des hydrocarbures de la fraction C_1-C_{15} et des gaz permanents. *Oil & Gas Science and Technology - Rev. IFP*, **25**, 741-751.

Durand, B., Espitalié, J., Oudin, J.L. (1970) Analyse géochimique de la matière organique extraite des roches sédimentaires. III. Accroissement de la rapidité du protocole opératoire par l'amélioration de l'appareillage. *Oil & Gas Science and Technology - Rev. IFP*, **25** 1268-1279.

Durand, B., Espitalié, J. (1971) Formation et évolution des hydrocarbures de C_1 à C_{15} et des gaz permanents dans les argiles du Toarcien du bassin de Paris. In: von Gaertner, H.R. and Wehner H. (Eds.), *Advances in Organic Geochemistry 1971*, Pergamon Press, Oxford, 455-468.

Durand, B., Espitalié, J., Nicaise, G., Combaz, A. (1972) Étude de la matière organique insoluble (kérogène) des argiles du Toarcien du Bassin de Paris. I - Étude par les procédés optiques, analyse élémentaire, étude en microscopie et diffraction électroniques. *Oil & Gas Science and Technology - Rev. IFP*, **27**, 865-884.

Durand, B., Espitalié, J. (1973) Évolution de la matière organique au cours de l'enfouissement des sédiments. *Comptes rendus de l'Académie des Sciences*, **276**, 2253-2256.

Durand, B., Espitalié, J. (1976) Geochemical studies on the organic matter from the Douala Basin (Cameroon) -II. Evolution of kerogen. *Geochimica et Cosmochimica Acta*, **40**, 801-808.

Durand, B. (Ed.) (1980) *Kerogen, Insoluble Organic Matter from Sedimentary Rocks*. Éditions Technip, Paris.

Durand, B. (1980) Sedimentary organic matter and kerogen. Definition and quantitative importance of kerogen, in Durand, B. (ed.), *Kerogen, Insoluble Organic Matter from Sedimentary Rocks*. Éditions Technip, Paris, 13-34.

Durand, B., Nicaise, G. (1980) Procedures of kerogen isolation, in Durand, B. (Ed.), *Kerogen, Insoluble Organic Matter from Sedimentary Rocks*. Éditions Technip, Paris, 35-53.

Durand, B., Monin, J.C. (1980) Elemental analysis of kerogens, in Durand, B. (Ed.), *Kerogen, Insoluble Organic Matter from Sedimentary Rocks*, Éditions Technip, Paris, 113-142.

Durand, B., Oudin, J.L. (1980) Exemple de migration des hydrocarbures dans une série deltaique: le delta de la Mahakam, Kalimantan, Indonésie. In: *10th World Petroleum Congress Proceedings vol. 2*, Heyden, London, 3-11.

Durand-Souron, C.(1980) Thermogravimetric analysis and associated techniques applied to kerogens. In: Durand, B. (Ed.), *Kerogen, Insoluble Organic Matter from Sedimentary Rocks*. Éditions Technip, Paris, pp.143-161.

Durand-Souron, C., Boulet, R., Durand, B. (1982) Formation of methane and hydrocarbons by pyrolysis of immature kerogens. *Geochimica et Cosmochimica Acta*, **46**, 1193-1202.

Ehmann, W.D., Koppenhaal, D.W., Hamrin Jr., C.E., Jones, W.C., Prasad, M.N., Tian, W.Z. (1986) Comparison of methods for the determination of organic oxygen in coals. *Fuel*, **65**, 1563-1570.

Espitalié, J., Durand, B., Roussel, J.C., Souron, C. (1973) Étude de la matière organique insoluble (kérogène) des argiles du Toarcien du Bassin de Paris. II - Étude en spectroscopie infrarouge, en analyse thermique différentielle et en analyse thermogravimétrique. *Oil & Gas Science and Technology - Rev. IFP*, **28**, 37-66.

Espitalié, J., Laporte, J.L., Madec, M., Marquis, F., Leplat, P., Paulet, J., Boutefeu, A. (1977a) Méthode rapide de caractérisation des roches mères, de leur potentiel pétrolier et de leur degré d'évolution. *Oil & Gas Science and Technology - Rev. IFP*, **32**, 23-42.

Espitalié, J., Madec, M., Tissot, B., Mennig, J.J., Leplat, P. (1977b) Source rock characterization method for petroleum exploration. *Offshore Technology Conference*, Paper OTC 2935, 439-444.

Espitalié, J., Senga Makadi, K., Trichet, J. (1984) Role of the mineral matrix during kerogen pyrolysis. *Advances in Organic Geochemistry 1983. Organic Geochemistry*, **6**, 365-382.

Faulon, J.L., Vandenbroucke, M., Drappier, J.M., Béhar, F., Romero, M.. (1990) 3D chemical model for geological macromolecules. *Organic Geochemistry*, **16**, 981-993.

Faulon, J.L. (1991) Prediction, elucidation et modélisation moléculaire: algorithmes et applications. *PhD Thesis*.

Fester, J.I., Robinson, W.E. (1966) Oxygen functional groups in Green River oil-shale kerogen and trona acids. In: Gould, R.F., (Ed.), *Coal Science, ACS Publications,* Washington D.C., 22-31.

Forsman, J.P. (1963) Geochemistry of kerogen, in Breger, I.A. (Ed.), *Organic Geochemistry, Earth Series,* monograph no. 16, Pergamon Press, Oxford, 148-182.

Forsman, J.P., Hunt, J.M. (1958) Insoluble organic matter (kerogen) in sedimentary rocks. *Geochimica et Cosmochimica Acta*, **15**, 170-182.

Fuchs, W., Sandhoff, A.G. (1942) Theory of coal pyrolysis. *Industrial and Engineering Chemistry,* **34**, 567-571.

Giraud, A. (1970) Application of pyrolysis and gas chromatography to geochemical characterization of kerogen in sedimentary rock. *AAPG Bulletin*, **54**, 439-455.

Given, P.H. (1960) The distribution of hydrogen in coals and its relation to coal structure. *Fuel*, **39**, 147-153 .

Hatcher, P.G. (1980) The origin, composition, chemical structure and diagenesis of humic substances, coals and kerogens as studied by nuclear magnetic resonance. *PhD Thesis*, University of Maryland.

Huc, A.Y. (1978) Geochimie organique des schistes bitumineux du Toarcien du bassin de Paris. *PhD Thesis*, University of Strasbourg.

Huc A.Y., Durand B., Roucache J., Vandenbroucke M., Pittion J.L. (1986) Comparison of three series of organic matter of continental origin. In: Leythaeuser, D., Rullkotter, J. (Eds.), Advances in organic geochemistry 1985. *Organic Geochemistry* **10,** 1/3, 65-72.

Huc, A.Y. (1990) Understanding organic facies: a key to improved quantitative petroleum evaluation of sedimentary basins. In: Huc, A.Y. (Ed.), *Deposition of Organic Facies, AAPG Studies in Geology, 30*, AAPG Publications, Tulsa, Oklahoma, pp. 1-11.

Khavari Khorasani, G., Michelsen, J.K. (1991) Geological and laboratory evidence for early generation of large amounts of liquid hydrocarbons from suberinite and subereous components. *Organic Geochemistry*, **17**, 849-863.

Klemme, H.D., Ulmishek, G.F. (1991) Effective petroleum source rocks of the world: stratigraphic distribution and controlling depositional factors. *AAPG Bulletin*, **75**, 1809-1851.

Kowalewski, I., Vandenbroucke, M., Huc, A.Y., Taylor, M.J., Faulon, J.L. (1996) Preliminary results on molecular modeling of asphaltenes using structure elucidation programs in conjunction with molecular simulation programs. *Energy & Fuels* **10**, 97-107.

Lafargue, E., Béhar, F. (1989) Application of a new preparative pyrolysis technique for the determination of source-rock types and oil/source-rock correlation. *Geochimica et Cosmochimica Acta*, **53**, 2973-2983.

Lafargue, E., Marquis, F., Pillot, D. (1998) Rock-Eval 6 applications in hydrocarbon exploration, production, and soil contamination studies. *Oil & Gas Science and Technology - Rev. IFP*, **53**, 421-437.

Larsen, J.W., Pan, C.S., Shawver, S. (1989) Effect of demineralization on the macromolecular structure of coals. *Energy & Fuels*, **3**, 557-561.

Leplat, P., Noël, R. (1972) Étude par pyrochromatographie en phase gazeuse et par réflectance de l'évolution de quelques kérogènes. In: von Gaertner, H.R. and Wehner H. (Eds.), *Advances in Organic Geochemistry 1971*, Pergamon Press, Oxford, 567-576.

Leplat, P., Noël, R. (1974) Contribution à l'étude de l'état de diagenèse des roches à kérogène par le dosage automatique rapide du CO_2 et du C_2H_4 libérés par pyrolyse de 400 à 900 °C. In: Tissot, B. and Bienner, F. (Eds.), *Advances in Organic Geochemistry 1973*, Éditions Technip, Paris, 107-115.

Lorant, F., Béhar, F. (2002) Late generation of methane from mature kerogens. *Energy & Fuels*, **16**, 412-427.

Louis, M. (1966) Études géochimiques sur les schistes cartons du Toarcien du Bassin de Paris, in Hobson, G.D. and Louis, M.C. (Eds.), *Advances in Organic Geochemistry 1964*, Pergamon Press, Oxford, 85-94.

Louis, M., Tissot, B. (1967) Influence de la température et de la pression sur la formation des hydrocarbures dans les argiles à kérogène. *7th World Petroleum Congress Proceedings*, **2**, Elsevier, Amsterdam, 47-60.

Maciel, G.E., Bartuska, V.J., Miknis, F.P. (1979) Characterization of the organic material in coal by proton-decoupled nuclear magnetic resonance with magic angle spinning. *Fuel*, **58**, 391-394.

Mc Iver, R.D. (1967) Composition of kerogen - clue to its role in the origin of petroleum. *7th World Petroleum Congress Proceedings*, **2**, Elsevier, Amsterdam, 25-36.

Marchand, A., Conard, J. (1980) Electron paramagnetic resonance in kerogen studies. In: Durand, B. (Ed.), *Kerogen, Insoluble Organic Matter from Sedimentary Rocks*. Éditions Technip, Paris, 243-270.

Monin, J.C., Durand, B., Vandenbroucke, M., Huc, A.Y. (1980) Experimental simulation of the natural transformation of kerogen, in Douglas, A.G. and Maxwell, J.R. (Eds.), *Advances in Organic Geochemistry 1979*, Pergamon Press, Oxford, 517-530.

Oberlin, A., Boulmier, J.L., Villey, M. (1980) Electron microscopic study of kerogen microtexture. Selected criteria for determining the evolution path and evolution stage of kerogen. In: Durand, B. (Ed.), *Kerogen, Insoluble Organic Matter from Sedimentary Rocks*. Éditions Technip, Paris, 191-241.

Oudin, J.L., Califet-Debyser, Y. (1969) Influence of depth and temperature on the structure and distribution of alkanes and aromatic hydrocarbons from rock extracts. In: *American Chemical Society Division of Petroleum Chemistry Preprints-New York 1969 Meeting*, ACS Publications, Washington D.C., 16-21.

Oudin, J.L. (1970) Analyse géochimique de la matière organique extraite des roches sédimentaires. I. Composés extractibles au chloroforme. *Oil & Gas Science and Technology - Rev. IFP*, **25**, 3-15.

Pelet, R. (1987) Vue d'ensemble sur les résultats du sondage Misedor. In: *Le sondage Misedor*, Editions Technip, Paris, 375-383.

Robinson, W.E. (1969) Isolation procedures for kerogens and associated soluble organic materials. In: Eglinton, G., Murphy, M.T.J. (Eds.). *Organic Geochemistry - Methods and Results*. Springer -Verlag, Berlin Heidelberg New York, 181-195.

Saxby, J.D. (1970) Isolation of kerogens in sediments by chemical methods. *Chemical Geology*, **6**, 173-184.

Shinn, J.H. (1984) From coal to single-stage and two-stage products: a reactive model of coal structure. *Fuel*, **63**, 1187-1196.

Sieskind, O., Ourisson, G., 1972. Hydrocarbures formés par le craquage thermocatalytique de l'acide stéarique en présence de montmorillonite. *Comptes rendus de l'Académie des Sciences*, **274**, 2186-2189.

Siskin, M., Scouten, C.G., Rose, K.D., Aczel, T., Colgrove, S.G., Pabst Jr., R.E. (1995) Detailed structural characterization of the organic material in Rundle Ramsay Crossing and Green River oil shales. In: Snape, C., (Ed.), *Composition, Geochemistry and Conversion of Oil Shales*, Kluwer Academic Publishers, Dordrecht, 143-158.

Solomon, P.R. (1981) New approaches in coal chemistry. In: *ACS Symposium Series 169*, Washington, 61-71.

Souron, C., Boulet, R., Espitalié, J. (1974) Etude par spectrométrie de masse de la décomposition thermique sous vide de kérogènes appartenant à deux lignées évolutives distinctes. *Oil & Gas Science and Technology - Rev. IFP*, **24**, 661-678.

Souron, C., Boulet, R., Espitalié, J. (1977) Étude par spectrométrie de masse de la décomposition thermique de roches sédimentaires contenant de la matière organique de deux types différents et comparaison avec les kérogènes correspondants. In: *Advances in Organic Geochemistry 1975*, Campos R. and Goni, Y. (Eds.), 797-820.

Spiro, C.L. (1981) Space-filling models for coal: a molecular description of coal plasticity. *Fuel*, **60**, 1121-1126.

Spiro, C.L., Kosky, P.G. (1982) Space-filling models for coal. 2. Extension to coals of various ranks. *Fuel*, **61**, 1080-1087.

Stankiewicz, B.A., Kruge, MA, Crelling, J.C., Salmon, G.L. (1994) Density gradient centrifugation: application to the separation of macerals of Type I, II and III sedimentary organic matter. *Energy & Fuels*, **8**, 1513-1521.

Stopes, M. (1935) On the petrology of banded bituminous coals. *Fuel*, **14**, 4-13.

Tissot, B., Califet-Debyser, Y., Deroo, G., Oudin, J.L. (1971) Origin and evolution of hydrocarbons in early Toarcian shales, Paris basin, France. *AAPG Bulletin*, **55**, 2177-2193.

Tissot, B. (1969) Premières données sur les mécanismes et la cinétique de la formation du pétrole dans les sédiments - Simulation d'un schéma réactionnel sur ordinateur. *Oil & Gas Science and Technology - Rev. IFP*, **24**, 470-501.

Tissot, B., Pelet, R. (1971) Nouvelles données sur les mécanismes de genèse et de migration du pétrole - Simulation mathématique et application à la prospection. In: *8th World Petroleum Congress Proceedings*, **2,** Applied Science Publishers, London, 35-46.

Tissot, B., Durand, B., Espitalié, J., Combaz, A. (1974) Influence of nature and diagenesis of organic matter in formation of petroleum. *AAPG Bulletin*, **58,** 499-506.

Tissot, B., Espitalié, J., 1975. L'évolution thermique des sédiments : applications d'une simulation mathématique. *Oil & Gas Science and Technology - Rev. IFP*, **30**, 743-777.

Tissot, B., Deroo, G, Hood, A. (1978) Geochemical study of the Uinta Basin: formation of petroleum from the Green River formation. *Geochimica et Cosmochimica Acta*, **42**, 1469-1485.

Tissot, B., Vandenbroucke, M. (1983) Geochemistry and pyrolysis of oil shales. In: Miknis, F.P., McKay, J.F. (eds.), *Geochemistry and Chemistry of Oil Shales, ACS Symposium Series 230*, 1-11.

Tissot, B.P., Welte, D.H., (1978) *Petroleum Formation and Occurrence*, Springer Verlag, Berlin.

Vandenbroucke, M., Albrecht, P., Durand, B. (1976) Geochemical studies on the organic matter from the Douala Basin (Cameroon) - III. Comparison with the early Toarcian shales, Paris basin, France. *Geochimica et Cosmochimica Acta*, **40**, 1241-1249.

Vandenbroucke, M., Durand, B., Hood, A. (1977) Thermal evolution experiments on a kerogen from the Green River Shales formation. Publication to *Advances in Organic Geochemistry 1977*, Moscow. *IFP report* no. 25168.

Vandenbroucke, M. (1980) Structure of kerogens as seen by investigations on soluble extracts. In: Durand, B. (Ed.), *Kerogen, Insoluble Organic Matter from Sedimentary Rocks*, Éditions Technip, Paris, 415-443.

Vandenbroucke, M., Durand, B., Oudin, J.L. (1983) Detecting migration phenomena in a geological series by means of C_1-C_{35} hydrocarbon amounts and distribution. In: Bjoroy at al. (Eds.), *Advances in Organic Geochemistry 1981*, Wiley & Sons, Chichester, 147-155.

Vandenbroucke, M., Béhar, F., Espitalié, J. (1988) Characterization of sedimentary organic matter by preparative pyrolysis: comparison with Rock-Eval pyrolysis and pyrolysis-gas chromatography techniques. *Energy & Fuels*, **2**, 252-258.

Vandenbroucke, M., Largeau, C. (2004) Kerogen formation, structure and evolution. Submitted to *Organic Geochemistry*.

van Krevelen, D.W. (1961) *Coal: Typology - Chemistry - Physics - Constitution*. Elsevier, The Netherlands.

Yen, T.F. (1976) Structural aspects of organic components in oil shales. In: Yen, T.F., Chilingarian, G.V. (Eds.) *Oil Shale, Developments in Petroleum Science*, 5, Elsevier, Amsterdam, 129-148.

Final manuscript received in March 2003

Oil & Gas Science and Technology – Rev. IFP, Vol. 58 (2003), No. 2, pp. 271-297
Copyright © 2003, Éditions Technip

From Organic Geochemistry to Statistical Thermodynamics: the Development of Simulation Methods for the Petroleum Industry

Ph. Ungerer[1,2]

1 Institut français du pétrole, 1 et 4, avenue de Bois-Préau, 92852 Rueil-Malmaison Cedex - France
2 Université de Paris-Sud, Laboratoire de chimie physique, UMR 8000 CNRS, Orsay - France
e-mail: philippe.ungerer@ifp.fr

Résumé — De la géochimie organique à la thermodynamique statistique : développement de méthodes de simulation pour l'industrie pétrolière — Le travail présenté constitue une synthèse des travaux auxquels l'auteur a contribué successivement dans trois domaines de recherche appliquée.

Le premier de ces domaines est la géochimie organique pétrolière, qui se donne pour objectif de comprendre la formation des gisements pétroliers à partir d'analyses détaillées de la composition des combustibles fossiles, de simulations expérimentales en laboratoire par pyrolyse et de reconstitutions de l'histoire géologique des bassins sédimentaires concernés. Ces reconstitutions, qui impliquent la simulation mathématique de nombreux phénomènes couplés comme le craquage thermique, l'écoulement des fluides en milieux poreux, la sédimentation, la compaction des roches sédimentaires et les équilibres entre phases, ont été l'objet des principales contributions de l'auteur dans ce domaine. Les résultats obtenus à l'*IFP* dans les années 1980, dans le cadre d'un travail d'équipe pluridisciplinaire, ont fourni les bases à de nombreuses applications industrielles dans le cadre d'une discipline nouvelle, les modèles de bassin.

Le deuxième de ces domaines est la thermodynamique pétrolière, où l'on s'attache à mesurer et à prédire les propriétés d'équilibre et de transport dans les procédés de production ou de traitement des pétroles bruts ou des gaz naturels. L'apport du travail présenté à ce domaine très vaste a été de caractériser des fluides pétroliers à haute pression, tant par la mise en évidence expérimentale des diagrammes de phase particuliers à certains fluides HP-HT que par la prédiction de leurs propriétés au moyen d'équations d'état. Il a permis aussi de comprendre plus précisément comment les équilibres de phase entre huile et gaz étaient susceptibles de modifier la composition des fluides pétroliers.

Le troisième domaine d'étude abordé est la thermodynamique statistique, dont les applications connaissent depuis peu une forte croissance avec la généralisation de la simulation moléculaire. Les méthodes de Monte Carlo ou de dynamique moléculaire apportent non seulement une compréhension à l'échelle microscopique, mais aussi une prédiction plus fiable des propriétés d'équilibre et de transport dans de nombreux cas où les modèles macroscopiques classiques sont limités par leur caractère empirique. En s'appuyant cette fois sur une collaboration étroite avec une équipe universitaire, il a été possible de mettre en œuvre des algorithmes améliorés et de développer un nouveau potentiel intermoléculaire afin de répondre aux besoins de l'industrie pétrolière. C'est ainsi que des applications effectives de la simulation moléculaire ont vu le jour dans le domaine du traitement des gaz riches en H_2S, des séparations par adsorption dans les zéolithes et du calcul de propriétés d'hydrocarbures non disponibles commercialement.

Bien que très différents au premier abord, ces trois domaines présentent des similitudes significatives comme leur articulation dominante autour de la physico-chimie, leur exigence en logiciels nouveaux ou l'optimisation empirique de paramètres dont on attend néanmoins un sens physique.

Abstract — From Organic Geochemistry to Statistical Thermodynamics: the Development of Simulation Methods for the Petroleum Industry — *This work is a synthesis of researches to which the author contributed successively in three areas of applied research.*

The first of these areas is petroleum organic geochemistry, which is aimed to understand the formation of petroleum reservoirs from detailed analyses of the composition of fossil fuels, experimental simulations in the laboratory by pyrolysis, and reconstructions of the geological history of the sedimentary basins concerned. These reconstructions, which imply the mathematical simulation of many coupled mechanisms such as thermal cracking, fluid flow in porous media, sedimentation, sedimentary rock compaction and inter-phase equilibria, were the subject of the author's main contributions in this field. The results obtained at IFP in the 1980s, as part of a multidisciplinary team project, provided the basis for numerous industrial applications in connection with the new discipline of basin models.

The second of these areas is petroleum thermodynamics, which involves measuring and predicting the equilibrium and transport properties in crude oil and natural gas production and processing. The contribution of the work presented in this vast area was to characterize petroleum fluids at high pressure, both by experimental identification of the phase diagrams specific to certain HP-HT fluids, and the prediction of their properties by means of equations of state. It also allowed a more accurate understanding of how the phase equilibria between oil and gas were likely to alter the composition of petroleum fluids.

The third area of study is statistical thermodynamics, whose applications have recently soared with the generalization of molecular simulation. Monte Carlo and molecular dynamics methods not only offer an understanding at microscopic scale, but also a more reliable prediction of the equilibrium and transport properties in many cases where classical macroscopic models are limited by their empirical character. By relying on this occasion on close collaboration with a university team, it was possible to implement improved algorithms and to develop a new intermolecular potential to meet the needs of the petroleum industry. Thus effective applications of molecular simulation materialized in the processing of H_2S *gases, separations by adsorption in zeolites, and the calculation of properties of commercially unavailable hydrocarbons.*

These three areas, which appear very different at first glance, display significant similarities, like their dominant articulation around physical chemistry, their need of new software, and the empirical optimization of parameters whereof a physical meaning is nonetheless expected.

INTRODUCTION

Fossil fuels—oil, coal and natural gas—represent the principal energy source for humanity today. The rate at which mankind exploits these resources, which have taken millions of years to come into being, may seem irrational, but their massive replacement by other energies is not seriously conceivable in the short term. The energy supply of the planet will therefore depend for a long time to come of the renewal of petroleum resources by the discovery and production of new reservoirs. This renewal is contingent on the continuous improvement of the techniques employed. In exploration, the days are past when the petroleum geologist was content to prospect in the field to pick or position a well, and he relies today on an in-depth knowledge of the mechanisms of formation of the oil reservoirs. In production, no longer are we content to let the oil gush from the production wells, and waterflooding and gas injection in the reservoirs is now systematic to optimize recovery. In refining, the distillation of crude oil is no longer sufficient, and has for decades been supplemented by processing methods and catalytic cracking to supply more efficient and less pollutant fuels.

This article, which summarizes the research projects to which the author contributed from 1980 to 2002 at *Institut français du pétrole* and the *University of Paris Sud*, offers an opportunity to dwell on some of these areas of applied research.

The first of these areas is petroleum geochemistry, which underwent a scientific revolution like all of earth science, after the advent of plate tectonics in the 1960s. This discipline, which participated in the tremendous impetus aimed at the time to quantify the behavior of the earth's crust in physical and chemical terms, had succeeded in demonstrating the biogenic origin of oil. The presence of many "geochemical fossils", hydrocarbons directly derived from molecules known in living beings, served to demonstrate that oil and natural gas were primarily produced by a slow thermal degradation of the sedimentary organic matter (Durand, 1980; Tissot and Welte, 1984).

In this field, the 1980s saw the development at *IFP* of ambitious numerical simulation methods to understand the formation of oil reservoirs. This work was aimed to improve the kinetic models of thermal cracking proposed some years earlier (Tissot and Espitalié, 1975), and also to propose an original approach to describe the transport processes responsible for its accumulation in reservoir rocks. These achievements were the focus of a multidisciplinary project integrating geosciences, chemistry, thermodynamics, mechanics and applied mathematics.

The second area of this recapitulation is the thermodynamics of petroleum fluids, and chiefly the experimental characterization of fluids at high pressure, which in petroleum production technology corresponds to the interval between 50 and 120 MPa. These investigations, propelled by the wish to produce very deep reservoirs, helped to identify phase diagrams of a specific type when a natural gas contains heavy *n*-alkanes which crystallize at low temperature. They also culminated with the measurement of volumetric properties and viscosities in the same conditions. As to modeling, we will above all have to address the difficulty of predicting the volumetric properties and phase equilibria of crude oils and natural gases. This inadequacy is not only explained by the empirical character of routine equations of state like Redlich-Kwong-Soave (Soave, 1972) and Peng-Robinson (Peng and Robinson, 1976), but also by the improper consideration of the fluid compositions: above 7 or 10 carbon atoms, engineers generally ignore the measured composition (Pedersen *et al.*, 1989). To improve the predictions of the properties of gas condensate, it was proposed to represent the heavy fraction by distributions of saturated and aromatic hydrocarbons. As we shall show, the results encouraged further work in this direction, which was unfortunately limited by the lack of experimental data on heavy hydrocarbons other than *n*-alkanes. This type of finding is also not specific to reservoir fluids: despite the progress offered by excess free energy models (Huron and Vidal, 1979), by contribution methods (Fredenslund *et al.*, 1977), by the application of the principle of corresponding states (Lee and Kesler, 1975) and by equations of state based on concepts of statistical thermodynamics (Chien *et al.*, 1983; Chapman *et al.*, 1989), thermodynamic models were often inaccurate. In many cases, the acquisition of experimental data enabling their adjustment was lengthy and costly: heavy hydrocarbons not commercially available, highly toxic components, very high pressures, new molecules, etc. In the specific example of multicomponent adsorption equilibria relevant to separation processes, classical thermodynamic models had to rely on selectivity measurements, which encumbered the search for better adsorbents (Ruthven, 1984).

These observations logically encouraged a trend towards a third area of research from 1997, molecular simulation. By taking account of the molecular structure and the different forms of intermolecular energy, simulation can offer more reliable predictions in extended areas. The continued progress of computation capacity enabled this discipline to occupy a leading place in physical chemistry (Allen and Tildesley, 1987; Frenkel and Smit, 1996). Thanks to the Gibbs ensemble method (Panagiotopoulos, 1987) it is now possible to calculate phase equilibria, while molecular dynamics provides access to transport properties. In terms of grand canonical simulation, adsorption processes can be simulated at a pertinent scale in microporous materials used to perform certain separations that are difficult in chemical engineering (Lachet *et al.*, 1997). For the effective use of molecular simulation in applied thermodynamics, many laboratories focus on algorithmic improvements to Monte Carlo methods employed to calculate the equilibrium properties (Smit *et al.*, 1995) or methods of molecular dynamics (Rowley, 1994; Dysthe *et al.*, 1999a; Dysthe *et al.*, 1999b; Dysthe *et al.*, 2000). In cooperation with the *University of Paris-Sud, IFP* attempted to contribute thereto by proposing an original method to calculate the bubble point of multicomponent mixtures and to calculate derivative properties like heat capacity or Joule-Thomson's coefficient. However, the accuracy of molecular simulation calculations always depends critically on the intermolecular potentials whose adjustment is relatively empirical. Continuing a project initiated by Toxvaerd (Toxvaerd, 1990; Toxvaerd, 1997), it was proposed to refine the determination of the anisotropic united atoms potential parameters for alkanes, on the basis of the thermodynamic properties which must be represented in chemical engineering. This work helped set the stage for a more general potential for other nonpolar hydrocarbons, and also for polar organic compounds, by also relying on the results of quantum chemistry.

1 NUMERICAL SIMULATION OF THE FORMATION OF OIL RESERVOIRS

1.1 Kinetics of Oil Formation

The pyrolysis of sedimentary organic matter with temperature programming under an inert gas stream (Espitalié *et al.*, 1985) helps to experimentally simulate the formation of oil and to simply identify the kinetic control of

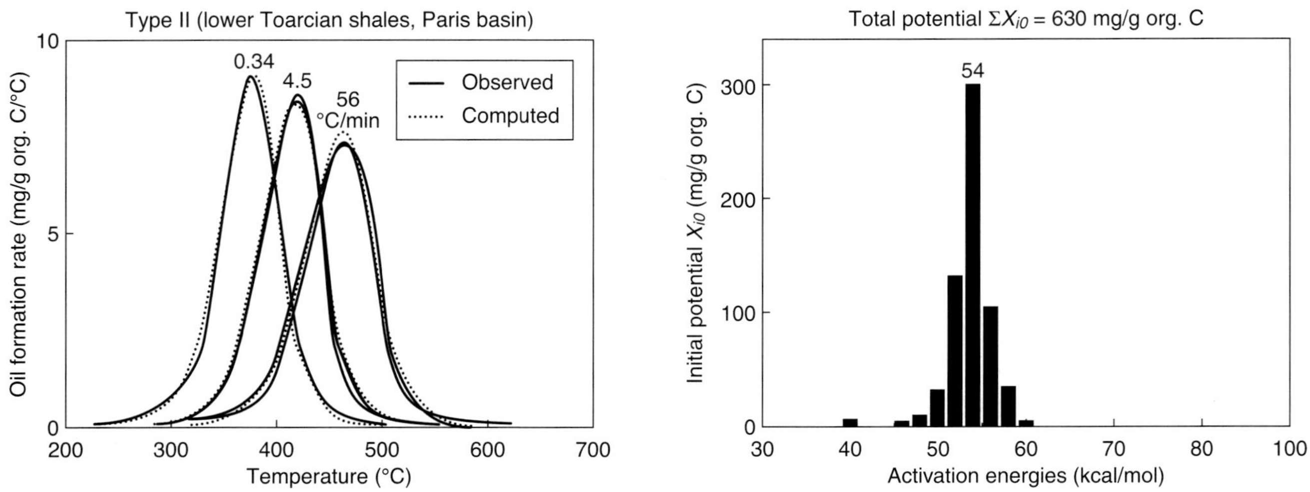

Figure 1

Pyrolysis curves at different heating rates of a sample of sedimentary organic matter taken from a source rock in the Paris basin, and histogram of activation energies of the kinetic model determined from these curves (Ungerer and Pelet, 1987).

this thermal cracking process. If pyrolysis is rapid (50°C/min) the maximum rate of liberation of hydrocarbon substances is observed around 500°C. If it takes place more slowly, at 0.3°C/min, this maximum release occurs around 400°C *(Fig. 1)*. Based on this type of observation, it is possible to develop a kinetic model involving several parallel first order reactions obeying the Arrhenius' law, where the main activation energies range from 50 to 60 kJ/mole (Ungerer *et al.*, 1985; Burnham *et al.*, 1987; Ungerer and Pelet, 1987; Ungerer, 1990). The determination of the parameters of the model (the weight of each reaction and their common preexponential factor) relies on an optimization method by a least squares criterion.

It is no doubt difficult to provide a solid justification for this model, proposed by Tissot (Tissot and Espitalié, 1975), rather than another. For example, the approach of Burnham at the *Lawrence Livermore Laboratory* (Burnham and Braun, 1990) led independently to slightly different parameters by using a continuous spectrum of parallel reactions rather than discrete reactions. On the other hand, there is no doubt that control of the parameters by experiments of very different durations is an excellent key to ensure that the problem of optimizing the twenty free parameters of the model is properly posed, in other words, that the solution is stable. It is only from the time that this stability is achieved that the model can be extrapolated to very slow evolutions like those taking place in the subsurface.

In sedimentary basins, where the burial rate is a few tens to a few thousands of meters per million years and where the temperature rises by 20 to 40°C per kilometer of depth, the rate of temperature rise is about 1 to 100°C per million years. The kinetic models fitted to the pyrolysis results predict that the maximum oil formation must occur at around 120-150°C in these conditions. It is in fact in sediments which have reached these temperatures that manifestations of thermal cracking are systematically observed, such as higher content of volatile hydrocarbons and a decrease in the hydrogen content of the residual solid organic matter—since the hydrocarbons formed are richer in hydrogen than the initial solid. The model also correctly predicts the shift of the pyrolysis peak towards the high temperatures which is used to characterize the thermal alteration of organic matter (Ungerer *et al.*, 1985). The validation of this approach required a patient analysis of natural samples not only involving *IFP* (Ungerer *et al.*, 1985; Ungerer and Pelet, 1987; Forbes *et al.*, 1991), but also the *Lawrence Livermore Laboratory* (Sweeney *et al.*, 1987) and the *KFA* team in Jülich (Welte *et al.*, 1997), among other leading laboratories active in this field. One of these validation examples is shown in Figure 2.

Admittedly, it is difficult to reconstruct to better than +/-10°C the temperature changes undergone by the rocks in their natural evolution, thereby limiting the accuracy of this validation. Despite these uncertainties, kinetic models describing the transformation of organic matter into oil are now intensively employed by the petroleum industry to delimit the zones of the sedimentary basins in which the oil has been formed.

In the industry, another major challenge is to predict the composition of the hydrocarbons formed, because it is much less advantageous economically to discover a gas reservoir than an oil reservoir, the second term designating a liquid form of petroleum in the reservoir conditions. To meet this

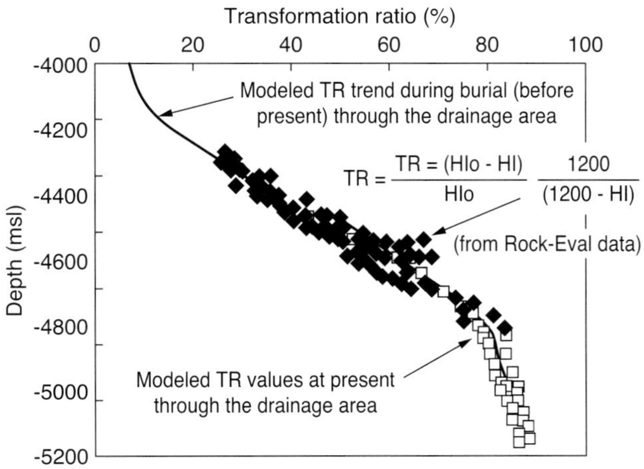

Figure 2

Trend as a function of depth of the transformation rate of organic matter into hydrocarbons, estimated from pyrolysis experiments on samples of wells of the Norwegian margin (rhombus) and calculated using a kinetic model like the one shown in Figure 1 (Forbes *et al.*, 1991).

demand, a first answer is to couple the inert gas flushing pyrolysis system with a cold trap system. Depending on the trap temperature, the curve of liberation of methane, hydrocarbons from C_1 to C_5, and C_1 to C_{15} can be recorded separately. Based on this additional experimental information, it was possible to make the kinetic model discussed above compositional, in other words, to supply at each stage the stoichiometry of the products formed in four main classes: methane, C_2-C_5, C_6-C_{15} and C_{15+} (Espitalié *et al.*, 1988). The studies conducted on organic matter of different origins *(Fig. 3)* helped to reveal systematic differences: coals of plant origin thus produce higher proportions of light hydrocarbons than matter of marine of lacustrine origin (Espitalié *et al.*, 1988; Welte *et al.*, 1997). However, the general predominance of the C_{15+} fraction in the reaction products clearly indicated that pyrolysis under a vector gas stream represented an open environment in which the hydrocarbons generated were preserved from the secondary cracking reactions that yield light hydrocarbons.

Although the compositional kinetic model thus developed was improved, it was unable to predict by itself the quantities of gas found in numerous basins. In consequence, the modeling of the secondary reactions in the conditions of the sedimentary basins (pressures of several hundred bar) proved indispensable.

1.2 Kinetics of Secondary Degradation Reactions

At the time when the development of basin models demanded better consideration of secondary reactions, it turned out that the pyrolysis of oils under pressure was also the subject of detailed investigations for the needs of enhanced recovery by *in situ* combustion (Behar *et al.*, 1988). From the theoretical standpoint, the excessively large number of elementary reactions involved made a detailed radical kinetic model inconceivable, and the kinetic models used in refining did not apply to complex systems such as crude oils. Furthermore, the few studies available attested that the influence of pressure was to significantly accelerate the cracking reactions up to 100 bar, and that the influence of

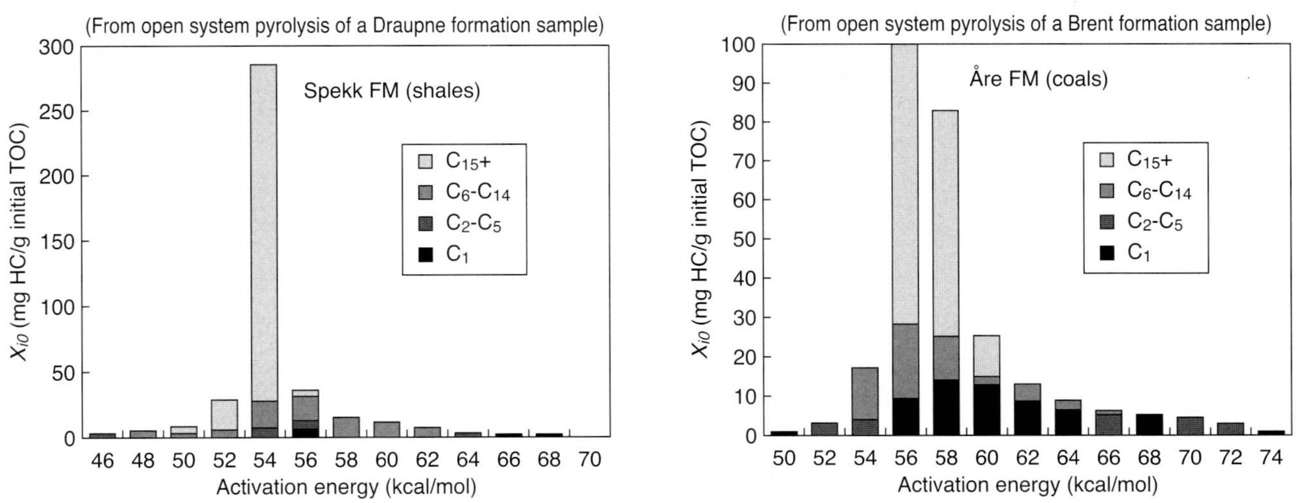

Figure 3

Histogram of activation energies used to model the composition of products generated by primary cracking by means of a parallel reaction kinetic model for two types of organic matter (Espitalié *et al.*, 1988).

pressure appeared to lessen beyond this limit. An empirical kinetic model was therefore proposed in which each of the unstable fractions (hydrocarbons other than methane and heavy products such as resins and asphaltenes) could be degraded by a first order reaction forming both the lighter and heavier fractions. In this system, the only final terms are methane and a polyaromatic carbonaceous residue called coke. This ultimate term also corresponds to the result of the thermodynamic stability analysis (Dayhoff *et al.*, 1967).

The optimization of the parameters of the secondary cracking model is difficult because they are numerous (activation energies, stoichiometric coefficients, pre-exponential factor). By employing a constrained optimization method, we can impose that the stoichiometric coefficients respect the conservation of carbon and hydrogen by assuming a constant elementary composition for a given fraction. But beyond the choice of the optimization method, it is crucial to conduct experiments covering a broad range of heating times and temperatures, without which certain parameters would be undetermined. In particular, it is important that the same degree of advancement of secondary cracking be simulated both by short term pyrolyses at high temperature and by longer pyrolyses at low temperature. Based on these principles, which demanded fruitful mutual feedback between models and experiments, it was possible to develop kinetic models for two typical examples of oils (saturated and aromatic) based on five fractions: the four fractions used in primary cracking (C_1, C_2-C_5, C_6-C_{15} and C_{15+}) and coke (Ungerer *et al.*, 1988). As for the primary cracking reactions, the extrapolation of the model to geological evolutions yielded results in good agreement with the highest temperatures at which the oils had been found in the sedimentary basins, or around 180°C. Subsequently, this kinetic model was improved on several occasions by increasing the number of fractions (Behar *et al.*, 1991) but by retaining the same general approach.

The coupling of primary cracking and secondary cracking models proved to accurately reflect the composition of the hydrocarbons generated by the main types of organic matter by pyrolysis in a closed medium, to the extent of the reproducibility of the experimental results (Ungerer, 1990). However, the application of this coupled model to sedimentary basins clashed with a major problem of principle, namely that neither pyrolysis in open medium, nor pyrolysis in closed medium, can be expected to fully simulate the thermal cracking mechanisms in a petroleum basin: the open medium would simulate the case of a system in which the hydrocarbons could travel to a colder reservoir as soon as they are formed, while the closed medium would simulate a basin in which these hydrocarbons cannot leave the place where they have been generated and would undergo the same thermal evolution as the solid organic matter. To simulate natural cases, which are intermediate, it is therefore necessary to consider the coupling between petroleum migration—name given to all its movements in the subsurface—and cracking mechanisms. By modeling the migration in a simplified manner, it was possible to compile a balance of the hydrocarbons expelled from the source rock in the drainage area of a known reservoir in the North Sea, and to confirm that their composition was compatible with that of the reservoir (Forbes *et al.*, 1991). This verification would ideally have demanded a dynamic flow model like the one discussed below, but it was not advanced enough to consider this.

Before concluding the discussion of kinetic models, it is important to state what is needed to have a more reliable model. As demonstrated by the work of Doué and Guiochon (Doué and Guiochon, 1968), the thermal cracking of alkanes at high pressure is characterized, inter alia, by the virtual absence of olefins in the reaction products, and is roughly explained by a radical mechanism. To examine how these results could be extended to cracking processes involving organic geochemistry, a project conducted alongside the *ENSIC* of Nancy attempted to completely simulate the cracking of light hydrocarbons (hexane and hexane-benzene mixture) under pressure (Kressmann *et al.*, 1990; Kressmann, 1991). Figure 4 shows the kinetic scheme obtained.

This work mainly offered the advantage of demonstrating the importance of bimolecular processes between the hexane molecules and the radicals *(Fig. 4)*, so that the apparent

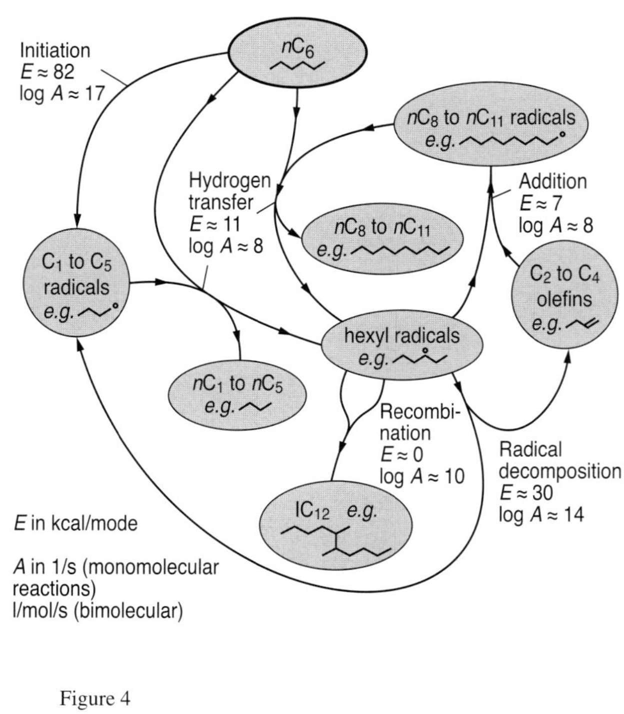

Figure 4

Radical cracking of hexane at 350°C (Kressmann *et al.*, 1990).

activation energy of the cracking process (50-60 kcal/mole) is much lower than that of monomolecular scission (over 80 kcal/mole). The importance of bimolecular reactions also questions the strong hypothesis of secondary cracking models, namely that the degradation rate of a fraction is independent of the concentrations of the other species. This being said, the radical mechanisms involved already include several hundred elementary processes for very simple systems with a low degree of advancement. Their resolution demanded solving systems of ordinary nonlinear and poorly conditioned differential equations, which demanded several hours of computer time on the computers available at the time (Cray II). The extension of the radical approach to real systems requires both the implementation of lumping methods in order to reduce the number of equations, as well as reliance on theoretical chemistry methods to calculate the kinetic parameters of the many elementary processes involved. This research path has advanced considerably since then (Duin *et al.*, 2001), but its application to complex systems is still an objective of present research paths.

1.3 Modeling of Oil Migration

The migration of oil and gas is a very general phenomenon since hydrocarbon accumulations are generally found in sedimentary rocks different from those in which oil is formed: the former, called reservoir rocks, are permeable rocks (sandstone, porous carbonates), while the latter, called source rocks, are generally very fine grained rocks (shale, marl, coal). Even in basins which are known to have generated large quantities of oil, an understanding of migration processes is essential to locate the costly exploration boreholes on the most favorable geological structures.

It was recognized very early that migration in permeable rocks was explained by a gravitational segregation and that this could be described by the classical formalism of flows in porous media. Similarly, it was acknowledged in the 1970s that the entrapment of hydrocarbons in the reservoirs was explained, depending on each case, by the impermeability of the overerlying rock—salt, anhydrite—or by capillary forces at the interface with the reservoir rock (Berg, 1975). On the contrary, the mechanisms at play in the source rocks were still the subject of debate in the early 1980s, because capillary forces were considered to pose an obstacle to migration in these rocks, whose pores were submicronic in size: some researchers felt that the hydrocarbons migrated in solution in the water expelled by the sediments during their compaction (Price, 1976), while others considered that molecular diffusion was the dominant mechanism (Leythaeuser and Yükler, 1982), and yet others suggested a two-phase flow in which the hydrocarbons formed a separate phase from the water (McAuliffe, 1980).

Following a critical analysis of the literature, it emerged that the main mechanism responsible for the expulsion of oil from the source rocks was the last of the three mentioned above, namely migration in a separate phase (Durand, 1983; Ungerer *et al.*, 1983). The underlying reasons can be summarized by simply stating that the solubility of liquid hydrocarbons in an aqueous phase is much lower and that molecular diffusion is a mechanism that would have excessively favored light hydrocarbons over the heavy fractions. On the contrary, expulsion in a separate phase better reflected the high electrical resistivity found in the source rocks, and a certain similarity of composition between the crude oils and the products remaining in the source rocks.

Once migration in a separate phase in source rocks such as reservoirs was accepted, it remained to find an adequate formalism to approach its modeling. This was a particularly ambitious task, because at the time, the only team that had tried to model the migration of oil at the scale of a basin had limited consideration to its movement in the reservoir rocks (Welte and Yükler, 1981). The approach pursued (Ungerer *et al.*, 1984) was characterized by consideration of the following processes:

- the geometric evolution of the basin was represented by a mobile mesh in which the rows of meshes are added successively to simulate the deposition process of new sediments;
- the thermal evolution of a level is inferred from the evolution of its depth, based on the geothermal gradient observed or from the solution of the Fourier's heat transfer equation;
- the formation of oil in the source rocks is described by the kinetic model of primary cracking discussed above;
- the fluid flow is described by a two-phase Darcy's law in which the permeability varies by several orders of magnitude according to the particle size distribution, porosity, hydrocarbon saturation (this term designating the fraction of pore volume occupied by the phase concerned) and the prevailing stress;
- compaction, *i.e.* the irreversible reduction of porosity of the sedimentary rocks, is represented by a law linking the porosity and the effective stress.

As complex as this formalism may appear, it was impossible to discard a single one of these elementary processes to represent the formation of oil reservoirs realistically. On the contrary, the model tended to sideline a vitally important mechanism with the phase equilibria between oil and gas, whose important implications during migration were the subject at the same time of the thesis of E. Nogaret at *LIMHP* (Nogaret, 1983).

To concretize this mathematical model by a numerical simulation, it was illusory to make a mere vertical unidimensional study as done by the authors dealing with the kinetics of oil formation. In fact, the formation of many

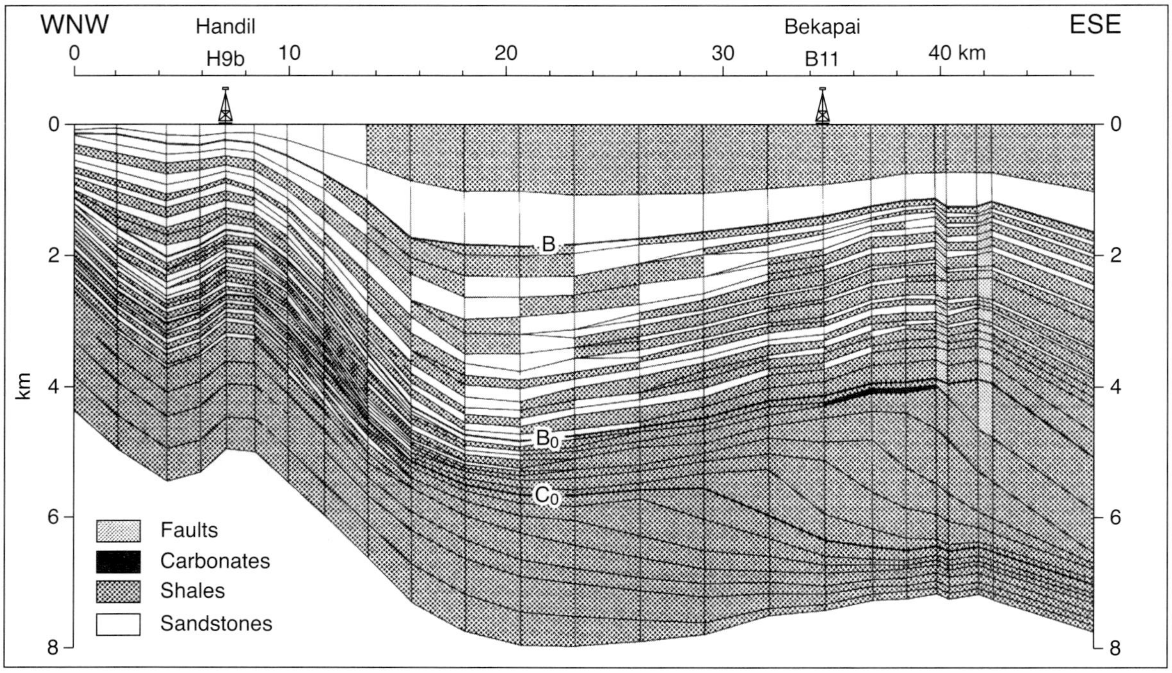

Figure 5

Mesh employed to simulate migration in the Mahakam river delta in Indonesia. The shale/sandstone intercalations are shown by rows of meshes touching the stratigraphic horizons (Ungerer *et al.*, 1990).

reservoirs cannot be explained without migration with a strong horizontal component. This is why the route of two-dimensional numerical simulation (vertical + horizontal) was forthwith selected, despite the numerical problems associated with the resolution of the system of equations. Given the total duration to be simulated, which ran into millions of years, it was in fact necessary to apply time steps of several thousand years to remain within reasonable computer time. This resulted in a numerical instability which demanded the development of implicit numerical schemes. Besides, the use of deformable and nonrectangular meshes was necessary to adapt to the limits of the geological discontinuities *(Fig. 5)*. This was an additional difficulty, because this type of mesh had barely been addressed in the literature.

Despite the shortcomings of the early simulations, the formalism proposed for migration rapidly proved capable of simulating all the mechanisms governing the formation of oil reservoirs in various cases: expulsion from source rocks, segregation by density in reservoir rocks, entrapment by and leakage through caprocks. Simultaneously, the interbedding of permeable and impermeable levels resulted in a very strong permeability anisotropy at the basin scale, which reconciled the requirement of a low vertical permeability—to explain the entrapment of the oil—and a high horizontal permeability, to supply reservoirs located far from the zones where the cracking was active. Beyond the petroleum aspects, the model accurately described the pressure and fluid flow regime at the scale of the basins. For example, the introduction of low permeabilities for shales (about one nanodarcy, or 10^{-21} m^2) helped to explain that despite the very large timescale of the mechanisms simulated, the pressure profiles often displayed very wide deviations from equilibrium at great depth *(Fig. 6)*.

Once these preliminary results were obtained, many improvements were made. A first alternative was to consider sedimentation mechanisms in greater detail and explicitly simulate heat transfer mechanisms, whose transient character is significant in certain basins (Doligez *et al.*, 1987). A second alternative concerned numerical resolution: test of a finite element method in connection with the thesis of V. Bouvier (Bouvier, 1989), implementation of a compositional scheme in that of I. Faille (Faille, 1992). At the same time, this coupled model of fluid migration and heat transfer saw its first application to real exploration cases, albeit for synthesis rather than prediction (Ungerer *et al.*, 1987) since it was limited to two dimensions *(Fig. 7)*.

The basic principles of the basin models set forth in the 1980's by the *IFP* team have not been fundamentally questioned, either in the field of thermal cracking or of migration. However, many improvements continued at the scientific level in teh 1990s: three dimensional modeling, explicit consideration of oil-gas phase equilibria during migration, representation of faults. At the industrial level, the use of basin models was progressively extended and generalized in oil exploration, and the *IFP* group is still one of the leading software suppliers in this area.

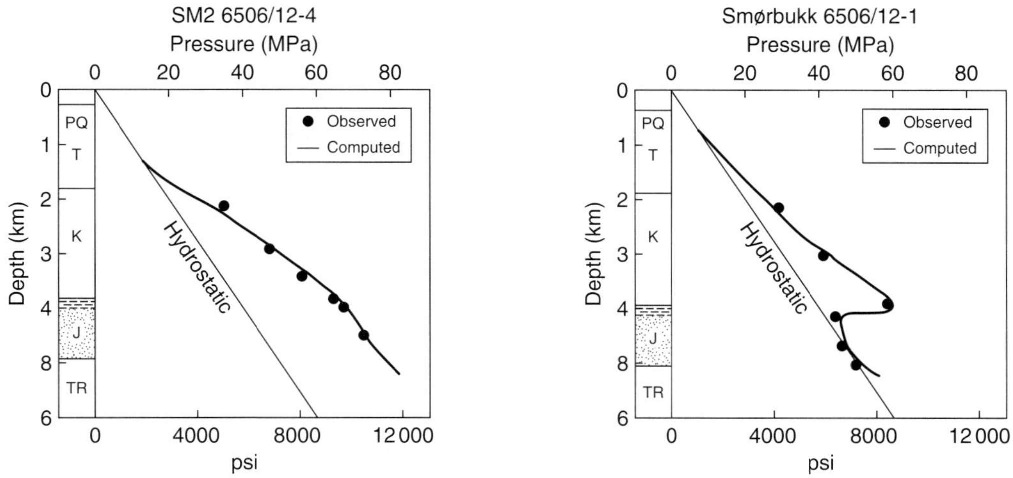

Figure 6

Variation in fluid pressure as a function of depth observed in two wells of the Norwegian margin (●) and calculated using the two-dimensional migration model (continuous line). The hydrostatic equilibrium profile is shown as a reference (Ungerer *et al.*, 1990).

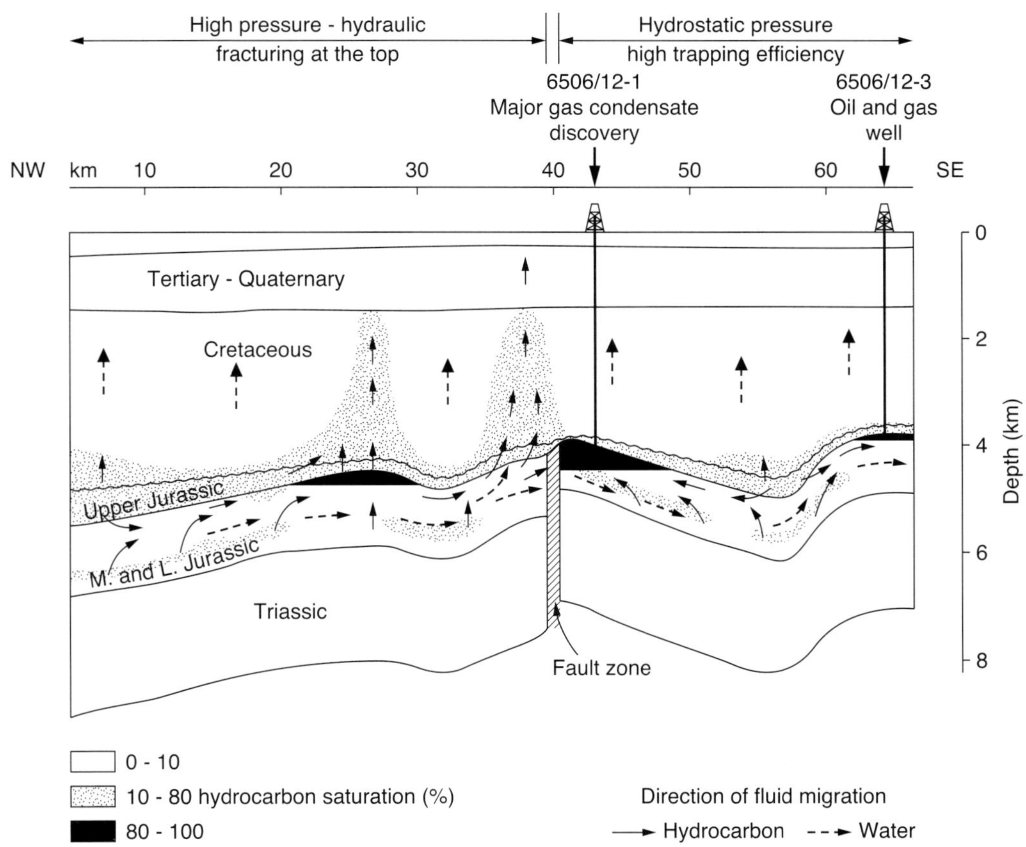

Figure 7

Geological section across a section of the Norwegian margin indicating the fluid migration directions calculated using a two-dimensional migration model. The hydrocarbon saturation is the fraction of pore volume occupied by the hydrocarbon phase (Ungerer *et al.*, 1990).

2 MEASUREMENT AND PREDICTION OF PROPERTIES OF CRUDE OIL AND NATURAL GAS AT HIGH PRESSURE

2.1 Phase Diagram of Natural Gases Containing Heavy Hydrocarbons

The natural gases discovered by *Elf* in the deepest boreholes of the North Sea and the Aquitaine basin in the late 1980s sometimes displayed unusual properties compared with the fluids normally encountered: particularly high dew point pressures, sometimes over 50 MPa, crystallization of paraffins at low temperature. These fluids also displayed a characteristic composition, rich in methane and containing heavy hydrocarbons up to C_{30} or more. To start producing these reservoirs, it was indispensable to have a clear understanding of the phase diagram of these fluids and to relate it to the composition. This raised specific experimental difficulties due to the high pressures and temperatures (190°C and 110 MPa). In cooperation with *Elf,* the phase diagrams of natural fluids *(Fig. 8)* and synthetic fluids *(Fig. 9)* were determined by means of high pressure cells (Ungerer *et al.*, 1995; Ungerer *et al.*, 1998).

The composition of the synthetic fluids had been selected in order to approximate that of natural fluids, methane being the majority compound (70% or more) with the remainder consisting of light liquid hydrocarbons. Their study revealed that the addition of heavy aromatic or paraffinic hydrocarbons in a small molar fraction (1% or less) resulted in the appearance of (gas + solid) and (gas + solid = liquid) domains at temperatures close to ambient. It also showed that the heavy hydrocarbons had the effect of considerably increasing the saturation pressures. These observations were not really surprising, because they fitted into the logic of the behavior of methane-heavy hydrocarbon binary mixtures investigated experimentally at the *University of Delft* (van der Kooi, 1981; Glaser, Peters *et al.*, 1985; Flöter, 1999), at *IFP* (Arnaud *et al.*, 1996; LeRoy *et al.*, 1997) and at the *CNRS* (Marteau *et al.*, 1998). For example, we found the negative slope of the curve of appearance of a solid phase at high pressure in the *(P, T),* diagram, as it had been observed on methane rich binary mixtures *(Fig. 8).*

Nevertheless, it was impossible to observe a diagram comprising a three-phase region in the *P, T* plane for a binary system due to the phase rule, whereas it was possible for our synthetic fluids since they had three components or more. Our experiments therefore provided a conceptual guide for the interpretation of the phase diagrams of natural fluids. These measurements were then used in the literature to validate thermodynamic models describing thermodynamic models describing crystallization in complex systems (Nichita *et al.*, 1999). In a different perpspective, the simultaneous use of experiments on model fluids and equations of state proved a very efficient approach to predict the distribution of light liquid hydrocarbons in oil and gas phases at depth. It was thus possible to provide organic geochemists with well-tested tools to understand subtle composition variations in oil reservoirs (Carpentier *et al.*, 1996).

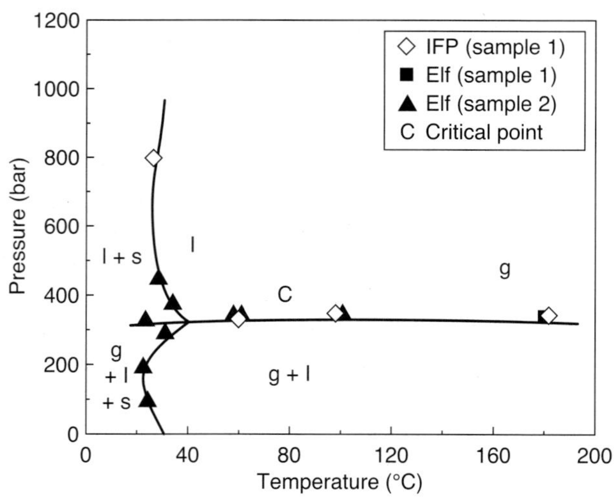

Figure 8

Phase diagram of a North Sea HP/HT gas condensate showing the appearance of a solid paraffin phase at low temperature.

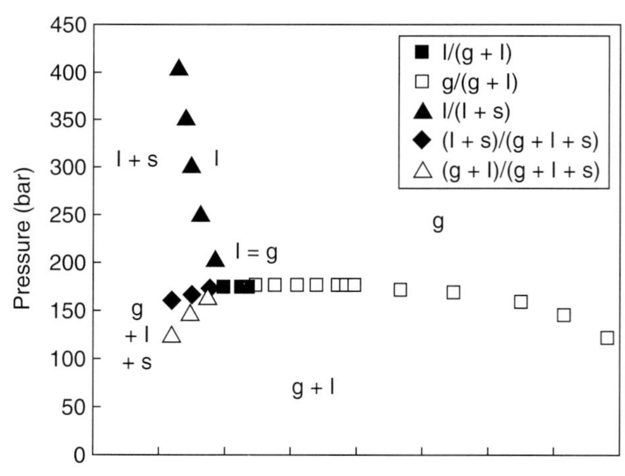

Figure 9

Phase diagram of a synthetic fluid consisting of methane (79%), pentane (20%) and α-methylnaphthalene (1%), showing the same type of behavior as gas condensate in the previous Figure 8 (Ungerer *et al.*, 1998).

2.2 Measurement and Correlation of Volumetric Properties at High Pressure

The prediction of volumetric properties by cubic equations of state is obviously insufficient, and to correct this defect, Péneloux *et al.* (1982) proposed the volume translation technique. The great advantage of this technique is to avoid modifying the representation of the liquid-vapor equilibria on which the cubic equations of state are fitted. Based on the literature data, it is possible to show that volume translation allows systematic errors to subsist (de Hemptinne and Ungerer, 1995). In particular, it was found that for *n*-alkanes from pentane to decane, the error committed at high pressure on the molar volume of a pure substance increases linearly with temperature. This finding led to the recommendation of a variable volume translation with temperature, whose parameters are adjusted to liquid volumes at 50 MPa (Ungerer and Batut, 1997; Sant'Ana *et al.*, 1999). Figure 10 gives an example of the improvement made. This approach helps avoid the inconsistencies observed at high pressure when volume translation is adjusted to accurately describe liquid volumes at saturation. It also offers the advantage of being compatible with the mixing rules classically used during the lumping of several pure substances into pseudocomponents.

In the case of a mixture, an adequate volume translation is not sufficient to clearly represent the molar volume, because the excess volume at constant pressure must also be represented. This quantity is generally small in the case of mixtures of liquids, for which it is less than 1% of the volume (Katzenski and Schneider, 1982; Ashcroft *et al.*, 1992). In the case of contrasted mixtures (gas-liquid) at high pressure, the measurements taken at *IFP* by J.F. Arnaud during his thesis project showed that the scale of the excess volume is significantly greater, raising the problem of its modeling. Volume translation has no influence on the excess volume in a cubic equation of state, but it is possible to introduce a quadratic mixing rule on the covolume to faithfully represent this quantity (Arnaud *et al.*, 1996). Measurements of excess volume thereby provide independent information of the phase equilibria measurements, the former being rather influenced by the covolume mixing rule and the latter by the attractive term of the equation of state.

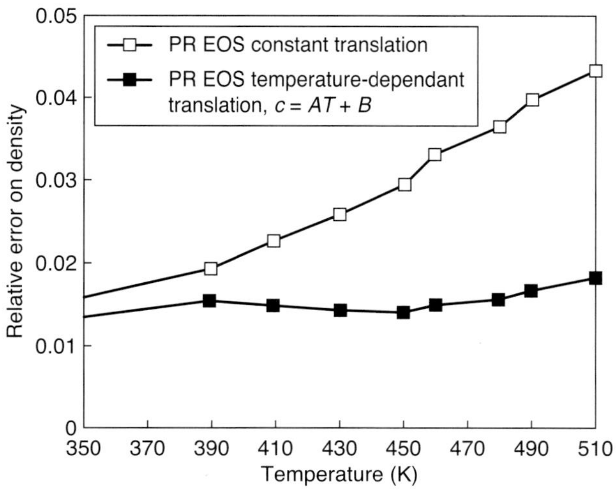

Figure 10

Relative error committed by the Peng-Robinson's equation on the density of liquid hexane at 50 MPa by means of a constant volume translation and a temperature dependent volume translation (Ungerer and Batut, 1997).

2.3 Representation of the Phase Behavior of Gas Condensates

A condensate gas is defined by the fact that a decrease in pressure at the reservoir temperature leads to the condensation of a liquid phase (Pedersen *et al.*, 1989). The composition of these gases is dominated by methane, of which the molar fraction is 60 to 80%, and by light hydrocarbon gases. The liquid hydrocarbons of these fluids are mainly hydrocarbons with C_5 to C_{10}, but may include heavier hydrocarbons: up to C_{20} or even C_{30}. As shown by the study of binary systems and synthetic mixtures discussed earlier, the phase behavior of methane rich mixtures is very sensitive to the composition of the heavy hydrocarbons. Hence it is not surprising that the prediction of the phase equilibria of gas condensate by equations of state is very sensitive to the representation of the heavy fraction of the gas. Unfortunately, the precise composition is generally unavailable, since chromatography cannot provide a molecular analysis above C_{10} due to the many isomers present (Durand *et al.*, 1989). It is hence routine practice to optimize the thermodynamic properties of the heavy fraction to best represent the liquid deposit curve. The resulting model is difficult to extrapolate to other temperatures.

To improve this procedure, distribution functions can be used to represent the hydrocarbons of the heavy fraction: this is the track explored by M. Sportisse in her thesis (Sportisse, 1996; Sportisse *et al.*, 1997). In this work, continuous distributions were employed to reduce the number of parameters to be optimized and not to reduce the number of equations of the phase equilibrium calculation. This approach helped to test several types of functions (exponential, gamma) to describe the heavy fraction by distributions of alkanes or aromatic hydrocarbons. It thus appeared that phase equilibria are just as sensitive to the chemical family as to the analytical form of the distribution. For example, the exclusive use of a distribution of *n*-alkanes leads to a systematic underestimation of the dew point pressure, and it

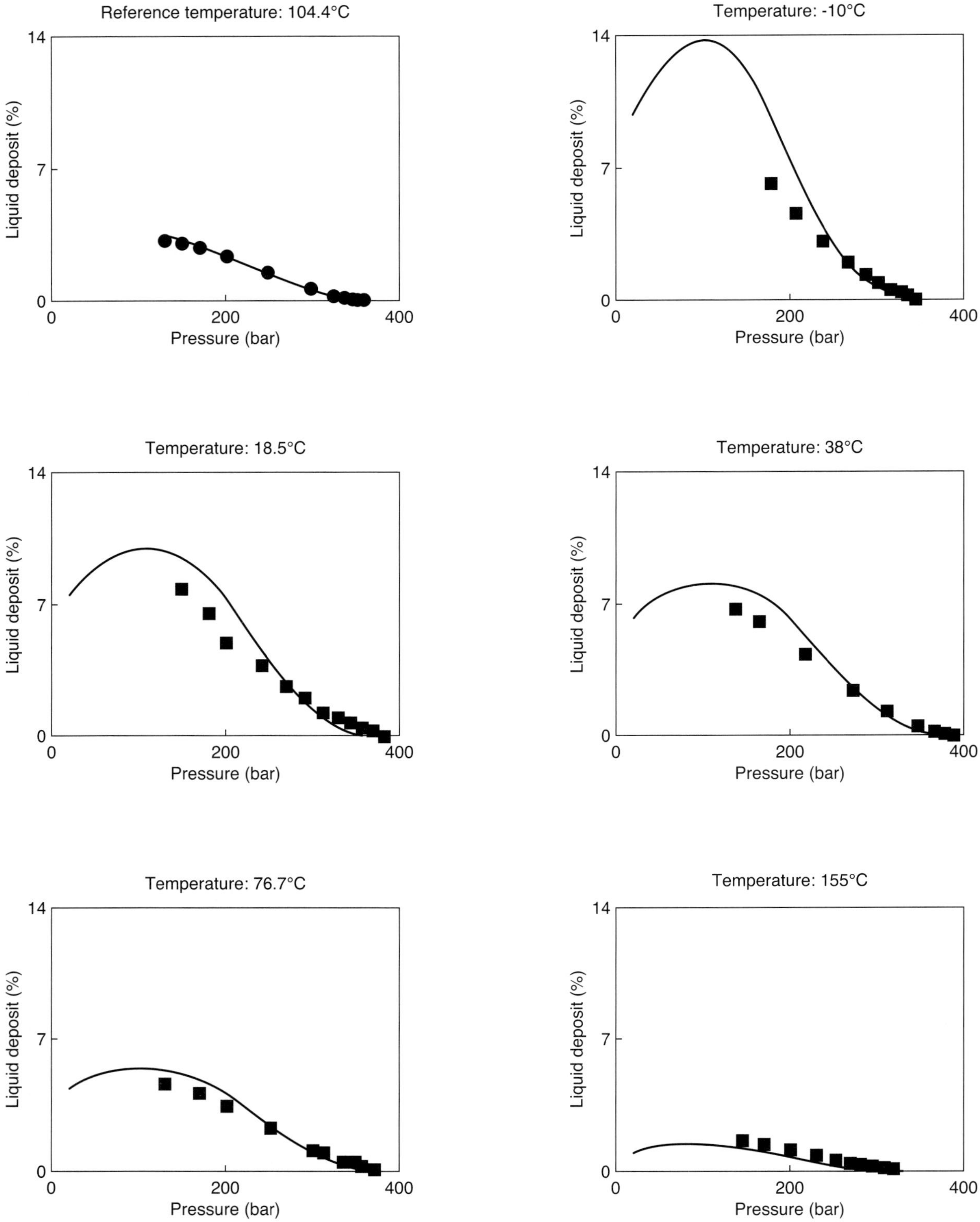

Figure 11

Curves of liquid deposition of a gas condensate observed (■) and calculated by representing the fraction by distribution functions (Sportisse *et al.*, 1997). The distribution of heavy hydrocarbons is determined here in order to represent the behavior observed at 104.4°C, and the curves corresponding to other temperatures are predicted without adjustment.

was necessary to include distributions of alkylbenzenes and polyaromatic hydrocarbons to correct this effect. For gases of which the detailed composition is known, the heavy fraction thus modeled proves to be fairly representative of the real composition. It also allows an improved prediction of the density of the condensate and a better extrapolation of the curves of liquid deposition towards the low temperatures *(Fig. 11)*.

3 PREDICTION OF THERMODYNAMIC PROPERTIES BY MOLECULAR SIMULATION

3.1 What Prospects for Petroleum Thermodynamics?

The petroleum industry would like to benefit from more predictive methods for a broad range of applications, including liquid-vapor equilibria, the crystallization of solid phases (paraffins, gas hydrates), the flocculation of asphaltenes, and adsorption, etc. However, progress is certainly difficult to achieve in this area, and the few results set forth in the previous chapter may appear modest when seen against the efforts made. Thermodynamicists are sometimes tempted to blame the lack of experimental data for a good share of these difficulties. It cannot be denied that in the field of crude oil and natural gas, the rarity of measured equilibrium properties of pure substances is a serious hindrance for hydrocarbons with more than ten carbon atoms other than *n*-alkanes, and this is no doubt one factor that limits potential progress. It is also true that the equilibrium data of binary mixtures at high pressure are not substantial enough, in the same way as crystallization measures under pressure, flocculation under pressure and adsorption selectivity on well-known adsorbents. The systematic acquisition of these data would be very expensive, and the industry is not prepared to finance this effort for profits that seem too far in the distant future.

Yet if the analysis is taken further, this situation could be explained as much by the inadequacy of present theories as by industry decisions.

Thus in the example of equations of state, it is questionable whether really decisive progress has been achieved since the work of Peng-Robinson and Soave. While many equations of state have improved the description of a particular system, up to now this has always occurred to the detriment of the generality of the equation. An eloquent example in this respect is the Lee-Kesler's method of corresponding states (Lee and Kesler, 1975), very accurate for components similar to the reference substances, but unrealistic for substances that are too different (Hemptinne and Ungerer, 1995). The SAFT equation also helps to clearly represent the equilibria of certain systems (Chapman *et al.*, 1989; Gil-Villegas *et al.*, 1997), but by remaining in the general scheme of linear molecules and, for each system, relying on a specific parameterization of the distribution function. Moreover, like any analytical equation of state, it does not reproduce the scaling laws near the critical point. On other equations of state, the addition of "crossover" terms helps to satisfy these scaling laws (Albright *et al.*, 1986; Rainwater and Williamson, 1986), but this procedure requires knowledge of the critical point, which one would in fact have preferred to calculate from the equation of state in the case of mixtures! Finally, another prohibitive weakness of existing theories is their difficulty in predicting differences in properties between near isomers (for example, between isoalkanes with the same number of branches). The consequence of these inadequacies is a very strong demand for experimental measurements to adjust model parameters to specific cases.

A more severe criticism could be leveled concerning the prediction of adsorption selectivities which the industry needs to design adsorption separation processes on crystalline microporous solids. The use of thermodynamic models similar to those employed in liquids or the reliance on statistical models (Ruthven, 1984; Tournier, 2000) does not help predict selectivities, but only to correlate them on the basis of experimental measurements.

Faced with these findings, it is likely that the community of thermodynamicists will have to completely change the problematics to provide general answers. Without much chance of being wrong, we can claim that this will entail reliance on theories wherein the molecular structure and the nature of the molecular interactions will be described in greater detail. Given the insufficient results of the models inspired by statistical thermodynamics, one can even claim that each molecule will have to be represented individually, as well as any changes in conformation, at least in delicate cases (adsorption, differences in properties between isomers). This is why the future of thermodynamic models is rather to be sought in terms of molecular simulation techniques, the subject of the last part of this article.

3.2 Computer Algorithms Based on Monte Carlo Methods

The first numerical simulations using Monte Carlo methods were nearly contemporary with the first appearance of computers, dating half a century back (Metropolis *et al.*, 1953). The first statistical bias algorithms (Rosenbluth and Rosenbluth, 1955) appeared soon thereafter, followed by the Widom insertion test which helps to calculate the chemical potential (Widom, 1963), mentioning only the best known of the pioneering work at the time. In the 1960-70s, we therefore had the theoretical concepts necessary to address complex systems, as attested by the still very real utility of the reference works written at the time (Münster, 1969; McQuarrie, 1976). However, the applications remained limited to small size molecules for many years.

Only in the 1980s did the continuous increase in computation capacities significantly broaden the field of possibilities of Monte Carlo methods. This is no doubt why many essential algorithms for calculating thermodynamic properties were developed quite late. Among them is the calculation of phase equilibria in the Gibbs ensemble (Panagiotopoulos, 1987) , the simulation of flexible alkanes by a configurational statistical bias (de Pablo *et al.*, 1992; Smit *et al.*, 1995), the simulation of water by an orientational bias (Cracknell *et al.*, 1990), the calculation of saturation vapor pressures at low temperature by the integration of the Clapeyron's equation (Kofke, 1993), the simulation of polymer materials by means of Monte Carlo movements acting on a limited segment of the molecule (Dodd *et al.*, 1993) and the calculation of phase equilibria not corresponding to a statistical ensemble *stricto sensu* thanks to pseudoensembles (Mehta and Kofke, 1994). These achievements allowed the development of the Gibbs code as part of a collaborative project between *IFP* and the *Physical Chemistry Laboratory* of the *University of Paris Sud* (Mackie *et al.*, 1997). Despite these breakthroughs, Monte Carlo methods cannot today meet all the needs of applied thermodynamics. This is largely due to the lack of appropriate intermolecular potentials, as we shall discuss below. But it is also due to the lack of algorithms in two directions: the calculation of phase equilibria under particular constraints and statistical bias methods adapted to the calculation of complex molecules.

3.2.1 Calculation of Phase Equilibria Under Particular Constraints

The Gibbs ensemble only partly covers the needs connected with phase equilibria, because the quantities imposed are the temperature, the total number of molecules, and a third quantity which is the volume or pressure. In view of the applications, algorithms are also needed in which the quantities are imposed. When classical thermodynamic models are used, it is for example routine to impose the composition of the liquid phase for a bubble point calculation or that of the vapor phase for a dew point calculation. Generally speaking, it is important to be able to plot in the *(P, T)* plane the limits of the domain wherein a complex mixture of a given composition becomes two-phase, *i.e.* its phase envelope (Michelsen, 1980). Among the other types of specifications are calculations at imposed enthalpy which serve to calculate the variation of temperature or pressure resulting from an isenthalpic expansion or from the input of a given quantity of energy. The direct calculation of azeotropes and critical points is another important need.

In this area, an original Monte Carlo algorithm has been proposed for the direct calculation of the bubble point (Ungerer *et al.*, 1999; Ungerer *et al.*, 2001) using two simulation boxes without interface but nonetheless coupled via changes in volume *(Fig. 12)*. This algorithm can be summarized by stating that it consists in calculating the probability of acceptance of a molecule transfer from one phase to the other as if we worked with the Gibbs' ensemble, but by refraining from updating the configuration of the liquid phase, whose composition thus remains rigorously constant. The vapor phase, in which the updates are effectively made, undergoes insertions and destructions comparable to those characterizing the grand canonical ensemble. Its mean composition and mean pressure supply the desired results in calculating the bubble point.

The results obtained were highly positive, both on monatomic molecules (Ungerer *et al.*, 1999) and on mixtures of alkanes (Ungerer *et al.*, 2001) *(Fig. 13)*.

Compared with the simulation techniques previously available (Vrabec and Fischer, 1995; Escobedo, 1998; Miyano, 1998), this work helped to avoid dependence on an iterative process to determine the liquid phase chemical potentials and hence offered undeniable progress. However, it appeared that the direct algorithm for calculating the

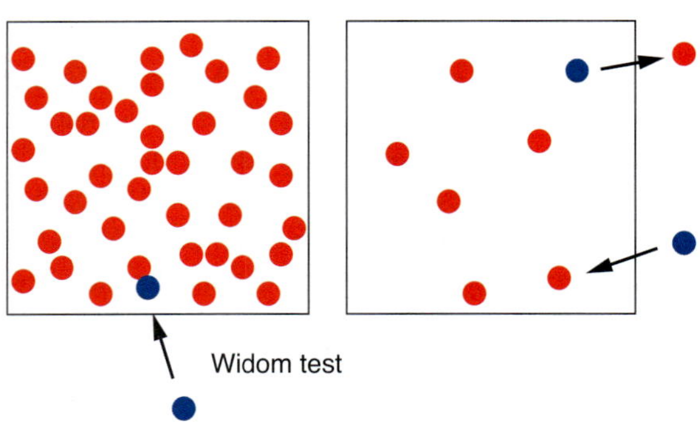

$$P_{acc}\,(\text{destruction}) = \min\left(1, \frac{(N_i^v + 1)\exp(-\beta\hat{\mu}_i^l)}{V^v\, W^{v-} \exp\left(\beta(\Delta U^{v-} - \Delta\bar{U}^{v-})\right)}\right)$$

$$P_{acc}\,(\text{insertion}) = \min\left(1, \frac{V^v\, W^{v-} \exp\left(\beta(\Delta U^{v+} - \Delta\bar{U}^{v+})\right)}{(N_i^v + 1)\exp(-\beta\hat{\mu}_i^l)}\right)$$

Figure 12

Algorithm of direct calculation of the bubble point at imposed temperature. The formulas shown for acceptance rates correspond to a configurational bias algorithm (Ungerer *et al.*, 2001).

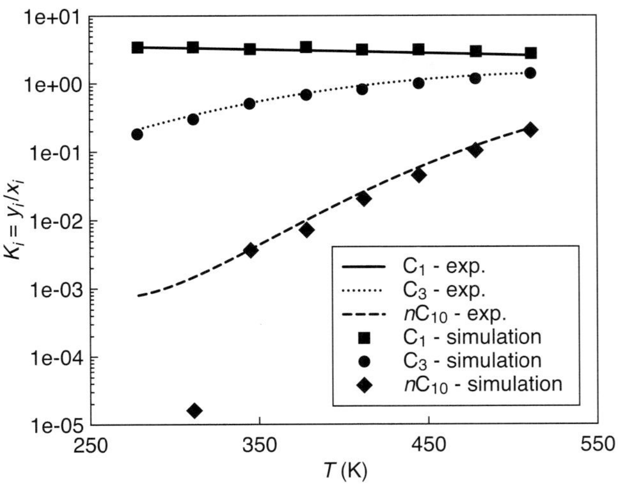

Figure 13

Liquid-vapor equilibrium constants in bubble point conditions calculated using the proposed algorithm (Ungerer *et al.*, 2001).

bubble point had been proposed independently at the *University of Wisconsin* by Escobedo under the term of *faked move* (Escobedo, 1999). This author having also proposed an alternative method adapted to the calculation of dew points—for which the direct algorithm is unstable—it was decided not to further pursue the development of this type of algorithm. In fact, Escobedo in the meantime proposed an algorithm for isenthalpic calculations (Escobedo and Chen, 2001) and the azeotrope calculation algorithms progressed (Pandit and Kofke, 1999). It would have been difficult to stand up to this competition. The sidelining of this research subject appears *a posteriori* to be a wise decision, especially since the determination of intermolecular potentials demanded major efforts.

However, this work provided many lessons. Inter alia, it served to identify the approximate nature of the factorization of the partition function as classically employed. Strictly speaking, the partial integration of the partition function reveals a factor whose dependence as a function of the conformation is presumed to be negligible in Monte Carlo techniques (Ungerer *et al.*, 2001). While this hypothesis seems justified for molecules with strong intramolecular potentials (Go and Sheraga, 1976) , it is not the case of long chain polymers, whose rigidity is weak at the scale of the chain.

The practical limitations of the Gibbs ensemble to a relatively restricted temperature interval (from the normal boiling point to close to the critical temperature) is another reason to employ complementary methods. For molecules of substantial size (C_{10} to C_{30}) the temperatures at which thermodynamic properties are to be calculated are in fact often lower than the boiling point. In the case of pure substances, thermodynamic integration has proved to be an ideal method for calculating the saturation vapor pressures at the same time as the densities and vaporization enthalpies (Ungerer *et al.*, 2000). To accelerate this type of calculation in the future, it is necessary to employ parallel tempering techniques (Yan and de Pablo, 1999) which, by exchanging configurations between simulations made at different temperatures, offer the advantage of effectively using the capacity of parallel machines, without necessarily demanding extreme speed of communication between processors. However, in using these methods, one must be sure to allow the configurations to decorrelate sufficiently before the exchanges.

3.2.2 Improved Statistical Bias

Thanks to configurational statistical bias techniques, it is possible to calculate the phase equilibria of linear or branched alkanes, which would be otherwise untractable even at the cost of very long computer time. However, the possibilities of simulation are extremely limited concerning large size molecules. Thus in the example of squalane—a C_{30} isoalkane—the use of configurational statistical bias in the Gibbs ensemble did not permit access to sufficiently low temperatures for comparison with experimental measurements (Neubauer *et al.*, 1999). To extend these calculations to temperatures closer to process requirements, it is undoubtedly necessary to improve the statistical bias algorithms.

In the case of long chain linear alkanes (C_{20} to C_{35}), E. Bourasseau, in his DEA diploma course at *IFP*, implanted a local rotational move that serves to relax the central portion of the chains, which is untouched by growth movements. This algorithm, in which a carbon atom rotates about the axis formed by its two neighbors, allows single-phase simulations at temperatures far below the boiling point (Bourasseau *et al.*, 2002). The advantage of this local move is that it applies to any molecule possessing a linear chain, regardless of its structure (alkylbenzenes, ethers, cycloalkanes). For very long chain alkanes (polyethylene), the introduction of a reptation move will be useful for improving effectiveness, but this move cannot be extended to all molecules.

In the case of isoalkanes with many branches, the local rotational movement is not sufficient to explore all the conformations of the middle chain because the branches all represent fixed points. To correctly simulate these compounds at low temperature, complementary Monte Carlo movements are necessary. One possibility is the pivot of a moiety of the molecule about one the bonds of the main chain, or a concerted rotational movement of several consecutive centers (Dodd *et al.*, 1993), or the simultaneous placement of two atoms at branches (Macedonia and Maginn, 1999).

Cyclic alkanes represent another family of molecules for which recent research has encountered difficulty in calculating the phase equilibria while taking account of the different possible internal conformations (Neubauer *et al.*, 1999). In his first thesis year, E. Bourasseau developed a particular transfer movement using a canonical reservoir of conformations, periodically updated by local rotational movements (Bourasseau *et al.*, 2002). To obtain a significant proportion of accepted movements, the transfer is made in two steps. In the first step, designed to find cavities of substantial size in the liquid, several insertion tests are performed on a Lennard-Jones' particle at places selected at random in the receiver phase, and one of them is selected for its Boltzmann's probability. In the second step, many insertion tests of the complete molecule are performed at the place selected in random orientations, the internal conformation being sorted at random in the reservoir. This statistical bias algorithm helped to reduce computer time by a factor of three per accepted transfer.

With a view to calculating larger and less flexible molecules, it is important to further improve the effectiveness of the transfers. How to do this? To start with, analyze the source of the effectiveness of the bias discussed above with respect to an unbiased transfer movement. Its success mainly stems from the fact that a cheap selection step (the test of several positions by a Lennard-Jones potential) helps to limit to a favorable position the energy calculation of the second step, which is more expensive because involving all the force centers of the molecule. In this way, the complete computation is avoided in sites where the overlap with the other molecules would in any case prevent the transfer from being accepted. In the case of inflexible molecules displaying a significant elongation, like polycyclic hydrocarbons (decaline, naphthalene), an interesting research direction would be to design a bias in three steps instead of two: the first step would be unchanged (insertion tests of Lennard-Jones' particles at several places), the second step would comprise insertion tests of a second Lennard-Jones particle at a fixed distance from the position selected for the first particle, and the third step would consist in testing several conformations and orientations of the molecule about the axis of the positions selected. In this movement, the aim of the first two steps would be to find elongated cavities in the liquid, so as to reserve the costly calculations of the final step for the most favorable sites.

A final point worth examining concerns the calculation of the chemical potential by insertion tests. Although the problem has been the subject of much work in the literature, no simple method appears to provide a definitive solution in the case of dense liquids (Kofke and Cummings, 1997). Reliance on progressive insertions is certainly possible, but perhaps difficult to insert in the normal course of a simulation, where other qualities than the chemical potential need to be calculated. A recent research direction consists in coupling a destruction test and an insertion test (Boulougouris *et al.*, 1999). This approach was successfully applied to calculate the solubility of alkanes in water, which is a difficult test (Boulougouris *et al.*, 2001). It could conceivably provide a fairly general solution to the direct calculation of the chemical potential at low temperature, when the Gibbs ensemble is inapplicable. In the case of pure substances, this could offer an alternative to thermodynamic integration.

3.3 Development of Improved Intermolecular Potentials

3.3.1 Why New Intermolecular Potentials?

Why is the development of intermolecular potentials still an important problem, despite the abundant reporting in the literature? There are many reasons for this, and they may be worth recalling. As a consequence of the present inability to accurately calculate the dispersion energy by *ab initio* calculation methods, the intermolecular potentials are of empirical form and their parameterization can assume various forms according to the final objective. Many potentials so far developed (Jorgensen and Madura, 1984; Cornell *et al.*, 1995; Sun, 1998) were not intended to represent the saturation vapor pressure, but only the density of the liquid phase and the enthalpy of vaporization in ambient conditions. Logically, these potentials offer a poor prediction of the saturation vapor pressure (Chen *et al.*, 1998). In fact, this quantity is essential for predicting phase equilibria because it guarantees the correctness of the chemical potential in a liquid phase for pure substances.

If we consider investigations aimed to represent liquid-vapor equilibria, the "all atoms" potentials in which each atom is explicitly represented (Chen and Siepmann, 1999) require very long computer time. The united atom potentials, in which the hydrogen atoms are not individually represented, offer the advantage of reducing computer time by about one order of magnitude, while allowing a good plot of the coexistence curve of pure substances. However, the united atom potentials *stricto sensu* which place the force center on the nucleus of the main atom of the group (usually carbon) are somewhat lacking in physical meaning and generality. This is demonstrated by the fact that in these potentials (Smit *et al.*, 1995; Martin and Siepmann, 1998; Nath *et al.*, 1998) the Lennard-Jones diameters attributed to the CH_3, CH_2 and CH groups are higher than that of methane (Möller *et al.*, 1992) and that in the case of iso-alkanes, the energy parameter attributed to the CH_3 group varies from 50 to 80 K depending on the position of the group and the size of the molecule (Martin and Siepmann, 1999; Nath and de Pablo, 2000). On the contrary, the anisotropic united atom potential proposed by Toxvaerd (Toxvaerd, 1990; Toxvaerd, 1997) seems to offer greater physical meaning, and this why the *Physical Chemistry Laboratory* decided to investigate this route from the start of its cooperation with *IFP*.

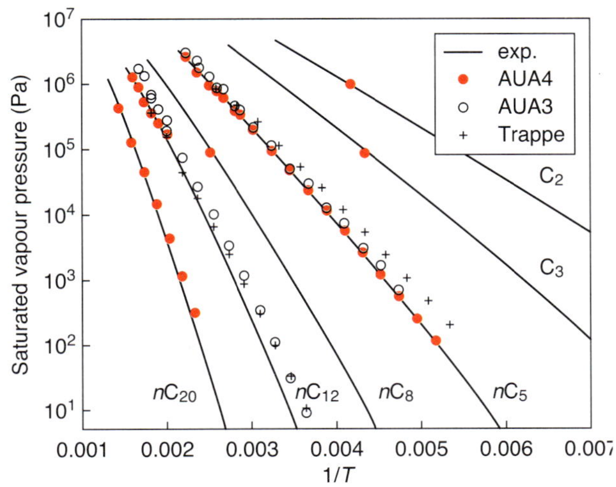

Figure 14

Calculation by simulation of the saturated vapor pressure of *n*-alkanes: comparison of TRAPPE united atom intermolecular potential (Martin and Siepmann, 1998), the AUA3 potential (Toxvaerd, 1997) and the optimized AUA4 potential (Ungerer *et al.*, 2000).

It was therefore proposed to optimize the parameters of the Toxvaerd's potential for *n*-alkanes, attempting to represent several properties in a wide range of temperatures: saturated vapor pressure, enthalpy of vaporization and liquid density *(Fig. 14)*.

Since then, E. Bourasseau has demonstrated that this potential correctly represented the properties of long chain alkanes (Bourasseau *et al.*, 2002b) and M. Lagache obtained good predictions of thermodynamic quantities derived for ethane and butane (Lagache *et al.*, 2001). This performance should be credited to the physical meaning of the parameters obtained by optimization. In particular, the fact that optimization can locate the positions of the force centers near the geometric center of the atoms of the group is highly encouraging. As to iso-alkanes, it has been demonstrated that the AUA potential optimized on isobutane could reasonably predict the difference in properties between *n*-heptane and its branched isomers.

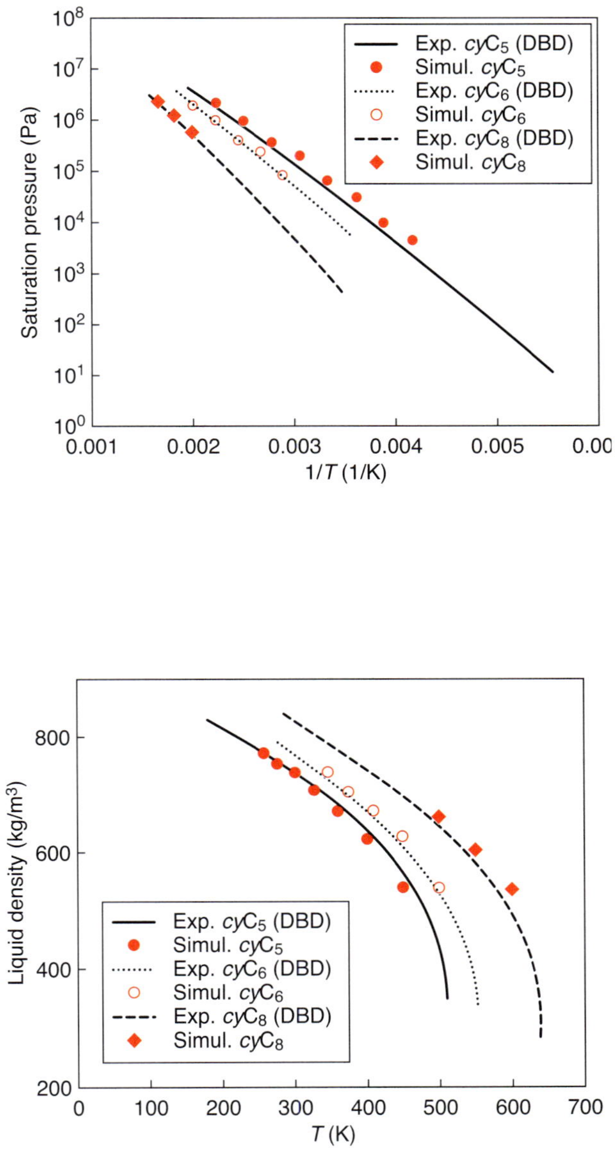

Figure 15

Results obtained by E. Bourasseau (Bourasseau, *et al.*, 2002a) using an AUA type intermolecular potential for cycloalkanes. The potential is fitted to the properties of cyclopentane and cyclohexane, but not to those of cyclooctane, which is also well represented.

3.3.2 Families of Components Considered

Since this work, E. Bourasseau continued the development of the AUA potential to adapt it to other hydrocarbon families. In the case of cycloalkanes, his work allowed the simultaneous representation of the properties of cyclopentane and cyclohexane with parameters of the CH_2 group very close to those of *n*-alkanes *(Fig. 15)*. The good representation of cyclooctane, which is not part of the molecules to which the potential was fitted, shows that this potential is more general for cyclic hydrocarbons than the united atom potentials *stricto sensu* (Errington and Panagiotopoulos, 1999; Neubauer *et al.*, 1999). In the case of olefins, the influence of the branches and the unsaturation position is clearly indicated by the AUA potential recently optimized by E. Bourasseau. Compared with the united atom potentials proposed by Spyriouni *et al.* (1999) or Wick *et al.* (2000) for olefins, this potential serves to represent more equilibrium properties (density, saturation pressures and enthalpies of vaporization) by means of parameters consistent with those employed for saturated hydrocarbons. A precise and transferable AUA potential has also been developed for aromatics, objects of close cooperation with A. Mackie's group of the *University of Tarragona*.

This research direction therefore offers highly promising prospects for mixed families encountered in petroleum fluids: alkylcycloalkanes, alkylaromatics, naphthenoaromatics, etc. Another interesting prospect is the development of potentials for organic molecules possessing polar functions, as initiated by J. Delhommelle successfully for organic sulfides and thiols (Delhommelle *et al.*, 2000). This approach consisted in relying on the results of quantum chemistry to determine the partial charges while preserving the anisotropic united atom potential of the alkanes for the nonpolar portion of the molecule. In cooperation with the *TotalFinaElf* group, S. Kranias carried out a work of the same type to represent other polar functions (ketones, aldehydes), in the course of publication. M. Lagache is currently working on a comparable development for the highly toxic organomercury compounds present in traces in some natural gases. Her work presents an additional difficulty insofar as the experimental data are less abundant and where the molecular structure of these compounds is poorly known. Hence there is every reason to believe that the family of potentials of anisotropic united atoms will be significantly enriched in the coming years.

3.3.3 Form of Interaction Potential and Mixing Rule

By thus proceeding into the development of intermolecular potentials, it is worth raising the question of the form of the interaction potential and the mixing rule. One could, for example, consider the replacement of the Lennard-Jones's potential by a Buckingham's potential, in which the exponential form of the repulsion is known to better represent the behavior of noble gases (Wu and Sadus, 2000). It is nonetheless possible that the larger number of parameters to be determined of the Buckingham's potential would mean indeterminations in the optimization process, and it is uncertain whether this potential will provide the same improvement for anisotropic united atoms as the one procured for monatomic gases.

In the same range of ideas, one could question whether the choice of the Lorentz-Berthelot's mixing rule is the best, insofar as the work done by J. Delhommelle on noble gases (Delhommelle and Millié, 2001) or by other researchers on more complex mixtures (Potoff *et al.*, 1999; Errington *et al.*, 2000) tend to show that the Kong and Waldmann-Hagler's rules could yield better results, at least for interactions between groups of very different size. These rules rely in particular on a geometric mean on the product $\varepsilon\sigma^6$ and not on ε, which conforms more to the theoretical form of the dispersion energy.

The tests performed on methane-alkane mixtures in connection with the Master degree work of E. Soldi-Lose in 2000 failed to identify any tangible effect of the mixing rule, so that it did not seem advisable to question the present choice so far. However, the recent study by A. Wender of $H_2S - CO_2$ mixtures showed that for temperatures slightly higher than the critical temperature of the mixture, its volumetric properties are particularly sensitive to the choice of the mixing rule. On this system, the results speak clearly in favor of the Kong and Waldmann-Hagler's rules, and could even help to decide between the two. An interesting research direction is hence to extend this type of comparison to other systems to determine the degree of generality of these conclusions.

3.3.4 Methodology of Optimization of Intermolecular Potentials

Another point worth examining is the optimization method developed during the work on *n*-alkanes, on which sufficient data is now available to consider improvements.

Undeniably, its strong point is the definition of the composite error function which integrates three quantities (saturation vapor pressure, enthalpy of vaporization and liquid density) at very different temperatures, on several pure substances. Provided allowance is made for necessary adjustments (for example, the consideration of volumetric data measured in supercritical conditions), this type of error function proves to be a powerful tool because it helps to use numerous independent data while demanding a moderate number of simulations. One could possibly consider entering derived quantities like those discussed in the following chapter in the error criteria. However, the slowness of convergence of these quantities obtained by fluctuation methods and the lesser availability of the corresponding experimental data, tend to imply that this would not be a major improvement to the optimization process.

On the contrary, the gradient method employed to determine the minimum of the error function is sometimes a weak point. Besides, the gradient method as proposed may display wide instabilities when the initial parameters are too far from the optimum. If this is the case, we cannot hope to find the optimum directly, and it must be accepted to decrease the error function step by step in the gradient direction. Moreover, the calculation of the gradient component by finite differences is a lengthy, inaccurate and difficult process. To avoid it, one can directly calculate the partial derivatives of a mean quantity X with respect to any parameter λ of the potential from the known fluctuation formula (Gray and Gubbins, 1984):

$$\frac{\partial\langle X\rangle}{\partial\lambda}=\left\langle\frac{\partial X}{\partial\lambda}\right\rangle-\beta\left\langle X\frac{\partial U}{\partial\lambda}\right\rangle+\beta\langle X\rangle\left\langle\frac{\partial U}{\partial\lambda}\right\rangle$$

where U is the total potential energy.

In exchange for the calculation of a few additional terms during the simulation, this direct calculation of the gradient components is likely to considerably accelerate the empirical portion of the development of intermolecular potentials. The first tests of this method conducted recently by E. Bourasseau effectively yielded absolutely positive results for representing the properties of olefins (Bourasseau *et al.*, 2003).

3.3.5 The Use of Anisotropic United Atoms for Adsorption

In terms of applications, adsorption in zeolites is a field in which the industry would be happy to have a united atom potential that displays transferability qualities comparable to those observed in liquid-vapor. On the scientific level, it is also interesting to examine whether the displacement of the force centers with respect to the carbon nuclei provides additional predictive power for simulating adsorption in zeolites bearing the compensation cationic charges. It is one of the major objectives assigned to the thesis P. Pascual, begun in autumn 2001 under the direction of A. Boutin on the adsorption of alkanes, to develop this new approach.

Based on the approach which yielded good results in liquid-vapor, one could propose to develop the zeolite-alkanes potential on the basis of adsorption isotherms covering a wide range of numbers of carbons and temperatures. In a first step, the consideration of zeolites devoid of extra-structural cations like silicalite would help to parameterize the oxygen-alkane and T-atom-alkane interactions. In the second step, the study of zeolites bearing cations in known crystallographic positions (like the NaY, KY and BaX faujasites) would help to derive the dispersion-repulsion parameters of cation-alkane interactions. And finally, if this succeeds, the approach could be extended to the adsorption of aromatics, olefins and cycloalkanes, so as to obtain a coherent interaction potential for adsorption and fluid phases. It is likely that this process will be confronted with a lack of data—if only of adsorption isotherms of pure substances measured at different temperatures—and it is therefore logical to propose an experimental program designed to fill these gaps.

3.4 Towards the Industrial Use of Monte Carlos Methods

A first concern of the *IFP* and *CNRS* investigations was to assess the accuracy with which it was possible to describe the properties of hydrocarbon gases at high pressure. The work of B. Neubauer had already established the feasibility for calculating one of the properties of natural gases containing several components, with highly encouraging results (Neubauer *et al.*, 1999). It was therefore tempting to grapple with a more difficult problem, namely the prediction of derivative thermodynamic quantities: heat capacity, thermal expansion coefficient, compressibility factor, Joule Thomson's coefficient. This topic is the first part of the thesis of M. Lagache, carried out at the LCP in connection with a Concerted Research Project involving six other French university laboratories and oil companies. To calculate the derivative quantities more directly, M. Lagache applied a fluctuation technique discussed briefly by Jorgensen (Jorgensen, 1986; Lagache *et al.*, 2001). It is in fact possible to calculate the derivative of a mean quantity X with respect to an intensive variable Y by considering the covariance of X and the extensive quantity Z associated with Y:

$$\frac{\partial X}{\partial Y} = \lambda\left(\langle XZ\rangle - \langle X\rangle\langle Z\rangle\right)$$

where λ is a factor depending on the variables considered.

Since Monte Carlo methods do not calculate the kinetic energy, it is necessary to use them only to calculate residual properties (deviation from ideal gas properties) and to calculate the ideal contributions from experimental measurements. Once this somewhat delicate division has been made, the results obtained are excellent, better in fact than the initial likelihoods. Without altering the intermolecular potentials which were nonetheless fitted to the first order derivatives of the free energy, it is in fact possible nearly quantitatively to represent the variations in quantities that are especially difficult to predict like the heat capacity at constant volume *(Fig. 16)*. Note that this test is particularly difficult since the C_v (ρ) curve at constant temperature shows a peak which the equations of state do not reproduce, even qualitatively. The correct representation of the peak of C_v by simulation offers hope that the slight divergence of this quantity at the critical point can also be reproduced if we use sufficiently large simulation boxes.

Another field of application of delicate phase equilibrium calculation is the behavior of acid gases, *i.e.* natural gases containing significant contents of H_2S and CO_2. In general, phase equilibria involving H_2S have been the subject of fewer experimental measurements as those exclusively involving hydrocarbons, due to the toxicity and corrosive nature of this gas. It is therefore extremely interesting to employ molecular simulation to predict these equilibria, as initiated by J. Delhommelle during his thesis work. To supplement these investigations, it was decided to simulate ternary phase diagrams in the water-methane-H_2S system, for which few data are available. In this study, it was important first to validate the choice of the intermolecular potentials by comparison with the properties of the pure substances of the mixture. It thus turned out that with the potentials considered for water and H_2S (Delhommelle, 2000), the properties of the pure substances were less well described when the polarization energy was taken into account. While this was not surprising in the case of water, since the TIP4P potential had been optimized without taking account of polarization

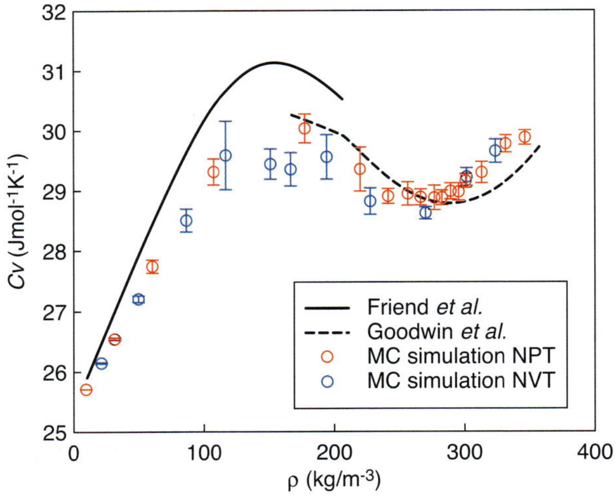

Figure 16

Calculation made by M. Lagache of the heat capacity at constant volume of methane from statistical fluctuations (T = 210 K , *i.e.* T/Tc = 1.1).

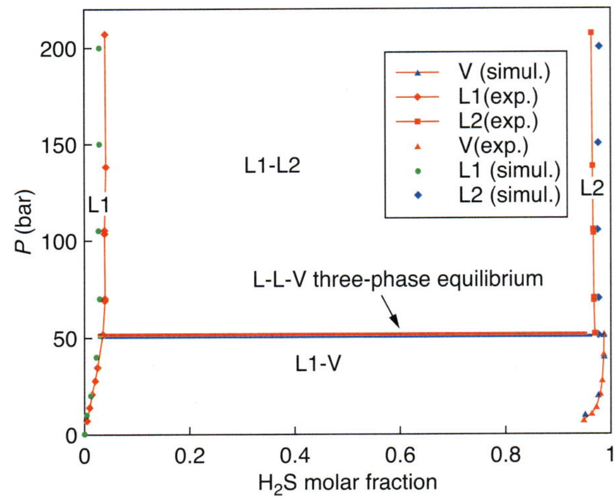

Figure 17

Phase diagram of the water-H_2S as a binary system at 343 K.

(Jorgensen *et al.*, 1983), this finding was difficult to explain in the case of the polarizable model of H_2S proposed by Delhommelle. Finally, nonpolarizable potentials were employed for H_2S (Kristof and J. Liszi, 1997) and for water (TIP4P). Once this stage of the choice of the potential was terminated, the analysis of the binary water-H_2S system *(Fig. 17)* provided the opportunity to employ a liquid-liquid-vapor equilibrium calculation at imposed total volume to calculate the three-phase equilibrium pressure. The prediction of the phase diagram is very good, and some details are extremely well rendered. The simulation helped to explain why the solubility of water is higher in the H_2S rich liquid phase than in the vapor phase: this behavior is probably connected with the self-combination of water molecules to form oligomers (dimers, trimers, etc.) in this phase *(Fig. 18)*. This mechanism, rather unexpected for solubility levels in the range of 2-3%, shows that we are not in the regime of infinite dilution for water.

Now that simulation has demonstrated its possibilities for acid gases, we can consider applying it more systematically for reinjection projects of these gases into the subsoil, after having separated them from the hydrocarbons: prediction of volumetric properties at high pressure, determination of the heat of mixing, phase equilibria of H_2S-hydrocarbon mixtures. The training student work of A. Wender recently completed at *IFP* helped to confirm these hopes by rapidly supplying a large number of high quality predictions. By resorting to molecular dynamics, it should be also possible to determine transport properties such as viscosity and thermal conductivity in the future.

Another type of application of Monte Carlo methods consists in predicting the properties of hydrocarbon components with more than ten carbon atoms, *i.e.* those making up most of diesels and kerosenes. It is in fact surprising to find that in this carbon range, only a few families, like *n*-alkanes, alkylbenzenes and condensed polyaromatics (naphthalene, phenanthrene, anthracene) have been the subject of sufficient measurements to accurately determine their equilibrium properties. In the absence of pure substances available commercially at reasonable prices, measurements are difficult to take on many isomers of little known families such as isoalkanes and cycloalkanes, with more than ten carbon atoms. The same applies to mixed compounds like alkylcyloalkanes and naphthenoaromatics, which are also well represented in crude oils (Petrov, 1984; Tissot and Welte, 1984). Since molecular simulation reasonably describes the differences in properties between isomers with the potentials recently developed and possesses good properties of extrapolation to high carbon numbers, it could supply more reliable predictions than the group contribution methods used today. This type of prediction proves feasible *(Fig. 19)* and is now seriously considered for better use of the detailed analytical results on petroleum products obtained by analytical methods coupling chromatography with mass spectrometry.

In the same range of ideas, the prediction of the behavior of polymers is a field in which molecular simulation can make a large contribution. For example, the solubility of gases at high pressure in amorphous polymer matrices is an important point that must be controlled for the design and

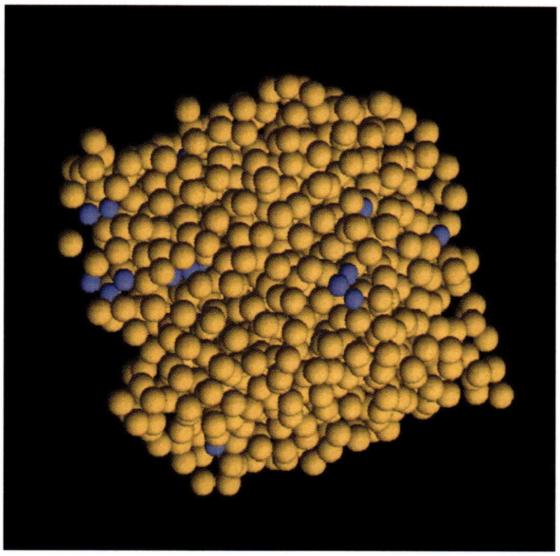

Figure 18

Example of configuration of a simulation box of the H_2S rich liquid phase at 343 K and 70 bar in equilibrium with an aqueous phase. The water molecules (mauve) are either isolated, or in the form of aggregates (dimers, trimers).

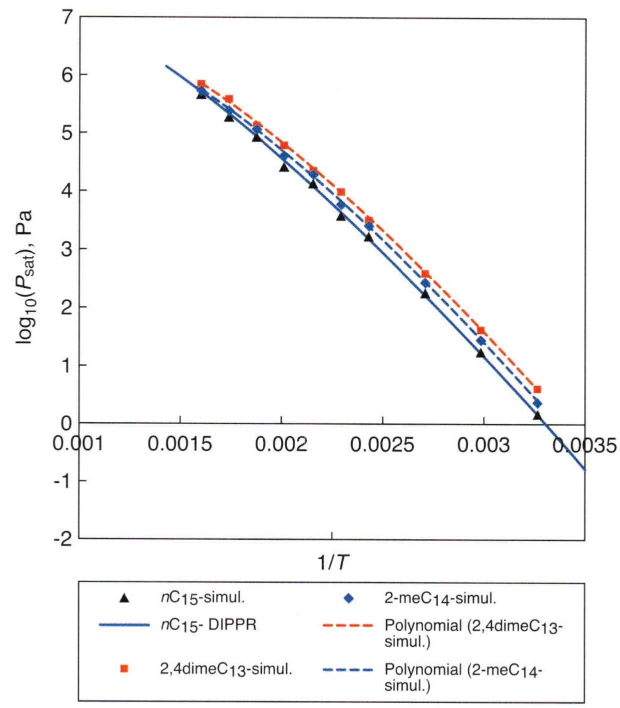

Figure 19

Prediction of the saturation vapor pressure of *n*-pentadecane and of two C_{15} alkanes by means of the AUA4 potential. The three simulations corresponding to the highest pressures were obtained directly by two-phase simulation (Gibbs ensemble), while the lower pressures were obtained by thermodynamic integration using single-phase simulations.

operation of flexible lines which used these materials considerably. The convergence of this type of system demands reliance on specific algorithms such as reptation, recoil growth (Consta *et al.*, 1999) and gradual insertion (Kaminski, 1994) to effectively sample all the conformations. It is to be hoped that the use of these algorithms will help to address industrial problems successfully.

A final industrial challenge, and certainly not the least, concerns the prediction of adsorption in zeolites. This prediction is essential to design adsorption separation processes used to separate isomers whose liquid-vapor equilibrium properties do not permit the economic use of conventional separation by distillation. Thus meta is separated from paraxylene, the latter being an important intermediate in petrochemicals. It is also by adsorption that iso-alkanes are separated from *n*-alkanes, the latter being undesirable in premium gasolines due to their poor octane number, and must therefore be recycled to the reactor, where they undergo conversion to iso-alkanes. These adsorption processes, as well as many other industrial processes using the same principle, are based on the specificity of molecule-adsorbent interaction at molecular scale. In this respect, it is not surprising that classic thermodynamic models fail to predict the adsorption selectivities in the absence of selectivity measurements on the absorbent concerned, even if these models incorporate concepts of statistical mechanics. This implies a real opportunity for the development of molecular simulation methods capable of supplying preliminary indications before experimental measurements become available.

Compared with cases of fluid phase equilibria, the simulation of adsorption equilibria is a more difficult problem in many respects. We are still far from having a family of potentials transferable from one zeolite to another and from one adsorbate to another: so far, the potentials used are in fact specific to a particular system. In his first thesis year at *LCP*, P. Pascual succeeded recently in producing a first improved anisotropic united atom potential. This potential displays transferability properties much better than those of classic united atom models (Vlugt *et al.*, 1998). Beyond this result, it must be kept in mind that the frequent presence of cations in zeolites makes electrostatic interactions and polarization more important, and the possibility that these cations could shift under the influence of adsorbed molecules highly complicates the calculations if these effects are to be considered. In fact, the simulation of adsorption in zeolites demands a bridging of two relatively distant disciplines, namely chemistry of solids and molecular simulation of phase equilibria. For the equilibria, the local electrostatic charges of the molecules in vacuum (Delhommelle *et al.*, 2000) may be determined by performing a quantum calculation of the electron structure (Hartree Fock or—Density Functional Theory—DFT). Similarly, an effective research direction could be to use methods of theoretical chemistry used in chemistry of solids (DFT with breakdown into plane waves) to calculate the local charges electrostatic charges in the zeolite pores. These methods are in fact the most widely used in the field of materials with a crystalline structure (Wimmer, 1993).

CONCLUSION

Although numerical simulation at the scale of a million years was met at its inception by the skepticism of many geologists marked by the naturalistic tradition of their field, this discipline has progressively gained credibility thanks to many regional studies that have helped to compare observations and simulation results. Designated today by the term of basin models, the tools derived from this approach are routinely used in industry and in the academic world. It is extremely satisfying to have been able to contribute to the advance of this discipline through basic research that is still recognized today. With the perspective of time, we can state that this success rewards the large risks assumed by confronting difficult problems. It must be acknowledged that our team had two big chances: to have had the backing of a hierarchy motivated by the development of basin models, and to have been operational at the right time to participate fully in this scientific adventure.

If we consider thermodynamics and molecular simulation, the situation is very different in many ways: the discipline is based on solid foundations, in which many highly competent teams have been active for many years and whose potential applications cover a broader field than the oil industry. Beyond these differences, there are some significant similarities with basin models: long computer time, the need for versatile softwares to analyze realistic systems, empirical optimization of parameters from which a physical meaning is nonetheless expected. Yet the most striking common feature is the difficulty that a new technology encounters at all levels to establish itself. In terms of applications, the confidence of industry can only be secured very gradually, once many examples have been favorably investigated, and this process is especially slow because the teams employing the new methods are made of a few people at the start! No doubt there is no panacea for success in establishing a new technology, except a strong dose of perseverance!

Having taken this look at the past and present, what does the future offer? In the field of basin models, the more detailed consideration of reaction mechanisms, based inter alia on a theoretical calculation of rate constants, is one path for progress that is highly interesting for better control of the kinetics of oil formation, including in cases where thermal alteration is not the only factor (thermoreduction of sulfates, for example). As to migration modeling, no doubt considerable work still needs to be done before realistic three-dimensional geometries (faults, etc.) can integrate flows involving equilibria between the water, oil and gas phases, not to speak of the physicochemical aspects which play a role in the isotopic composition of the fluids.

In the field of applied thermodynamics, interesting prospects exist in the use of new techniques, including inter alia, vibrating wire viscodensimetry used at the *University of Clermont-Ferrand* for fluids of high pressure. The measurement of low saturation vapor pressures with very small quantities of samples developed at the *University of Lyon*, the systematic analysis of petroleum fluids by GC-MS coupling. Yet it is no doubt in modeling, and particularly in molecular simulation, that the greatest prospects for progress exist. Generally speaking, the algorithms and intermolecular potentials developed in recent years have only been partly used for industrial purposes, and their practical application alone is a very fruitful pathway. And still concerning the petroleum industry and chemical engineering, three research directions can be distinguished: the calculation of properties of pure toxic substances, unstable or unavailable commercially; the simulation of complex systems in the fluid state (equilibria involving polar compounds, gas condensate, supercritical fluids) and amorphous state (polymers); and adsorption equilibria in industrial adsorbents.

Beyond the industrial application of already validated techniques, the improvement of algorithms and molecular simulation softwares also offers rich prospects. Considering Monte Carlo methods and equilibrium properties, worth noting is the establishment of parallel tempering techniques, which will ultimately help to improve a factor that clearly limits the uses of molecular simulation, namely the productivity of the engineer's work. Many other improvements are worth testing, such as the non-Boltzman sampling and the gradual insertion technique. With the explicit simulation of interface, it will be possible to predict interfacial tensions and achieve a detailed understanding of the role of surfactants (Goujon *et al.*, 2001; Chen *et al.*, 2002). In a nearby area, network calculations help to consider much larger systems (a few tens of nanometers) thereby helping to simulate micellar systems that are inaccessible with classical atomistic techniques. The simulation of crystalline molecular solids, such as crystals of paraffins or hydrates, could also become an attractive tool.

It would be incomplete to fail to mention the techniques of molecular dynamics among these prospects. Whether for free fluids or for confined fluids, the prediction of the transport properties by classic models is much less satisfactory than that of equilibrium properties. In this area, simulation is already capable of surprisingly accurate predictions when relaxation times are not too long. By contrast, the prediction of transport properties remains an unresolved problem when relaxation is very slow—except for the case in which the path is known and where kinetic Monte Carlo techniques applies. Increased computer power will probably not suffice this problem, and its solution rather entails the development of new algorithms.

It will be useful to terminate by stressing the importance that molecular simulation can and must assume. In quantum chemistry, the success of this discipline is such that some quantities are already obtained more reliably by calculation than by experiment. However, this must not obscure the fact that for most applications, simulation techniques are still based on empirical parameters and approximate numerical solutions. This is particularly the case of statistical thermodynamics discussed here. This means that simulation remains dependent on the experimental approach and can only substitute for it partially, after intensive validation tests. Nor can simulation substitute for classical models, which will remain indispensable for rapidly calculating properties in process simulators where the calculations must be repeated millions if not billions of times. On the contrary, it is legitimate that simulation claims a more important place in research and development programs of the petroleum industry and chemical industries. It is striking in this respect to observe the difference in involvement of the chemical engineering laboratories in Europe and the United States, where modeling at the molecular scale is clearly felt as the most promising "frontier", in scientific as well as industrial terms. The Old Continent still has time to seize the opportunities, but it cannot procrastinate.

REFERENCES

Albright, P.C., Sengers, J.V., Nicoll, J.F. and Ley-Koo, M. (1986) A Crossover Description for the Thermodynamic Properties of Fluids in the Critical Region. *Int. J. of Thermophysics,* **7**, 75.

Allen, M.P. and Tildesley, D.J. (1987) *Computer Simulation of Liquids*, Oxford, Oxford Science Publications.

Arnaud, J.F., Ungerer, P., Béhar, E., Moracchini G. and Sanchez, J. (1996) Excess Volumes and Saturation Pressures for the System Methane + *n*-Tetracosane at 374 K. Representation by Improved EOS Mixing Rules. *Fluid Phase Equilibria,* **124**, 177-207.

Ashcroft, S.J., Booker, D.R. and Turner, J.C.R. (1992) Volumetric Behaviour of Mixtures of Crude Oils and Light Hydrocarbons. *J. of the Institute of Energy,* **65**, 131-136.

Béhar, F., Ungerer, P., Audibert, A. and Villalba, M. (1988) Experimental Study and Kinetic Modelling of Crude Oil Pyrolysis in Relation to Thermal Recovery Processes. *4th UNITAR/UNDP International Conference on Heavy Crude and Tar Sands*, Edmonton, Alberta Oil Sands Technology Abd Research Authority.

Béhar, F., Ungerer, P., Kressmann, S. and Rudkiewicz, J.L. (1991) Thermal Evolution of Crude Oils in Sedimentary Basins: Experimental Simulation in a Confined System and Kinetic Modelling. *Oil & Gas Science and Technology,* **46**, 2, 151-181.

Berg, R.R. (1975) Capillary Pressure in Stratigraphic Traps. American Association of Petroleum, *Geologists Bulletin,* **59**, 939-956.

Boulougouris, G.C., Economou, I.G. and Theodorou, D.N. (1999) On the Calculation of the Chemical Potential Using the Particle Deletion Scheme. *Molecular Physic,* **96**, 905-913.

Boulougouris, G.N., Economou, I.G. and Theodorou, D.N. (2001) Calculation of the Chemical Potential of Chain Molecules Using the Staged Particle Deletion Scheme. *J. Chem. Phys.*, **115**, 8231.

Bourasseau, E., Ungerer, P. and Boutin, A. (2002a) Prediction of Equilibrium Properties of Cyclic Alkanes by Monte Carlo Simulation - New Anisotropic United Atoms Potential - New Transfer Bias Method. *J. Phys. Chem.*, B, **106**, 5483.

Bourasseau, E., Ungerer, P., Boutin, A. and Fuchs A.H., (2002b) Monte Carlo Simulation of Branched Alkanes and Long Chain *n*-Alkanes with Anisotropic United Atoms Intermolecular Potential. *Molecular Simulation,* **28**, 317-336.

Bourasseau, E., Haboudou, M.A., A., Boutin, Fuchs, H. and Ungerer, P. (2003) New Optimization Method for Intermolecular Potentials - Optimization of a New Anisotropic United Atoms Potential for Olefins - Prediction of Equilibrium Properties. *J. Chem. Phys.*, in press.

Bouvier, V. (1989) Modélisation bidimensionnelle des phénomènes de transport dans les bassins sédimentaires par la méthode des éléments finis. *Thèse de doctorat en hydrologie et hydrogéologie quantitatives,* Paris, École nationale supérieure des Mines.

Burnham, A.K., Braun, R.L. and Gregg, H.R. (1987) Comparison of Methods for Measuring Kerogen Pyrolysis Rates and Fitting Kinetic Parameters. *Journal of Energy and Fuels,* **1**, 452-458.

Burnham, A.K. and Braun, R.L. (1990) Detailed Model of Petroleum Formation, Destruction, and Expulsion from Lacustrine and Marine Source Rocks. *Organic Geochemistry,* **16**, 27-40.

Carpentier, B., Ungerer, P., Kowalewski, I. Magnier, C., Courcy, J.P. and Huc, A.Y. (1996) Molecular and Isotopic Fractionation of Light Hydrocarbons Between Oil and Gas Phases. *Organic Geochemistry,* **24,** 12, 1115-1139.

Chapman, W.G., Gubbins, K.E., Jackson, G. and Radosz, M. (1989) SAFT: Equation of State Solution Model for Associating Fluids. *Fluid Phase Equilibria,* **52**, 31-38.

Chen, B., Martin, M.G. and Siepmann, J.I. (1998) Thermodynamic Properties of the Williams, OPLS-AA, and MMFF94 All-Atom Force Fields for Normal Alkanes. *J. Phys. Chem.,* B, **102**, 2578-2586.

Chen, B. and Siepmann, I.J. (1999) Transferable Potentials for Phase Equilibria: 3. Explicit - Hydrogen Description of Normal Alkanes. *Journal of Physical Chemistry,* **103**, 5370-5379.

Chen, B., Siepmann, J.I. and Klein, M.L. (2002) Vapor-Liquid Interfacial Properties of Mutually Saturated Water/1-Butanol Solutions. *J. Am. Chem. Soc.,* **124**, 12232.

Chien, C.H., Greencorn, R.A. and Chao, K.C. (1983) Chain-of-Rotators Equation of State. *American Institute of Chemical Engineering Journal,* **24**, 849-860.

Consta, S., Wilding, N.B., Frenkel D. and Alexandraowicz, Z. (1999) Recoil Growth: an Efficient Simulation Method for Multi-Polymer Systems. *J. Chem. Phys.,* **110**, 3220.

Cornell, W.D. and Cieplak, P. *et al.* (1995) A Second Generation Force Field for the Simulation of Proteins, Nucleic Acids and Organic Molecules. *J. Am. Chem. Soc.,* **117**, 5179.

Cracknell, R.F., Nicholson, D. and Parsonage, N.G. (1990) Rotational Insertion Bias: a Novel Method for Simulating Dense Phases of Structured Particles, with Particular Application to Water. *Mol. Phys.,* **71**, 931.

Dayhoff, M.O., Lippincott, E.R., Eck, R.V. and Nagarajan, G. (1967) Thermodynamic Equilibrium in Prebiological Atmospheres of C, H, O, N, P, S, and Cl. *NASA,* 249.

Delhommelle, J. (2000) Etablissement de potentiels d'interaction pour la simulation moléculaire - Application à la prédiction des équilibres liquide-vapeur de mélanges binaires alcane-molécule multipolaire. *Thesis,* Orsay, Paris XI, 169.

Delhommelle, J., Tschirwitz, C., Ungerer, P., Granucci, G., Millié, P, Pattou, D. and Fuchs, A.H. (2000) Derivation of an Optimized Potential Model for Phase Equilibria (OPPE) for Sulfides and Thiols. *Journal of Physical Chemistry,* B, **104**, 4745-4753.

Delhommelle, J. and Millié, Ph. (2001) Inadequacy of the Lorentz-Berthelot Combining Rules for Accurate Predictions of Equilibrium Properties by Molecular Simulation. *Molecular Physics,* **99**, 619-625.

Dodd, L.R., Boone, T.D. and Theodorou, D.N. (1993) A Concerted Rotation Algorithm for Atomistic Monte Carlo Simulation of Polymer Melts and Glasses. *Mol. Phys.,* **78**, 961-996.

Doligez, B., Ungerer, P., Chénet, P.Y., Burrus, J., Bessis, F. and Bessereau, G. (1987) Numerical Modelling of Sedimentation, Heat Transfer, Hydrocarbon Formation and Fluid Migration in the Viking Graben, North Sea. *Petroleum Geology of NorthWest Europe*. J.B. and K. Glennie, Graham and Trotman, 1039-1048.

Doué, F. and Guiochon, G. (1968) Étude théorique et expérimentale de la cinétique de décomposition chimique du *n*-hexadécane, de son mécanisme et de la composition du mélange de produits obtenus. *Journal de chimie physique et de physico-chimie biologique,* **64**, 395-409.

Duin, A.C.T. van, Dasgupta, S., Lorant, F. and Goddard, W.A. (2001) A Reactive Force Field for Hydrocarbons. *Journal of Physical Chemistry,* A, **105**, 9396-9409.

Durand, B., Ed. (1980) *Kerogen, Insoluble Organic Matter from Sedimentary Rocks*, Éditions Technip, Paris.

Durand, B. (1983) Present Trends in Organic Geochemistry in Research on Migration of Hydrocarbons. *10th International Meeting on Organic Geochemistry*, Bergen, Wiley.

Durand, J.P., Fafet, A. and Barreau, A. (1989) Direct and Automatic Capillary GC Analysis for Molecular Weight Determination and Distribution in Crude Oils and Condensates up to C_{20}. *Journal of High Resolution Chromatography,* **12**, 230-233.

Dysthe, D., Fuchs, A.H. and Rousseau, B. (1999) Fluid Transport Properties by Equilibrium Molecular Dynamics. I. Methodology at Extreme Fluid States. *Journal of Chemical Physics,* **110**, 4047.

Dysthe, D., Fuchs, A.H. and Rousseau, B. (1999) Fluid Transport Properties by Equilibrium Molecular Dynamics. II. Multi-component Systems. *Journal of Chemical Physics,* **110**, 4060.

Dysthe, D., Fuchs, A.H. and Rousseau, B. (2000) Fluid Transport Properties by Equilibrium Molecular Dynamics. III. Evaluation of United atom Interaction Potential Models for Pure Alkanes. *Journal of Chemical Physics,* **112**, 7581.

Errington, J.R. and Panagiotopoulos, A.Z. (1999) New Inter-molecular Potential Models for Benzene and Cyclohexane. *J. Chem. Phys.*, **111**, 9731.

Errington, J.R., Boulougouris, G.C., Economou, I.G., Panagio-topoulos, A.Z. and Theodorou, D.N. (2000) Molecular Simulation of Phase Equilibria for Water - *n*-Butane and Water - *n*-Hexane Mixtures. *J. Phys. Chem.,* B, **104**, 4958.

Escobedo, F.A. (1999) Tracing Coexistence Lines in Multi-component Fluid Mixtures by Molecular Simulation. *Journal of Chemical Physics,* **110**, 11999-12010.

Escobedo, F.A. (1998) Novel Pseudoensembles for Simulation of Multicomponent Phase Equilibria. *J. Chem. Phys.,* **108**, 8761-8772.

Escobedo, F.A. and Chen, Z. (2001) Simulation of Isoenthalps and Joule-Thomson Inversion Curves of Pure Fluids and Mixtures. *Molecular Simulation,* **26**, 395-416.

Espitalié, J., Deroo, G. and Marquis, F. (1985) La pyrolyse Rock-Eval et ses applications (1). *Oil & Gas Science and Technology,* **40**, 563-579.

Espitalié, J., Ungerer, P., Irwin, I. and Marquis, F. (1988) Primary Cracking of Kerogens. Experimenting and Modelling C_1, C_2-C_5, C_6-C_{15} and C_{15+} Classes of Hydrocarbons Formed. *Organic Geochemistry,* **13**, 893-899.

Faille, I. (1992) Modélisation bidimensionnelle de la genèse et la migration des hydrocarbures dans un bassin sédimentaire. *Thèse de doctorat en mathématiques appliquées*, Grenoble, université Joseph Fourier.

Flöter, E. (1999) Hyperbaric Reservoir Fluids - Phase Behaviour of Model Systems. *Thèse de doctorat*, Delft, Technische Universiteit Delft.

Forbes, P., Ungerer, P., Kuhfuss, A., Riis, F. and Eggen, S. (1991) Compositional Modeling of Petroleum Generation and Expulsion: Trial Application to a Local Mass Balance in the Smorbukk Sor Field, Haltenbanken Area, Norway. *American Association of Petroleum Geologists Bulletin,* **75,** 5, 873-893.

Fredenslund, A., Gmehling, J. and Rasmussen, P. (1977*) Vapour-Liquid Equilibrium Using UNIFAC*, New York, Elsevier.

Frenkel, D. and Smit, B. (1996) *Understanding Molecular Simulation,* San Diego, Academic Press.

Gil-Villegas, A., Galindo, A., Whitehead, P.J., Mills, S.J. and Jackson, G. (1997) Statistical Associating Fluid Theory for Chain Molecules with Attractive Potentials of Variable Range. *J. Chem. Phys.,* **106**, 4168-4186.

Glaser, M., Peters, C.J., Kooi, H.J.V. de and Lichtenthaler, R.W. (1985) Phase Equilibria of the System Methane + *n*-Hexadecane. *J. Chem. Thermodynamics,* **17**, 803-815.

Go, N. and Sheraga, H.A. (1976) On the Use of Classical Statistical Mechanics in the Treatment of Polymer Chain Conformation. *Macromolecules,* **9**, 535.

Goujon, F., Malfreyt, P., Boutin, A. and Fuchs, A.H. (2001) Vapour-Liquid Phase Equilibria of *n*-Alkanes by Direct Monte Carlo Simulations. *Molecular Simulation,* **27**, 99-114.

Gray, C.G. and Gubbins, K.E. (1984) *Theory of Molecular Fluids*, Oxford, Oxford Science.

Hemptinne, J.C. de and Ungerer, P. (1995) Accuracy of the Volumetric Predictions of Some Important Equations of State for Hydrocarbons, Including a Modified Version of the Lee-Kesler Method. *Fluid Phase Equilibria,* **106**, 81-109.

Huron, M.J. and Vidal, J. (1979) New Mixing Rules in Simple Equations of State for Representing Vapor-Liquid Equilibria of Strongly Non-Ideal Mixtures. *Fluid Phase Equilibria,* **1**, 247-265.

Jorgensen, W.L., Chandrasekhar, J. and Madura, J.D. (1983) Comparison of Simple Potential Functions for Simulating Liquid Water. *J. Chem. Phys.,* **79**, 926-935.

Jorgensen, W.L. and Madura, J.D. (1984) Optimized Inter-Molecular Potential Functions for Liquid Hydrocarbons. *Journal of the American Chemical Society,* **106**, 6638.

Jorgensen, W.L. (1986) Optimized Intermolecular Potential Functions for Liquid Alcohols. *J. Phys. Chem.*, **90**, 1276-1284.

Kaminsky, R.D. (1994) Monte Carlo Evaluation of Ensemble Averages Involving Particle Number Variations in Dense Fluid Systems. *J. Chem. Phys.*, **101**, 4986.

Katzenski, G. and Schneider, G.M. (1982) Excess Volumes of Liquid *n*-Alkane Binaries from 10 to 200 MPa at 298.15K. *J. Chem. Thermodynamics*, **14**, 801-802.

Kofke, D.A. (1993) Direct Evaluation of Phase Coexistence by Molecular Simulation via Integration Along the Saturation Line. *Journal of Chemical Physics*, **98**, 4149.

Kofke, D.A. and Cummings, P.T. (1997) Quantitative Comparison and Optimization of Methods for Evaluating the Chemical Potential by Molecular Simulation. *Molecular Physics*, **92**, 973.

Kooi, H.J. van de (1981) Metingen en berekningen aan het systeem methaan-n-eicosaan. *PhD Thesis*, Delft Technical University.

Kressmann, S., Dominé, F., Enguehard, F., Béhar, F., Ungerer, P. Billaud, F. and Goffe, B. (1990) Secondary Cracking of Crude Oils: Experimental Simulation and Kinetic Interpretation on Model Compounds. *Organic Geochemistry*, **16**.

Kressmann, S. (1991) Craquage thermique de mélanges d'hydrocarbures à haute pression : étude cinétique expérimentale et modélisation numérique, implications pour la géochimie pétrolière. *Thèse*, Université Pierre et Marie Curie (Paris VI).

Kristof, T. and Liszi, J. (1997) Effective Intermolecular Potential for Fluid Hydrogen Sulfide. *J. Phys. Chem.*, B, **101**, 5480-5483.

Lachet, V., Boutin, A., Tavitian, B. and Fuchs, A.H. (1997) Grand Canonical Monte Carlo Simulations of Adsorption of Xylene Molecules in Faujasite Zeolites. *Faraday Discuss.*, **106**, 307.

Lagache, M., Ungerer, P., Boutin, A. and Fuchs, A.H. (2001) Prediction of Thermodynamic Derivative Properties of Fluids by Monte Carlo Simulation. *Physical Chemistry Chemical Physics*, **3**, 4333-4339.

Lee, B.I. and Kesler, M.G. (1975) A Generalized Thermodynamic Correlation Based on Three-Parameter Corresponding States. *AIChE Journal*, **21**, 510.

LeRoy, S., Béhar, E. and Ungerer, P. (1997) Vapour-Liquid Equilibrium Data for Synthetic Hydrocarbon Mixtures. Application to Modelling of Migration from Source to Reservoir Rocks. *Fluid Phase Equilibria*, **135**, 63-82.

Leythaeuser, D. and Yükler, A. (1982) Role of Diffusion in Primary Migration of Hydrocarbons. *American Association of Petroleum Geologists Bulletin*, **66**, 408-429.

Macedonia, M.D. and Maginn, E.J. (1999) A Biased Grand Canonical Monte Carlo Method for Simulating Adsorption Using All-Atom and Branched United Atom Models. *Mol. Phys.*, **96**, 1375-1390.

Mackie, A., Tavitian, B., Boutin, A. and Fuchs, A.H. (1997) Vapour-Liquid Phase Equilibria Predictions of Methane-Alkane Mixtures by Monte Carlo Simulation. *Molecular Simulation*, **1**, 19.

Marteau, P., Tobaly, P., Ruffier-Meray, V. and Hemptinne, J.C. de (1998) High Pressure Phase Diagrams of Methane- + Squalane and Methane + Hexatriacontane Mixtures. *J. Chem. Eng. Data*, **43**, 362-366.

Martin, M.G. and Siepmann, I.J. (1998) Transferable Models for Phase Equilibria 1. United-Atom Description of *n*-Alkanes. *Journal of Physical Chemistry*, B, **102**, 2569.

Martin, M.G. and Siepmann, I.J. (1999) Novel Configurational-Bias Monte Carlo Method for Branched Molecules. Transferable Ppotentials for Phase Equilibria. 2. United-Atoms Description of Branched Alkanes. *Journal of Physical Chemistry*, B, **103**, 4508.

McAuliffe, C.D. (1980) Physical and Chemical Constraints on Petroleum Migration. *Studies in Geology 10. Tulsa, American Association of Petroleum Geologists*, 89-108.

McQuarrie, D.A. (1976) *Statistical Mechanics*, Harper and Collins.

Mehta, M. and Kofke, D.A. (1994) Coexistence Diagrams of Mixtures by Molecular Simulation. *Chemical Engineering Science*, **49**, 2633-2645.

Metropolis, N., Rosenbluth, A.W., Rosenbluth, M.N., Teller, A.H. and Teller, E. (1953) Equation of State Calculations by Fast Computing Machines. *J. Chem. Phys.*, **21**, 1087.

Michelsen, M.L. (1980) Calculation of Phase Envelopes and Critical Points for Multicomponent Mixtures. *Fluid Phase Equilibria*, **4**, 1-10.

Miyano, Y. (1998) Vapor-Liquid Equilibria from Molecular Simulations Using the Algorithm in Equation of State Calculations. *Fluid Phase Equilibria*, **144**, 137-144.

Möller, D., Oprzynski, J., Müller, A. and J. Fischer (1992) Prediction of Thermodynamic Properties of Fluid Mixtures by Molecular Dynamics Simulations: Methane - Ethane. *Mol. Phys.*, **75**, 363.

Münster, A. (1969) *Statistical Thermodynamics*, Berlin, Springer.

Nath, S.A., Escobedo, F.A. and Pablo, J.J. de (1998) On the Simulation of Vapour-Liquid Equilibria for Alkanes. *Journal of Chemical Physics*, **108**, 9905.

Nath, S.K. and Pablo, J.J. de (2000) Simulation of Vapour-Liquid Equilibria for Branched Alkanes. *Molecular Physics*, **98**, 231-238.

Neubauer, B., Boutin, A., Tavitian, B. and Fuchs, A.H. (1999) Gibbs Ensemble Simulation of Vapour-Liquid Equilibria of Cyclic Alkanes. *Molecular Physics*, **97**, 769-776.

Neubauer, B., Delhommelle, J., Boutin, A., Tavitian, B. and Fuchs, A.H. (1999) Monte Carlo Simulations of Squalane in the Gibbs Ensemble. *Fluid Phase Equilibria*, **155**, 167-176.

Neubauer, B., Tavitian, B., Boutin, A. and Ungerer, P. (1999) Molecular Simulations on Volumetric Properties of Natural Gas. *Fluid Phase Equilibria*, **161**, 45-62.

Nichita, D., Goual, L. and Firoozabadi, A. (1999) Wax Precipitation in Gas Condensate Mixtures. *SPE Annual Conference*, Houston.

Nogaret, E. (1983) Solubilité des hydrocarbures dans le gaz naturel comprimé: application à la migration du pétrole dans les bassins sédimentaires. *Thèse*, École nationale supérieure des Mines de Paris.

Pablo, J.J. de, Laso, M. and Suter, U.W. (1992) Simulation of Polyethylene Above and Below the Melting Point. *J. Chem. Phys.*, **96**, 2395.

Panagiotopoulos, A.Z. (1987) Direct Determination of Phase Coexistence Properties of Fluids by Monte Carlo Simulation in a New Ensemble. *Molecular Physics*, **61**, 813-826.

Pandit, S.P. and Kofke, D.A. (1999) Evaluation of a Locus of Azeotropes by Molecular Simulation. *AIChE Journal*, **45**, 2237.

Pedersen, K.S., Fredenslund, A. and Thomassen, P. (1989) *Properties of Oils and Natural Gases*, Gulf.

Péneloux, A., Rauzy, E. and Freze, R. (1982) A Consistent Correction for Redlich-Kwong-Soave Volumes. *Fluid Phase Equilibria*, **8**, 7-23.

Peng, D.Y. and Robinson, D.B. (1976) A New Two-Constant Equation of State. *Industrial Chemical Engineering Fundamentals*, **15**, 59-64.

Petrov, A.A. (1984) *Petroleum Hydrocarbons*, Berlin, Springer.

Potoff, J.J., Errington, J.R. and Panagiotopoulos, A.Z. (1999) Molecular Simulation of Phase Equilibria for Mixtures of Polar and Non-Polar Components. *Mol. Phys.*, **97**, 1073.

Price, L. (1976) Aqueous Solubility of Petroleum Applied to its Origin and Pprimary Migration. *American Association of Petroleum Geologists Bulletin,* **60**, 213-244.

Rainwater, J.C. and Williamson, F.R. (1986) Vapor-Liquid Equilibrium of Near-Critical Binary Alkane Mixtures. *Int. J. of Thermophysics,* **7**, 65-75.

Rosenbluth, M.N. and Rosenbluth, A.W. (1955) Monte Carlo Calculation of the Average Extension of Molecular Chains. *J. Chem. Phys.,* **23**, 356.

Rowley, R.L. (1994) *Statistical Mechanics for Thermophysical Property Prediction,* Prentice Hall.

Ruthven, D.M. (1984) *Principles of Adsorption and Adsorption Processes.* New York, Wiley.

Sant'Ana, H.B. de, Ungerer, P. and Hemptinne, J.C. de (1999) Evaluation of an Improved Volume Translation for the Prediction of Hydrocarbon Volumetric Properties. *Fluid Phase Equilibria,* **154**,193-204.

Smit, B., Karaborni, S. and Siepmann, I.J. (1995) Computer Simulation of Vapor-Liquid Phase Equilibria of *n*-Alkanes. *Journal of Chemical Physics,* **102**, 2126.

Soave, G. (1972) Equilibrium Constants from a Modified Redlich-Kwong Equation of State. *Chemical Engineering Science,* **27**, 1197-1203.

Sportisse, M. (1996) Modélisation de propriétés thermodynamiques des gaz à condensat par représentation de la fraction lourde à l'aide de fonctions de distribution, *Thèse* Université de la Méditerrannée Aix-Marseille II, 197.

Sportisse, M., Barreau, A. and Ungerer, P. (1997) Modeling of Gas Condensates Properties Using Continuous Distribution Functions for the Characterization of the Heavy Fraction. *Fluid Phase Equilibria,* **139**, 255-276.

Spyriouni, T., Economou, I.G. and Theodorou, D.N. (1999) Molecular Simulation od Alpha-olefins Using a New United - Atom Potential Model: Vapor-Liquid Equilibria of Pure Compounds and Mixtures. *J. Am. Chem. Soc.*, **121**, 3407-3413.

Sun, H. (1998) COMPASS: an Ab Initio Force-Field Optimized for Condensed Phase Applications - Overview with Details on Alkane and Benzene Compounds. *J. Phys. Chem.*, B, **102**, 7338-7364.

Sweeney, J., Burnham, A.K. and Braun, R.L. (1987) A Model of Hydrocarbon Maturation from Type I Kerogen: Application to the Uinta basin, Utah. *American Association of Petroleum Geologists Bulletin,* **71**, 967-985.

Tissot, B.P. and Espitalié, J. (1975) L'évolution de la matière organique des sédiments: application d'une simulation mathématique. *Oil & Gas Science and Technology,* **30**, 743-777.

Tissot, B.P. and Welte, D.H. (1984) *Petroleum Formation and Occurrence*, Berlin, Springer.

Tournier, H. (2000) Étude expérimentale and modélisation des équilibres d'adsorption compétitive d'aromatiques en phase liquide sur des faujasites X et Y. Dijon, *Thèse* université de Bourgogne**,** 256.

Toxvaerd, S. (1990) Molecular Dynamics Calculation of the Equation of State of Alkanes. *Journal of Chemical Physics,* **93**, 4290.

Toxvaerd, S. (1997) Equation of State for Alkanes II. *Journal of Chemical Physic,* **107**, 5197.

Ungerer, P., Béhar, E. and Discamps, D., Eds. (1983) Tentative Calculation of the Overall Volume Expansion of Organic Matter During Hydrocarbon Genesis from Geochemistry Data. Implications for Primary Migration. *Advances in Organic Geochemistry,* 1981, Wiley.

Ungerer, P., Bessis, F., Chénet, P.Y., Durand, B., Nogaret, E., Chiarelli, A., Oudin, B. and Perrin, J.F. (1984) Geological and Geochemical Models in Oil Exploration; Principles and Practical Examples. Petroleum Geochemistry and Basin Evaluation. R.J. Murris and G. Demaison eds, *American Association of Petroleum Geologists*, **35,** 53-77.

Ungerer, P., Espitalié, J., Marquis, F. and Durand, B. (1985) Use of Kinetic Models of Organic Matter Evolution for the Reconstruction of Paleotemperatures. Application to the Case of the Gironville Well (France). *Thermal Modeling of Sedimentary Basins*, Carcans, Éditions Technip.

Ungerer, P., Bessereau, G., Junca, J. and Rabiller, P. (1987) Application d'un modèle de migration des hydrocarbures à l'évaluation des permis. *12th World Petroleum Congress*, Houston.

Ungerer, P. and Pelet, R. (1987) Extrapolation of the Kinetics of Oil and Gas Formation from Laboratory Experiments to Sedimentary Basins. *Nature,* **327,** 6117, 52-54.

Ungerer, P., Béhar, F., Villalba, M., Heum, O.R. and Audibert, A. (1988) Kinetic Modelling of Oil Cracking. *Organic Geochemistry,* **13**, 857-868.

Ungerer, P. (1990) State of the Art of Research in Kinetic Modelling of Oil Formation and Expulsion. *Organic Geochemistry,* **16**, 1-25.

Ungerer, P., Burrus, J., Doligez, B., Chénet, P.Y. and Bessis, F. (1990) Basin Evaluation by Integrated Two-Dimensional Modeling of Heat Transfer, Fluid Flow, Hydrocarbon Generation, and Migration. *American Association of Petroleum Geologists Bulletin,* **74,** 3, 309-335.

Ungerer, P., Burrus, J., Doligez, B., Chénet, P.Y. and Bessis, F. (1991) Évaluation des bassins par modélisation intégrée en deux dimensions des transferts thermiques, de l'écoulement des fluides, de la genèse et de la migration des hydrocarbures. *Revue de l'Institut français du pétrole,* **46,** 1, 3-39.

Ungerer, P., Faissat, B., Leibovici, C., Zhou, H., Béhar, E. Moracchini, G. and Courcy, J.P. (1995) High Pressure - High Temperature Reservoir Fluids: Investigation of Synthetic Condensate Gases Containing a Solid Hydrocarbon. *Fluid Phase Equilibria,* **111**, 287-311.

Ungerer, P. and Batut, C. (1997) Prédiction des propriétés volumétriques des hydrocarbures par une translation de volume améliorée. *Oil & Gas Science and Technology,* **52,** 6, 609-623.

Ungerer, P., Batut, C., Moracchini, G., Sant'Ana, H.B. de, Carrier, J. and Jensen, D.M. (1998) Measurement and Prediction of Volumetric and Transport Properties of Reservoir Fluids at High Pressure. *Oil & Gas Science and Technology,* **53,** 3, 265-281.

Ungerer, P., Boutin, A., and Fuchs, A.H. (1999) Direct Calculation of Bubble Points by Monte Carlo Simulation. *Molecular Physics,* **97**, 523-539.

Ungerer, P., Beauvais, C., Delhommelle, J., Boutin, A., Rousseau, B. and Fuchs, A.H. (2000) Optimization of the Anisotropic United Atoms Intermolecular Potential for *n*-Alkanes. *Journal of Chemical Physics,* **112**, 5499-5510.

Ungerer, P., Boutin, A. and Fuchs, A.H. (2001) Direct Calculation of Bubble Points for Alkane Mixtures by Monte Carlo Simulation. *Molecular Physics,* **99**, 1423-1434.

Vlugt, T.J.H., Krishna, R. and Smit, B. (1999) Molecular Simulations of Adsorption Isotherms for Linear and Branched

Alkanes and their Mixtures in Silicalite. *J. Phys. Chem.*, B, **103**, 1102-1118.

Vrabec, J. and Fischer, J. (1995) Vapor-Liquid Equilibria of Mixtures from the NPT Plus Test Particle Method. *Molecular Physics,* **85**, 781.

Welte, D.H. and Yükler, A. (1981) Petroleum Origin and Accumulation in Basin Evolution - A Quantitative Model. *American Association of Petroleum Geologists Bulletin,* **65**, 1387-1396.

Welte, D.H., Horsfield, B. and Baker, D.R. (1997) *Petroleum and Basin Evolution*, Berlin, Springer.

Widom, B. (1963) Some Topics in the Theory of Fluids. *J. Chem. Phys.*, **39**, 2808-2812.

Wick, C.D., Martin, M.G. and Siepmann, J.I. (2000) Transferable Potentials for Phase Equilibria. 4. United-Atom Description of Linear and Branched Alkenes and Alkylbenzenes. *J. Phys. Chem.*, B, **104**, 8008.

Wimmer, E. (1993) Computational Materials Design: a Perspective for Atomistic Approaches. *J. Computer-aided Materials Design,* **1**, 215-242.

Wu, G.W. and Sadus, R.J. (2000) Molecular Simulation of the High-Pressure Phase Equilibria of Binary Atomic Fluid Mixtures Using the Exponential-6 Intermolecular Potential. *Fluid Phase Equilibria,* **170**, 269-284.

Yan, Q. and de Pablo, J.J. (1999) Hyper-Parallel Tempering Monte Carlo: Application to the Lennard-Jones Fluid and the Restricted Primitive Model. *J. Chem. Phys.,* **111**, 9509.

Final manuscript received in January 2003

Oil & Gas Science and Technology – Rev. IFP, Vol. 58 (2003), No. 2, pp. 299-311
Copyright © 2003, Éditions Technip

Gas Isotopes Tracing: an Important Tool for Hydrocarbons Exploration

A. Prinzhofer[1] and A. Battani[1]

1 Institut français du pétrole, 1 et 4, avenue de Bois-Préau, 92852 Rueil-Malmaison Cedex - France
e-mail: alain.prinzhofer@ifp.fr - anne.battani@ifp.fr

Résumé — Les isotopes du gaz utilisés comme traceurs : un outil important pour l'exploration des hydrocarbures — Le gaz naturel peut être considéré aujourd'hui à la fois comme un traceur fiable pour la connaissance des hydrocarbures liquides qui peuvent lui être associés, et comme un enjeu économique en lui même. La mesure isotopique combinée des isotopes stables et des gaz rares donne des contraintes majeures pour reconstruire l'histoire géologique des hydrocarbures, depuis leur genèse jusqu'à leur accumulation. Des progrès récents dans la mesure des isotopes stables du carbone du gaz naturel (du méthane au butane et du dioxyde de carbone), liés principalement à l'utilisation de GC-C-IRMS (Gas Chromatograph-Combustion-Isotopic ratios mass spectrometer) ont permis de retracer de nombreux processus physico-chimiques qui affectent le gaz naturel, au lieu d'utiliser ces signatures comme de simples empreintes digitales d'origines, comme c'était l'usage auparavant. Ces reconstructions fournissent des informations capitales à la fois sur leur origine et sur l'évolution dynamique des fluides hydrocarbures entre la roche mère et l'accumulation dans les réservoirs. L'association de cette méthodologie avec l'utilisation d'autre traceurs naturels augmente notre connaissance de l'histoire des hydrocarbures dans les bassins sédimentaires. Entre autres méthodologies potentielles, les isotopes des gaz rares peuvent apparaître comme l'ultime outil de diagnostic, leur totale inertie chimique nous permettant de les utiliser tant comme des traceurs fiables de sources, que comme des indicateurs quantitatifs de processus physiques (état des phases, migration et fuites au travers des formations géologiques). Qui plus est, comme certains isotopes (^4He, ^{40}Ar par exemple) sont produit par radioactivité naturelle, il représentent des horloges géologiques, donnant accès à la quantification des temps de résidence des hydrocarbures dans les réservoirs. Plusieurs applications nouvelles sont présentées dans cet article, associant les isotopes stables avec ceux des gaz rares : une nouvelle méthode permettant de distinguer l'origine bactérienne ou thermogénique du gaz, et la caractérisation de paramètres en relation avec la genèse du gaz thermogénique (craquage primaire ou craquage secondaire, degré d'ouverture du système, relations entre les signatures isotopiques et la biodégradation). Enfin, nous présentons une tentative de quantification de la proportion d'hydrocarbures ayant fui hors d'un réservoir, et des valeurs relatives des temps de résidence des hydrocarbures dans un système de réservoirs hétérogènes.

Abstract — Gas Isotopes Tracing: an Important Tool for Hydrocarbons Exploration — *Gas may be considered today both as a reliable tracer for the understanding of associated liquid hydrocarbons and as an economic target. Isotopic measurements of stables isotopes and noble gases give important clues to reconstruct the geological history of hydrocarbons from their generation to their accumulation. Recent analytical advances in carbon isotopes of natural gases (methane to butane and carbon dioxide) due mainly to the development of GC-C-IRMS (Gas Chromatograph-Combustion-Isotopic ratios mass spectrometer) have allowed to reconstruct some of the physico-chemical processes which affect natural gas chemical signatures, instead of using these signatures as simple fingerprinting of origins as it was the* *case some time ago. These reconstructions provide important information on both the origins and the dynamic behavior of hydrocarbon fluids between the source rocks and the accumulations in reservoirs. Correlating this methodology with other natural tracers increases the understanding of hydrocarbon history in sedimentary basins. Among other new potential methodologies, noble gas coupled to stable isotopes are the new frontier tool, as their chemical inertia enables their use as precise tracers of sources and of associated physical processes (state of the phases, migration and leakage). Moreover, because some isotopes (^4He, ^{40}Ar for example) are produced by natural radioactivity, they may very well represent geological "clocks", giving potentially a quantification of the residence times of hydrocarbons in a reservoir. Several new applications of this association of stable isotopes and noble gases are presented in this paper: a new way for distinguishing bacterial and thermogenic gas origins, and the characterization of the parameters related to the genesis of thermogenic gas (primary versus secondary cracking, openness of the system, relations between gas isotope signatures and biodegradation). At last is presented a tentative quantification of the proportion of hydrocarbon gases leaked from a reservoir, and a quantitative relative residence times of hydrocarbons in heterogeneous reservoirs.*

INTRODUCTION

The exploration of natural gas became during the last decade as important as the liquid hydrocarbons quest. Although a large part of the R&D work performed for petroleum exploration could be used for the understanding of gas accumulations, it rapidly appeared that some specific problems were to be addressed specifically for the understanding of gas history. Therefore, some geochemical tools could be especially valuable for gas characterization. Natural gas is a relatively simple mixture when compared to oil, as it contains fewer gas molecules in major proportions: the hydrocarbon part consists of methane, ethane, propane, normal-butane and isobutane (the isomers of pentane are already highly fractionated between the liquid and gas phase at a separator, and cannot be generally used as reliable data). The three most common nonhydrocarbon gases are CO_2, N_2 and H_2S. Their concentration ranges from 0 to almost 100%, and they are considered a hazard in gas exploration. Among the other molecules in trace amounts, the whole family of noble gases (He, Ne, Ar, Kr, Xe and Rn) is always present in natural gas, and other compounds as hydrogen or mercury have been sometimes detected. Apart from the chemical proportions of these different molecules, the use of isotopic ratios has been rightfully adapted to oil and gas interpretations. Part of the reason is that of the GC-C-IRMS technique (coupling of a gas chromatograph with a mass spectrometer devoted to stable isotopes measurements) changed drastically the amount and quality of data obtainable from gas samples. Ten years ago, the whole analysis of carbon isotopes of one gas sample took several days, whereas it is now possible on a routine basis in half an hour. Moreover, this online technique allows to analyze much smaller volumes, rendering possible today the analysis of the carbon isotopic signature of an hydrocarbon molecule with an abundance as low as 100 ppm, which was not feasible before. Likewise, very recent technological development of the same kind of online analysis for the hydrogen isotope composition of hydrocarbons will probably give a new geochemical leap for the study of hydrocarbon gases in the next decade. The implementation of noble gases and respective isotope studies are still in a process of new development and research for the purpose of hydrocarbon exploration. However, the wide knowledge of the geochemical behavior of these natural tracers, associated with some pioneer work in oil and gas characterization (Zartmann *et al.*, 1961; Bosch and Mazor, 1988; Ballentine and O'Nions, 1994; Ballentine and Sherwood-Lollar, 2002) indicates that their generalized use is probably of paramount importance for a better understanding of petroleum system.

The reconstruction of the history of a petroleum system involves several steps. Some of them are already well developed with the association of organic geochemistry and basin modeling. However, several other aspects still need a better quantitative estimate: reservoir continuity, distance and rate of migration of the hydrocarbons from the source rocks to the reservoir, and from the petroleum system to the atmosphere, age and residence time of accumulated hydrocarbons, origins and sources of natural gas, prediction of the risk of nonhydrocarbon gases as CO_2 and N_2. A more accurate attention given to the chemical and isotopic signatures of natural gases (hydrocarbons and non-hydrocarbons) may have other industrial implications, for example in the field of CO_2 sequestration, or for the estimation of potential productions of helium as a by-product of gas accumulations. We present in this paper some examples of these new methodologies for several case studies, highlighting the performances for the prediction of quality/volume during exploration, by using stable isotopes associated with noble gas isotopes.

Stable isotopes are natural tracers related to the source signature of gas samples, and fractionated by the chemical and physical processes which affect in a slightly different way the different isotopes of a same chemical compound, with the general rule that lighter isotopes present a higher chemical reactivity, and are more mobile than the heavier ones. Their actual isotopic signature is therefore a superposition of a lot of chemical and physical informations, rendering sometimes difficult their interpretation.

On the other hand, noble gas molecules have the unique property to be absolutely inert chemically, and to allow the tracing of sources and of physical processes of segregation during the history of a petroleum system without any interference with chemical processes (Ballentine and O'Nions, 1994). The second important advantage of noble gases is that they may be classified in two main groups. Some of them are fossil, created during the formation of the solar system, with their main reservoir being the atmosphere, and completely conservative as they are not chemically reactive. These molecules enter hydrocarbon fluids through their solubilization in water in the zones of water recharge, and are then exchanged and fractionated between crustal fluids (gas, oil and water). The other group comprises species produced through natural nuclear reactions, and may accumulate through time, offering potential natural clocks inside the crustal fluids. More in-depth will be discussed in Chapter 4 about radiogenic/nucleogenic production. Interestingly, their large range of molecular mass (from 3 for helium to 136 for xenon) gives a large spectrum of physical properties (solubilities between phases, diffusion coefficients, etc.). These various properties, well constrained through numerous academic studies, may and are potentially used quantitatively for petroleum system evaluation.

Among the numerous applications which are being developed associating carbon isotopes and noble gas geochemistry, one may quote:

- The source of nonhydrocarbon gases like CO_2 and N_2. Indeed, the stable isotopes of these two compounds generally do not give a unique answer for their origins, whereas the isotopic signature of helium $^3He/^4He$ indicates the possible proportion of these molecules coming from the mantle, *versus* those accumulated directly from a continental crust (Battani, 1999; Battani *et al.*, 2000).
- The distinction between a bacterial and a thermogenic origin of natural gas may be confirmed by a completely novel methodology using the apparent production ratio of two radiogenic isotopes of noble gases, ^4He and ^{40}Ar.
- Segregation during the migration of natural gas may be monitored through the carbon isotopes of methane (Prinzhofer and Pernaton, 1997; Prinzhofer *et al.*, 2000a). However, these signatures are strongly dependent upon the genesis of the gas (bacterial *versus* thermogenic) and on the maturity of the source, obscuring the part of the fractionation due to gas migration through porous rocks. On the contrary, because of their inertia, isotopic and chemical ratios of noble gases are fractionated only by physical processes, *i.e.* in this case the interaction between fluids and fractionation during diffusion. The projection of such data in migration maps at the basin scale or at the reservoir scale are coherent with stable isotopes but also much less "noisy".
- As the fossil isotopes of noble gases enter the gas phase from the water phase, their absolute and relative concentrations are directly linked to the amount of water equilibrated with the gas accumulation. This direct estimate of the importance of water contact with hydrocarbons has obvious direct implications for any estimation of a risk of biodegradation or hydrodynamism in a petroleum reservoir.
- The amount of radiogenic noble gases is linked to several key parameters, the amount of parent isotope, the efficiency of accumulation in the hydrocarbon phase and the time. This last parameter is extremely important, as it allows to test for the first time direct chronological information for hydrocarbons accumulations.
- Finally, apart from the application to petroleum exploration, and because of their chemical inertia, noble gases may be used to monitor multi-layered or compartmentalized fields production, the relative proportions produced from each block of the field *versus* time. This kind of new geochemical monitoring of a multi-play gas field may give a precise and less expensive information about the evolution of the production.

We shall now provide more details of some examples of these different applications.

1 BACTERIAL METHANOGENESIS *VERSUS* THERMAL CRACKING: CHARACTERIZATION OF GAS ORIGINS AND NEW AMBIGUITIES

The origins of methanogenesis in hydrocarbon reservoirs have been extensively studied through the last decades (Colombo *et al.*, 1970; Stahl, 1977; Bernard *et al.*, 1977; Schoell, 1980, 1983; Galimov, 1988; Faber *et al.*, 1992; Prinzhofer and Huc, 1995; Prinzhofer *et al.*, 2000a; Prinzhofer *et al.*, 2000b). Even if some unconventional methane generation has been noted to occur through marginal inorganic processes (Charlou and Donval, 1993; Charlou *et al.*, 1998; Guo *et al.*, 1997; Wang *et al.*, 1997; Szatmari, 1989), the main origins of hydrocarbons are organic. They are traditionally subdivided in two classes of processes: one process occurs during the thermal degradation of organic matter, and is associated with the generation of large amounts of heavier gas molecules and liquid hydrocarbons; the other generation process occurs through bacterial methanogenesis with insignificant amounts of other hydrocarbon molecules. The distinction between these two genetic processes is still subject to discussion in several case studies. Bacterial contamination is an important issue when characterizing the gas origin and migration with stable isotopes, as the correct interpretation of the data is sometimes delicate: in the last two decades, it was deemed easy to distinguish gas generated by thermal cracking from methanogenesis using the carbon and hydrogen isotopes (Bernard *et al.*, 1977; Schoell, 1983: Faber *et al.*, 1992). These authors demonstrated that bacterial genesis of gas gave gas enriched in methane, isotopically light compared to a thermal generation. Several diagrams were proposed to illustrate the distinction between a bacterial field of generation and a thermogenic one (in Figure 1 is represented the first of these diagrams, from Bernard *et al.*, 1977). Gases with intermediate compositions were interpreted by these authors as mixes between a bacterial and a thermogenic end-member. The supplementary information obtained with the hydrogen isotopic ratio of methane enables to add a characterization of the biogenic path which occurred, *i.e.* the fermentation of acetates or the reduction of CO_2 (Schoell, 1983; Faber *et al.*, 1992).

However, recent studies have shown that bacterial oxidation of bacterial gas may give a residual gas showing a thermogenic signature, whereas diffusive processes during migration may give a bacterial signature to a thermogenic gas (Prinzhofer and Pernaton, 1997). In Figure 2 are plotted the results of several migration experiments (where the gas migration occurs by diffusion, through a slice of shale or sandstone, saturated or not with water) where it is possible to obtain a purely bacterial signature for methane, or a signature supposedly corresponding to a mixing, when the gas is actually a purely thermogenic gas. This indicates that

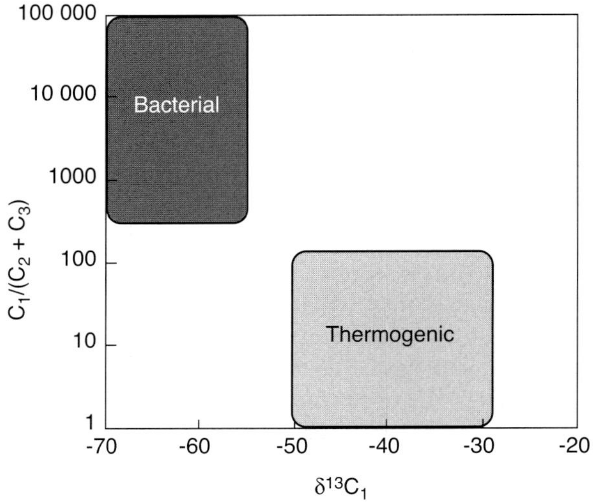

Figure 1

Distinction of the two different organic geneses of natural gas (bacterial *versus* thermogenic). This distinction is based on the dryness of bacterial gas, and on the fact that the carbon isotopic ratio of methane is lighter for bacterial gas than for thermogenic gas. From Bernard *et al.* (1977), modified.

Figure 1 represents the genetic signature of the gases, but that post-genetic processes as chemical and bacterial alteration, or physical migration may change drastically these original fingerprints.

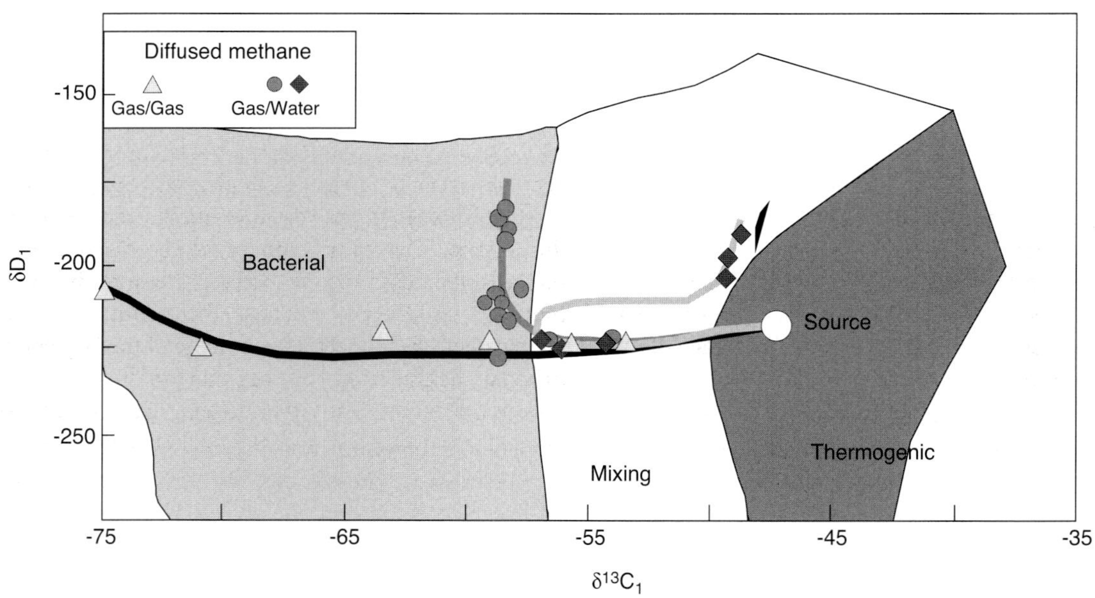

Figure 2

Experimental results of methane migration through shales, saturated or not with water. The migration occurs by water solubilization/diffusion, and the data are presented in a hydrogen *versus* carbon isotopic ratios. The ranges of genetic signatures corresponding to bacterial and thermogenic methane are also represented (from Faber *et al.*, 1992), showing that a purely thermogenic methane may present a bacterial signature (or a mixed one), due to its migration through a porous rock. The series of data correspond to a migration through water saturated shales, but with different petrophysical properties. From Prinzhofer and Pernaton (1997), modified.

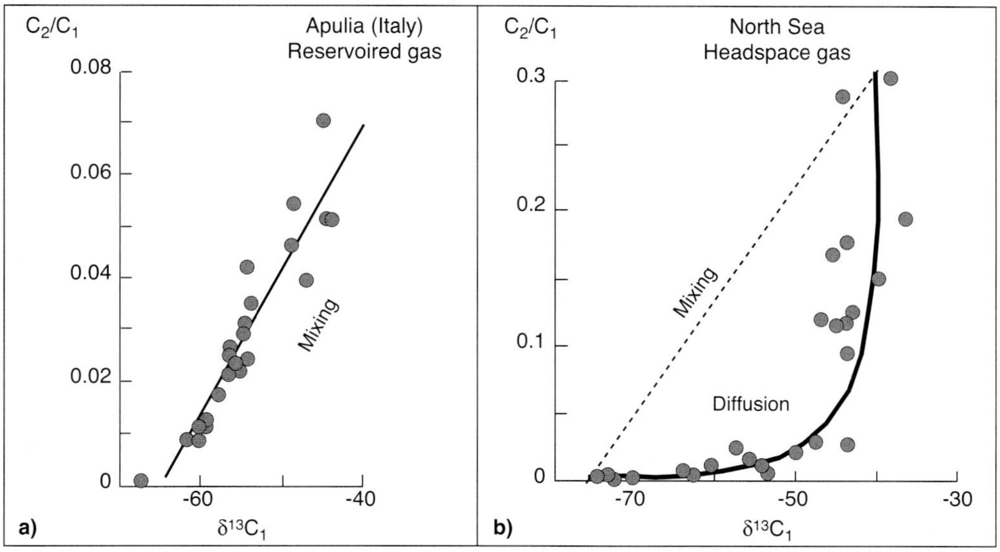

Figure 3

Two examples of gas series in the mixing diagram C_2/C_1 *versus* $\delta^{13}C$ of methane. a) gas data from Apulia (Ricchiuto and Schoell, 1988), showing a straight line interpreted as a mixture between a thermogenic and a bacterial end-member. b) gases from headspace gases collected during a drilling of a North Sea field. The curved trend is incompatible with a mixing between two end-members, and is interpreted as the result of a chemical and isotopic fractionation during the leakage of a thermogenic fluid accumulated in a reservoir located below. From Prinzhofer and Pernaton (1997), modified.

In some cases, it is then possible to distinguish these processes using the principle of the mixing diagrams. Prinzhofer and Pernaton (1997) nonetheless demonstrated that for gases with isotopically light methane (supposedly contaminated by bacterial activity), a mixing process could not explain the observed geochemical trends, and that a fractionation linked to post-genetic processes as migration and adsorption/desorption needed to be invoked. Furthermore, in some other case studies, even with mixing diagrams, the obtained results remained ambiguous. Figure 3 shows data from Apulia (Italy), which are consistent with a mixing scenario between bacterial and thermogenic end-members, and data from headspace gases incompatible with a mixing, and interpreted as due to a fractionation during leakage of the gas out of the reservoir.

The authors have developed very recently a totally independent way of distinguishing these two gas origins (Prinzhofer, 2001), using the ratios of two radiogenic noble gases, ^{4}He and $^{40}Ar^{*}$ (radiogenic part of total ^{40}Ar), as they are not affected by chemical secondary processes. Additionnally, physical processes should not fractionate helium and argon in the range of the observed variations. As these molecules are generated from the mineral network, their ratio in a gas phase depends only on the production ratio in the rocks and the physical fractionation during their transfer from the solid to the gas phase. As helium diffuses more easily than argon, the ratio $^{4}He/^{40}Ar^{*}$ is expected higher in bacterial gas, as the generation temperature is low enough to forbid the major part of argon to leave the solid (Elliot *et al.*, 1993). On the contrary, the higher generation temperatures of thermogenic gas promote a very effective transfer of both molecules, and the $^{4}He/^{40}Ar^{*}$ ratio measured in these gases is always close to the average crustal production ratio. Several well-documented natural examples, with unambiguous origins of the gases, validate this new way of characterizing gas generation *(Fig. 4)*. However, it becomes clear with such examples that the gap supposedly existing between the bacterial generation and the thermal one diminishes. The estimation of newly discovered hidden biomass, the possibility for some bacteria to live in extreme thermal conditions pushes the limits of bacterial methanogenesis probably in the range of the beginning of the thermal degradation of organic matter. The ancient observation of an "early thermogenic gas" which would be dry and non associated with oils may be reconciled with new observations concerning the behavior of reduced carbon cycles in a sedimentary column.

2 ESTIMATION OF THE MATURITY OF A GAS: PRIMARY *VERSUS* SECONDARY CRACKING. POSSIBLE SECONDARY BIODEGRADATION

As the proportions and isotope ratios of methane are highly subjected to post-genetic processes, a new attempt for characterizing the maturity of the gas was done using the

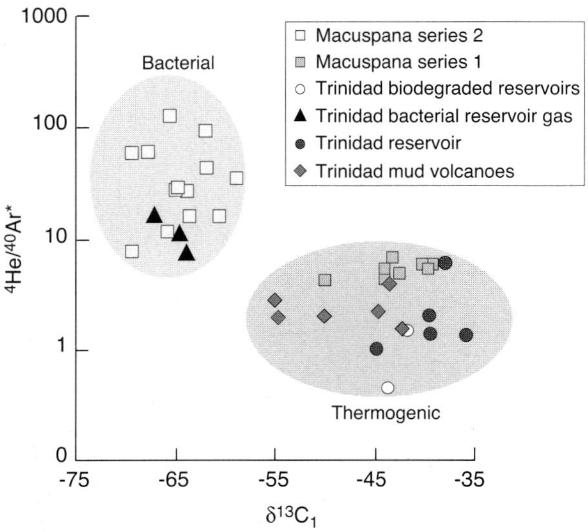

Figure 4

Distinction of thermogenic and bacterial origin of natural gas, combining the use of carbon isotopes and noble gases. The gases presented in this diagram have unambiguous bacterial or thermogenic origins, without any mixing or post-genetic fractionation. Bacterial gases have light carbon isotopic signatures, and high apparent production ratios $^4He/^{40}Ar^*$ compared to the average crustal value. Data from Prinzhofer *et al.* (2000c) and unpublished data.

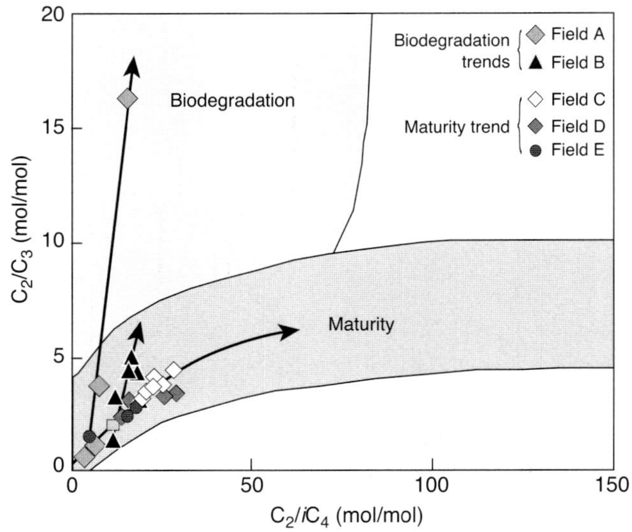

Figure 5

Diagram C_2/C_3 *versus* C_2/iC_4 (Prinzhofer *et al.*, 2000b), distinguishing a maturity trend from a biodegradation trend. Some natural gas examples are shown exhibiting both processes. Field A, gases have a lower maturity than field B, both exhibiting a trend of biodegradation respectively.

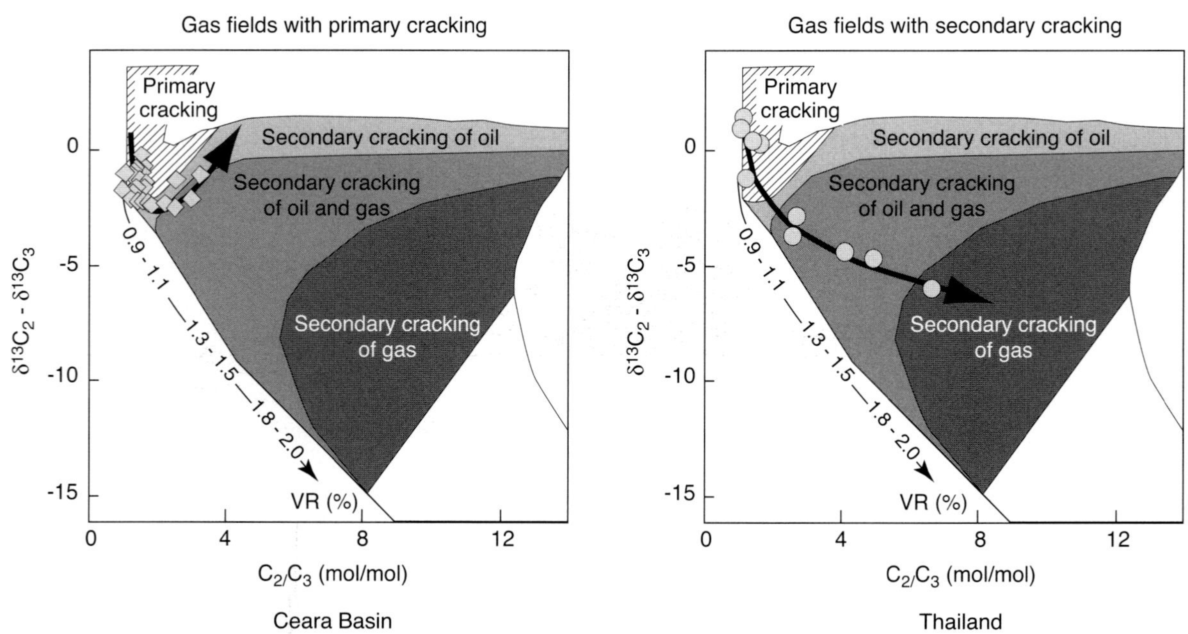

Figure 6

$\delta^{13}C_2$- $\delta^{13}C_3$ *versus* C_2/C_3 diagram (Lorant *et al.*, 1998), with two examples of gas series from the Ceara Basin, Brazil (Prinzhofer *et al.*, 2000b) and from Thailand. The first case presents a trend of maturity of an open system during the primary cracking of kerogen, the second case presents a wider range of maturity in a more closed system, ranging from the primary cracking to secondary cracking of oil and gas.

chemical and isotopic signatures of the C_2-C_4 fraction of the gases (Lorant *et al.*, 1998; Prinzhofer *et al.*, 2000a; Prinzhofer *et al.*, 2000b). Two main diagrams were established, the first one (C_2/C_3 *versus* C_2/iC_4) allowing to distinguish any gas biodegradation *versus* a maturity trend. Its principle is that for a maturity trend, iC_4 proportions decrease more rapidly than C_3 proportions. On the contrary, during a biodegradation process, propane (and nC_4) are more efficiently altered than iC_4. Figure 5 gives some examples of gas signatures plotted in this diagram. It is seen that for a given basin, gases from some reservoirs exhibit a trend of maturity, whereas for some other hydrocarbon accumulations, a trend of biodegradation may be defined.

An other diagram using both the chemical and isotopic signatures of ethane and propane was developed in order to better characterize the maturity of the gas. For correctly interpreting data, an other important parameter to take into account is the degree of openness of the system, as the maturity trends depend drastically on the retention of the hydrocarbons in the generation area or not. Figure 6 presents gas data from two different basins, one corresponding to gas generated during the primary cracking of the kerogen, the other one where a large proportion of the gases comes from the secondary cracking of the liquid hydrocarbons. This efficiency of accumulation is primordial when characterizing the composition of accumulated fluids. Indeed, as shown in Figure 7, for the same source rock and for the same maturation history, a reservoir accumulating all the produced hydrocarbons will be filled with oil and gas, whereas a reservoir accumulating only in the last part of the generated hydrocarbons may be filled only with dry gas.

3 GAS AS A TRACER OF MIGRATION: STABLE ISOTOPES *VERSUS* NOBLE GAS SIGNATURES

Segregation during the migration of natural gas may be monitored through the carbon isotopes of methane (Prinzhofer and Pernaton, 1997; Prinzhofer *et al.*, 2000a). As presented earlier, the carbon isotopes of methane fractionate during migration with an enrichment in the light isotope ^{12}C. This observation obscures the diagnostic on the origins of gases, but may help in the same time the reconstruction of the directions of migration. However, as said before, these signatures are strongly dependent on the kind of genesis of the gas (bacterial *versus* thermogenic), on the isotopic signature of the source and of its maturity, complexing the fractionation due to gas migration through porous rocks.

On the contrary, because of their chemical inertia, isotopic and chemical ratios of noble gases are fractionated only by physical processes, *i.e.* in this case the interaction between fluids and fractionation during diffusion. The ratios between two fossil isotopes of noble gases (of two different chemical species) are controlled by the atmospheric ratio, the relative

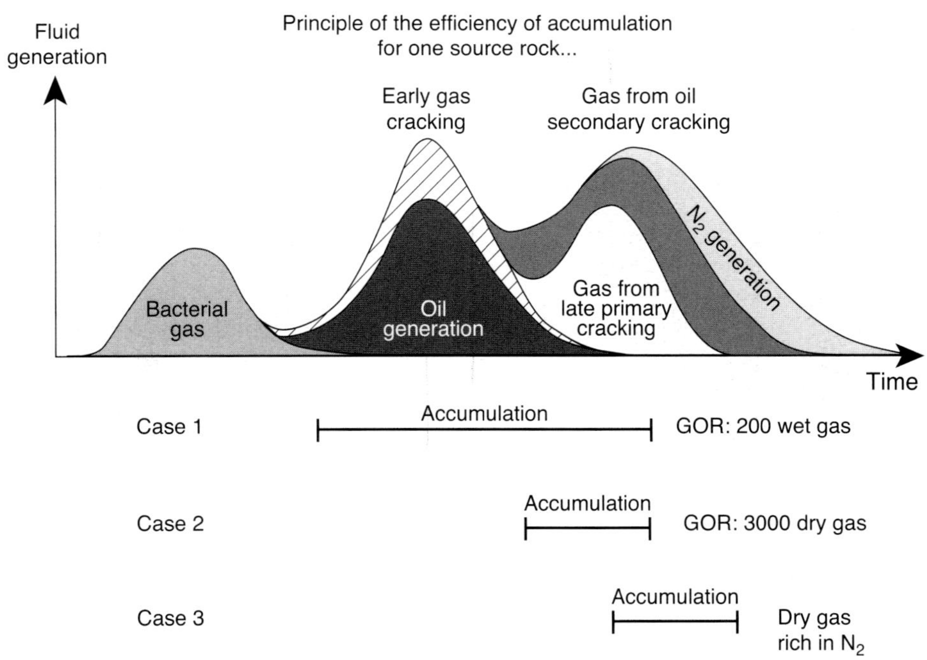

Figure 7

General scheme of hydrocarbon and N_2 generation from organic matter. The range of accumulation time has a direct influence on the composition of reservoired hydrocarbons, a late accumulation being enriched in gas as compared to a complete accumulation. For overmature source rocks, a late accumulation may lead to a gas highly enriched in nitrogen.

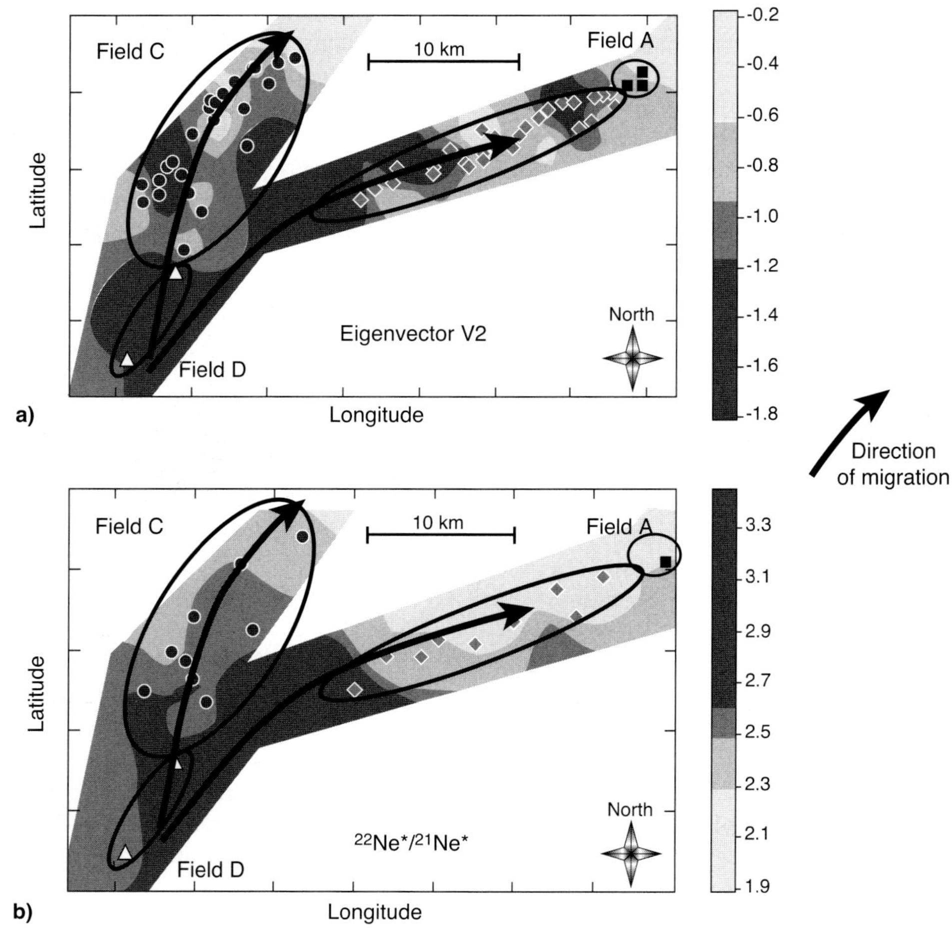

Figure 8

Migration maps in a series of reservoirs of a Brazilian basin (Prinzhofer *et al.*, 2000d), using (a) the PCA method based on the proportion and the carbon isotopic ratios of hydrocarbons (Prinzhofer *et al.*, 2000a) and (b) the isotopic ratios of the nucleogenic part of ^{21}Ne and ^{22}Ne (Prinzhofer *et al.*, 2000d). Both tracers indicate a direction of migration from the southwest to the northeast.

solubilities of noble gas in water in the aquifer recharges, and to fractionation processes due to migration in sedimentary rocks. For radiogenic and nucleogenic isotopes, their relative ratios are controlled by the average crustal production ratio and by the transfer function from the rocks to the hydrocarbons. It is then possible to characterize the direction of migration with the same physical principle, which implies that a light molecule moves faster than a heavy one. The fractionation may occur for both fossil and radiogenic isotopes, indicating a late fractionation. On the contrary, if only the fossil isotopes are fractionated, this means that the differentiation occurred before the integration of radiogenic isotopes into the hydrocarbons. In Figure 8 are presented two maps of fractionation due to migration. Figure 8a represents a fractionation linked mainly to the carbon isotopic ratios of methane, quantified with the second eigenvector of a PCA, mainly linked with the carbon isotope composition of methane (Prinzhofer *et al.*, 2000a). An enrichment in ^{12}C from the southwest to the northeast indicates that the pod of generation is likely located southwest of the reservoirs. When comparing such results with a ratio of two nucleogenic isotopes ^{21}Ne* and ^{22}Ne* *(Fig. 8b)*, the same relative enrichment in light isotopes is visible to the northeast of the studied area, confirming the direction of accumulation of hydrocarbons.

In other cases, mixture with bacterial methane will obscure the fractionation of carbon isotopes due to migration, while noble gases will not be affected by this hydrocarbon mixing process, and will give a consistent direction of migration. Figure 9 represents in the basin of Macuspana (Mexico) the fractionation of ratios of fossil and radiogenic isotopes respectively (Prinzhofer *et al.*, 2000c). A consistent direction of migration from south to north is visible, although the values of δ^{13}C of methane do not give a clear clue about migration because of an important mixing between a bacterial and a thermogenic end-member (Prinzhofer *et al.*, 2000c). Figure 9 c and 9d show that the projection of the iso- δ^{13}C of methane and ethane are similar, although some shift does

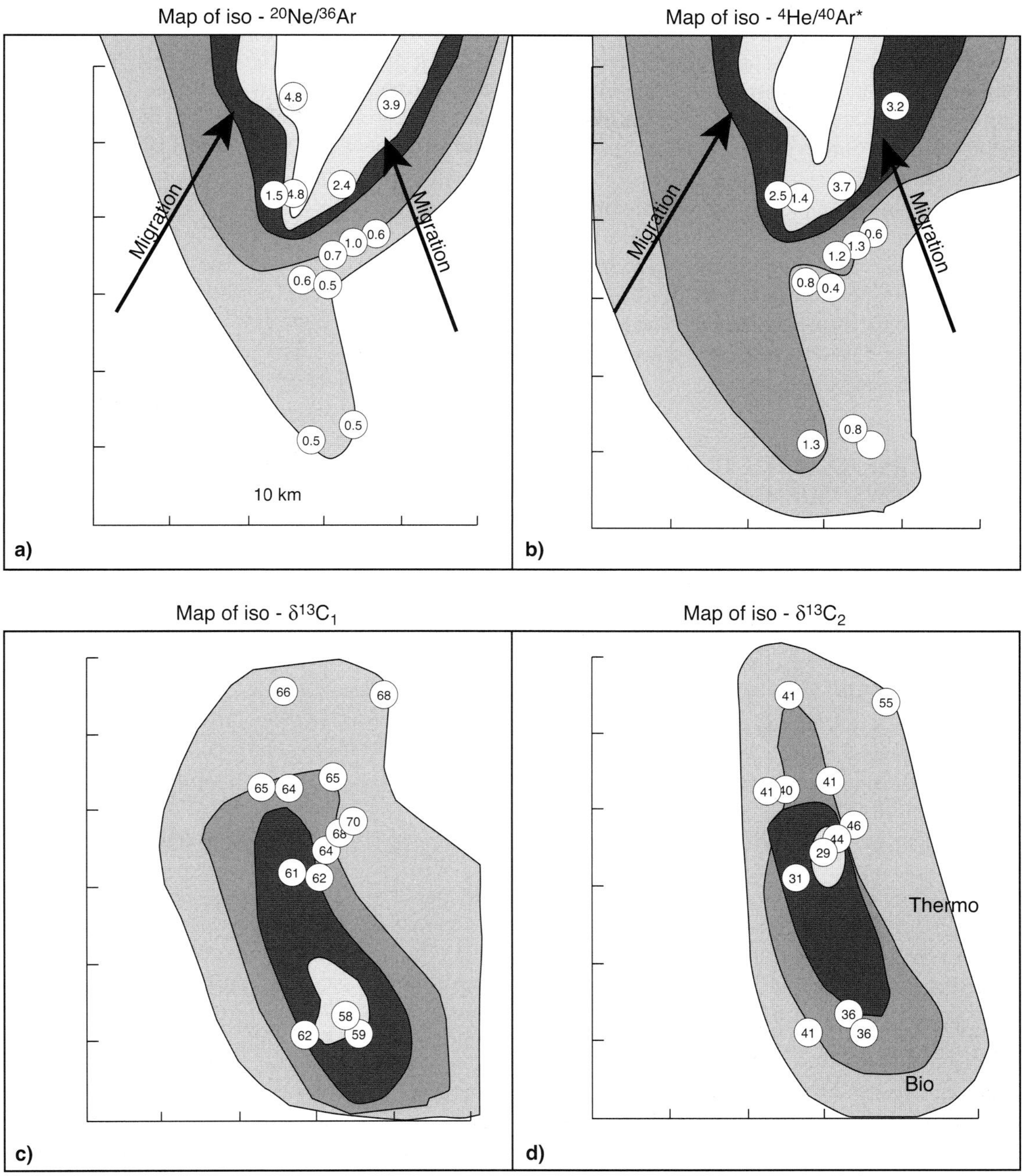

Figure 9

Maps of several gas geochemical parameters in the Macuspana Basin, Mexico (Prinzhofer *et al.*, 2000c).

a) map of the iso-values of the ratio of two fossil noble gas isotopes $^{20}Ne/^{36}Ar$.

b) map of the iso-values of the ratio of two radiogenic noble gas isotopes $^{4}He/^{40}Ar^{*}$.

c) map of the iso-values of the carbon isotopic ratio of methane.

d) map of the iso-values of the carbon isotopic ratio of ethane.

The four maps indicate a consistent direction of migration from the south to the north.

persist between the positions of the less negative $\delta^{13}C$. These images were interpreted (Prinzhofer *et al.*, 2000c) as due to a mixing between a bacterial gas impregnated with a small amount of thermogenic gas, whose proportion increases from the edge to the center of the studied area. The shift between the two maximum of $\delta^{13}C$ between methane and ethane is further interpreted as due to a higher fractionation of methane because of longer migration. As the heaviest methane is shifted south of the studied area, this would indicate a heavier residue of leakage located south, *i.e.* a migration from the south to the north, in perfect agreement with the fractionations of the noble gas isotopes. The combined approach of $\delta^{13}C$ and noble gas demonstrates the importance of gain of information from all these gaseous species.

4 NEW CHALLENGES: RESIDENCE TIME OF HYDROCARBONS, QUANTIFICATION OF LEAKAGE OUT OF THE RESERVOIRS

As was assessed in the introduction, noble gases (He, Ne, Ar, Kr, Xe) contain isotopes generated through natural nuclear reactions occurring in the minerals. Some isotopes are radiogenic, *i.e.* generated by the radioactive decay of a radioactive isotope. The others are called nucleogenic, and their generation occurs through nuclear reactions as (α,n) or (n,α) on nuclei of Li, Mg, F, O, etc. The absolute amount of generated stable radiogenic and nucleogenic isotopes is increasing linearly with time, for compatible time ranges involved in petroleum systems (several hundred of million years). For a given radiogenic or nucleogenic isotope, a part of it comes from generation occurring before the history of the petroleum system, and presents an atmospheric proportion when compared to the same isotopes of the same elements. This makes possible the distinction, for a given concentration of a so-called radiogenic or nucleogenic isotope, of a "fossil" part and of a "radiogenic" part, if an other isotope of the same element is only fossil. Taking the example of argon, as ^{36}Ar is only fossil, it is possible to calculate the concentration of fossil ^{40}Ar ($^{40}Ar_o$), knowing the atmospheric ratio $^{40}Ar/^{36}Ar$ equal to 295.5:

$$^{40}Ar_o = 295.5 \cdot {}^{36}Ar \quad (1)$$

and the concentration of the radiogenic ^{40}Ar ($^{40}Ar^*$):

$$^{40}Ar^* = {}^{40}Ar - 40Ar_o = {}^{40}Ar\,(1\text{-}295.5 \cdot {}^{36}Ar) \quad (2)$$

As both isotopes, fossil and radiogenic/nucleogenic, are rigorously the same physical element (*i.e.* will not suffer any physical or chemical fractionation as they are physically undistinguishable), the ratio of radiogenic (or nucleogenic) over fossil concentration is linearly linked to the residence time of the hydrocarbon fluid. The coefficient of proportionality is not constrained yet, but for a given geological situation, as for example, several fields that produce with the same geological interval and with hydrocarbons originating from the same source rocks, it is possible to quantify in a relative way the residence times of hydrocarbon accumulations. Figure 10a shows that for the same example than illustrated and discussed on Figure 8, the gases from the two fields A and B have similar ratios in a diagram $^{40}Ar^*$ *versus* $^{40}Ar_o$, and their different values define a straight line passing through the origin. The gases from fields C and D define an other straight line with a slope presenting a larger value. The ratio between the two slopes equals 1.34. Looking at other nucleogenic isotopes as ^{21}Ne and ^{22}Ne for the same gas wells, the same kind of diagram is visible, with ratios between the two linear correlation trends of 1.30 and 1.44 respectively *(Figs 10b and c)*. The good consistency between these three values of slope ratios confirms the quantitative estimate of the relative residence time of hydrocarbons in these reservoirs, the hydrocarbons of reservoirs C and D having a residence time approximately 35% longer than in the reservoirs A and B.

Another new powerful application of noble gas isotopes consists on using them as inert tracers of leakage from the reservoirs through caprocks or aquifers. In fact, it is common sense to say that helium leaks much more easily than other gases because of its size (except hydrogen). As the noble gas family corresponds to a series of gas molecules from helium to xenon with increasing size, the latest being a large molecule which does not migrates easily, it is possible to trace a leakage phenomenon looking either at concentrations of different noble gas compounds, or at ratios between two noble gas compounds of different size, the smaller moving out more rapidly than the larger. An increase of a "leaking" process should result in a decrease of the helium concentrations, and an increase of the ratios (heavy compound/light compound). Figure 11 shows for the same four fields of Figures 8 and 10, some correlation diagrams, plotting a ratio of two fossil noble gas isotopes *versus* the concentration of ^4He. As the concentrations of ^4He decrease, all the ratios heavy/light compounds increase. The two trends for each diagram correspond to the two families of fields, A-B and C-D respectively. It has been noticed that the absolute concentrations of argon are relatively constant in these families, whereas the concentrations of helium and neon are decreasing *versus* the importance of accumulation leakage, while the concentrations of krypton and xenon are increasing. This indicates that the leakage factor for the hydrocarbons is in the same order than the leakage of argon, as the argon concentrations —constant—correspond to the ratios of the absolute amount of argon over the whole gas amount. It is then possible to quantify the proportions of hydrocarbons lost through leakage for each gas sample, fitting the curves of Figure 11 with different diffusivities for helium, neon, argon, krypton and xenon (for the noble gas ratios), and using the same diffusivity for argon and for hydrocarbons

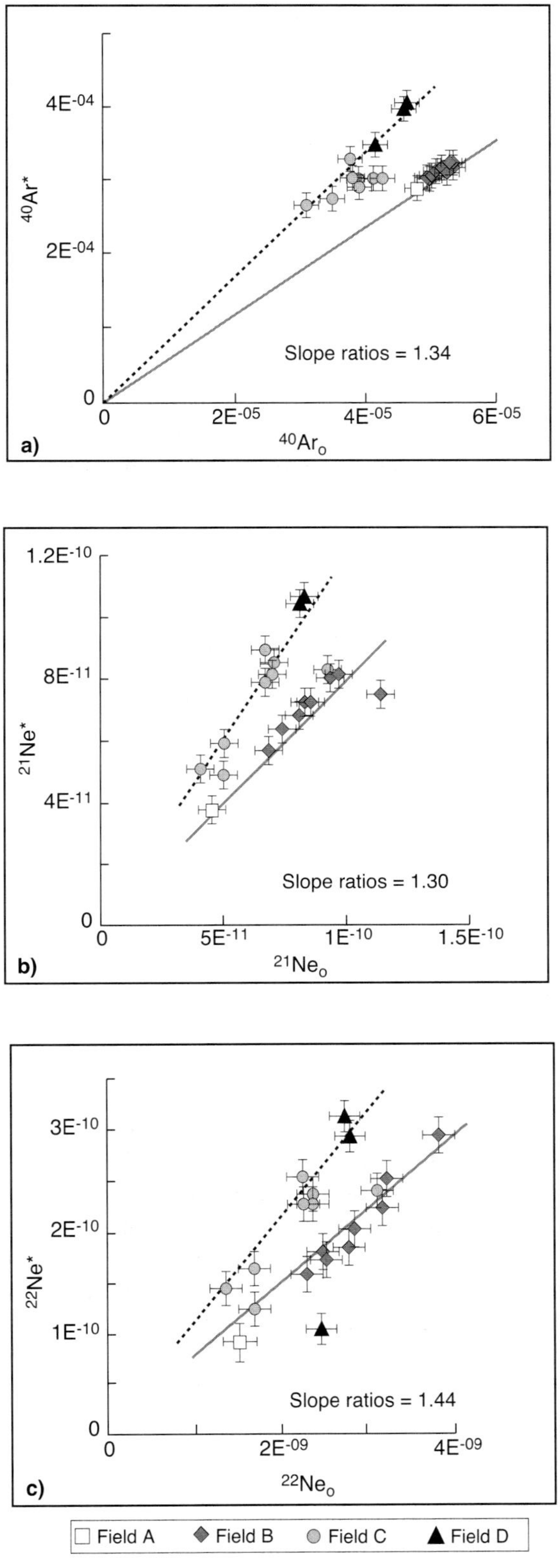

Figure 10

Correlation between the fossil and the radiogenic/nucleogenic part of the isotopes ^{40}Ar, ^{21}Ne and ^{22}Ne for the four fields of Figure 8. The ratios X^*/X_o are proportional to the residence time of the hydrocarbons.

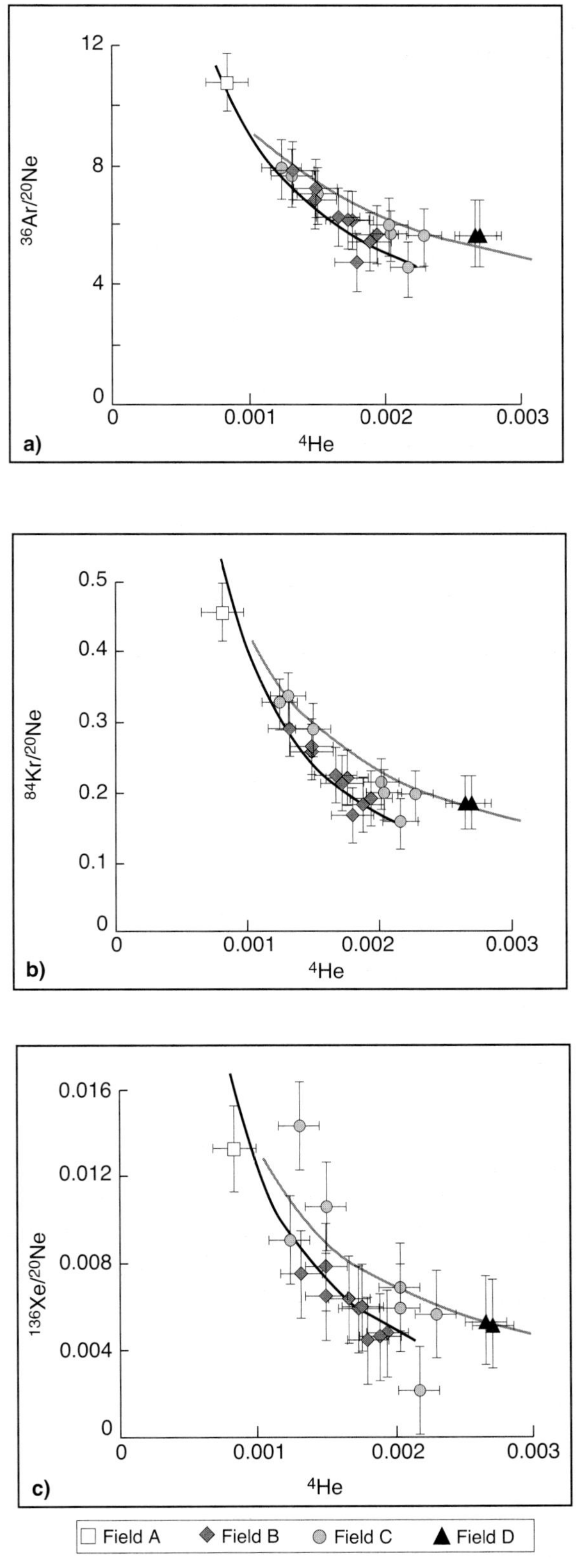

Figure 11

Correlations between the ratios of two fossil isotopes of noble gases and the concentrations of the radiogenic ^{4}He for the four fields from Figure 8. The observed trends are interpreted as due to leakage, which decreases the ^{4}He concentrations and increases the ratios heavy/light isotopes.

for the helium concentrations. The result of this calculation is presented in the map of Figure 12. The proportions of displaced hydrocarbons range between 16 and 80%, and their repartition indicates that fields B and D exhibit smaller leakage than fields A and C.

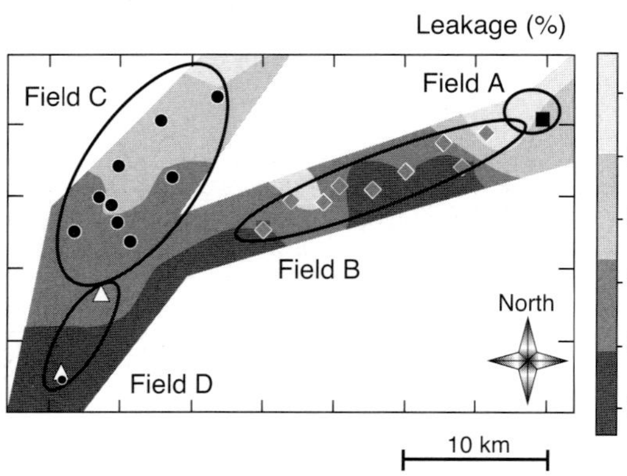

Figure 12

Map of the quantified proportion of hydrocarbons lost by leakage for the four fields of Figure 8.

CONCLUSIONS

From the recent developments in the geochemistry of natural gases, it appears that the use of carbon isotopes, extended to the whole range of gas molecules, gives an evaluation of the physico-chemical processes affecting hydrocarbons in sedimentary basins, including the source of the hydrocarbons, their genesis and their post-genetic history as their migration, accumulation, leakage and chemical alteration (bacterial and/or thermal). The recent development of noble gas isotopes tracing of these fluids gives a more accurate view of some of these processes, as the type of genesis and the secondary and tertiary migration of hydrocarbons. Some very new quantitative informations about the proportions of displaced hydrocarbons, and the residence time of hydrocarbons in reservoirs is still under development, giving to gas geochemistry new challenges for future applications outside oil and gas exploration, as for example the monitoring of hydrocarbon production, the monitoring of CO_2 sequestration in aquifers or in depleted hydrocarbon reservoirs.

REFERENCES

Ballentine, C.J. and O'Nions, R.K. (1994) The Use of Natural He, Ne and Ar Isotopes to Study Hydrocarbon-Related Fluid Provenance, Migration and Mass Balance in Sedimentary Basins. From Parnell J. (ed.) 1994, *Geofluids: Origin, Migration and Evolution of Fluids in Sedimentary Basins. Geological Society, Special Publication*, 78, 347-361.

Ballentine, C.J. and Sherwood-Lollar, B.S. (2002) Regional Groundwater Focusing of Nitrogen and Noble Gases into the Hufoton-Panhandle Giant Field, USA. Geochim. *Cosmochim. Acta*, **66**, 14, 2483-2497.

Battani, A. (1999) Utilisation des gaz rares He, Ne et Ar pour l'exploration pétrolière et gazière : exemple des accumulations du Pakistan et de Macuspana (Mexique). *PhD*, Paris University.

Battani, A., Sarda, Ph. and Prinzhofer, A. (2000) Geochemical Study of Pakistani Natural Gas Accumulations Combining Major Elements and Rare Gas Tracing. *Earth and Planet. Sci. Lett.*, **181**, 229-249.

Bernard, B.B., Brooks, J.M. and Sackett, W.M. (1977) A Geochemical Model for Characterization of Hydrocarbon Gas Sources in Marine Sediments. In *Offshore Technology Conference*, May 1977, OTC 2934, 435-438.

Bosch, A. and Mazor, E. (1988) Natural Gas Association with Water and Oil as Depicted by Atmospheric Noble Gases: Case Study from the Southeastern Mediterranean Coastal Plain. *Earth Planet. Sci. Letters*, **87**, 338-346.

Charlou, J.L. and Donval, J.P. (1993) Hydrothermal Methane Venting between 12°N and 26°N along the Mid-Atlantic Ridge. *Journal of Geophys. Research*, **98**, B6, 9625-9642.

Charlou, J.L., Fouquet, Y., Bougault, H., Donval, J.P., Étoubleau, J., Jean-Baptiste, Ph., Dapoigny, A., Appriou, P. and Rona, P.A. (1998) Intense CH_4 Plumes Generated by Serpentinization of Ultramafic Rocks at the Intersection of the 15°20'N Fracture Zone and the Mid-Atlantic Ridge. *Geochim. Cosmochim. Acta*, **62**, 13, 2323-2333.

Colombo, U., Gazzarrini, F., Gonfiantini, R., Kneuper, G., Teichmüller, M. and Teichmüller, R. (1970) Carbon Isotope Study on Methane from German Coal Deposits. Hobson, G.D., Speers, G.C. (Eds.), *Advances in Organic Geochemistry*, 1966, Pergamon Press, Oxford, 1-26.

Elliot, T., Ballentine, C.J., O'Nions, R.K. and Ricchiuto, T. (1993) Carbon, Helium, Neon and Argon Isotopes in a Po Basin (Northern Italy) Natural Gas Field. *Chemical Geology*, **106**, 429-440.

Faber, E., Stahl, W.J. and Whiticar, M.J. (1992) *Distinction of Bacterial and Thermogenic Hydrocarbon Gases: Bacterial Gas*, R. Vially (Ed.), 63-74.

Galimov, E.M. (1988) Sources and Mechanisms of Formation of Gaseous Hydrocarbons in Sedimentary Rocks. In *Origins of Methane in the Earth, Chem. Geol. (Special Issue)*, Schoell, M. (Ed.), **71**, 77-95.

Guo, Zhanqian, Wang, Xianbin and Liu Wenlong (1997) Reservoir-Forming of Abiotic Gas in Songliao Basin. *Science in China*, Series D, **40**, 6, 621-626.

Lorant, F., Prinzhofer, A., Béhar, F. and Huc, A.Y. (1998) Isotopic (^{13}C) and Molecular Constraints on the Formation and the Accumulation of Thermogenic Hydrocarbon Gases. *Chemical Geology*, **147**, 249-264.

Prinzhofer, A. and Huc, A.Y. (1995) Genetic and Post-Genetic Molecular and Isotopic Fractionations in Natural Gases, *Chemical Geology*, **126**, 3-4, 281-290.

Prinzhofer, A. and Pernaton, E. (1997) Isotopically Light Methane in Natural Gases: Bacterial Imprint or Segregative Migration? *Chemical Geology*, **142**, 193-200.

Prinzhofer, A., Mello, M.R. and Takaki, T. (2000a) Geochemical Characterization of Natural Gas: a Physical Multivariable Approach and its Applications in Maturity and Migration Estimates. *AAPG Bull.*, **84**, 8, 1152-1172.

Prinzhofer, A., Mello, M.R., Da Silva Freitas, L.C. and Takaki, T. (2000b) New Geochemical Characterization of Natural Gas and its Use in Oil and Gas Evaluation. In *Petroleum systems of South*

Atlantic Margins, *AAPG Memoir 73*, M.R. Mello and B.J. Katz (Eds.), 107-119.

Prinzhofer, A., Guzman-Vega, M.A., Battani, A and Escudero, M. (2000c) Gas Geochemistry of the Macuspana Basin, Mexico: Thermogenic Accumulations in Sediments Impregnated by Bacterial Gas. *Marine and Petroleum Geology*, **17**, 1029-1040.

Prinzhofer, A., Vaz Dos Santos Neto, E., Magnier, C., Takaki, T. and Mello, M.R. (2000d) Geochemical Signatures of Gas and Oils in an Amazonian Accumulation. A dynamic Reconstruction of Hydrocarbon Filling and Leakage of a Reservoir. *7th ALAGO Meeting*, 26-28 October 2000 (extended abstract).

Prinzhofer, A. (2001) Distinguishing Thermogenic and Biogenic Gases: the Past (Fingerprinting), the Present (Stable Isotopes and Geological Processes) and the Future (Noble Gas Isotope Signatures). *AAPG Hedberg conference*, Veracruz, Mexico, 4-7 Nov. 2001 (extended abstract).

Ricchiuto, T. and Schoell, M. (1988) Origin of Natural Gases in the Apulian Basin in South Italy: a Case history of Mixing of Gases of Deep and Shallow Origin. In *Advances in Organic Geochemistry*, Matavalli, L. Novelli, L. (Eds.), 311-318.

Schoell, M. (1980) The Hydrogen and Carbon Isotopic Composition of Methane from Natural Gases of Various Origins. *Geochimica Cosmochimica Acta*, **44**, 649-661.

Schoell, M. (1983) Genetic Characterization of Natural Gases, *AAPG Bulletin*, **67**, 12, 2225-2238.

Stahl, W.J. (1977) Carbon and Nitrogen Isotopes in Hydrocarbon Research and Exploration. *Chemical Geology*, **20**, 121-149.

Szatmari, P. (1989) Petroleum Formation by Fisher-Tropsch Synthesis in Plate Tectonics. *AAPG Bulletin*, **73**, 989-998.

Wang Xianbin, Li Chunyuan, Chen Jianfa, Xia Xinyu, Guo Zhanqian and Xie Hongsen (1997) On Abiogenic Natural Gas. *Chinese Science Bulletin*, **42**, 16, 1327-1337.

Zartman, R.E., Wasserburg, G.J. and Reynolds, J.H. (1961) Helium, Argon and Carbon in some Natural Gases. *Journ. of Geophys. Resarch*, **66**, 277-306.

Final manuscript received in February 2003

Oil & Gas Science and Technology – Rev. IFP, Vol. 58 (2003), No. 2, pp. 313-324
Copyright © 2003, Éditions Technip

Basin Modeling in Complex Area: Examples from Eastern Venezuelan and Canadian Foothills

F. Schneider[1]

1 Institut français du pétrole, 1 et 4, avenue de Bois-Préau, 92852 Rueil-Malmaison Cedex - France
e-mail: frederic.schneider@ifp.fr

Résumé — Modélisation de bassin en zones complexes : exemples des piémonts vénézuéliens et canadiens — L'objectif de la modélisation de bassin est de reconstruire l'histoire géologique d'un bassin sédimentaire et de ses systèmes pétroliers afin de pouvoir prédire l'existence d'accumulations. Les modèles de bassin prennent en compte la compaction, les transferts de chaleur, la génération des hydrocarbures et la circulation des fluides. Cependant, les modèles de bassin classiques ne sont pas utilisables dans les zones de piémont où la géométrie est complexe. C'est pour répondre à ce besoin que le prototype Ceres a été développé. Ce prototype est capable de simuler dans une section 2D des écoulements triphasiques pour un bassin dont la géométrie évolue au cours du temps sous l'effet de la sédimentation, de la compaction, des érosions, de la tectonique salifère et des déplacements des blocs le long des failles.

Une étude est généralement constituée de trois étapes principales. La première consiste à construire la section à l'époque actuelle. Ceci est généralement effectué à partir de l'interprétation sismique, des données de puits, des observations d'affleurements et des analyses de carottes. Durant cette étape, la section est généralement équilibrée en utilisant un logiciel tel que Locace.

La seconde étape consiste à la restauration de la section. La section à l'époque actuelle est restaurée progressivement dans le passé pour chacune des couches définies jusqu'à ce que le substratum soit atteint. Pour équilibrer les sections au cours du temps et pour contraindre les parties érodées, il est parfois nécessaire d'utiliser un logiciel cinématique direct comme Thrustpack. Ce prototype nous permet alors de construire des sections intermédiaires consistantes d'un point de vue cinématique.

La troisième étape est consacrée aux simulations directes. Pour résoudre les équations utilisées dans les modèles de bassins, nous avons dû développer des méthodes numériques originales qui mettent en œuvre des techniques de décomposition de domaines.

La méthodologie précédente est le fruit de tests effectués sur une coupe en Alberta (Canada) et une autre au Venezuela. Si l'on considère l'origine des fluides dans les grès d'âge Oligocène de la structure de El Furrial ou bien dans les dolomies d'âge Mississippien des piémonts canadiens, on reconnaît un épisode commun d'expulsion de fluides le long de la stratification. Avant la phase de flexure et de chevauchement, les fluides sont essentiellement expulsés vers la surface sous l'effet de la compaction. Ensuite, à cause de la flexure et des chevauchements, les fluides sont poussés le long de la stratification avec des vitesses qui atteignent plusieurs dizaines de kilomètres par million d'années. Cet épisode d'expulsion de fluide s'arrête à un endroit donné quand les failles deviennent actives en cet endroit.

Abstract* — *Basin Modeling in Complex Area: Examples from Eastern Venezuelan and Canadian Foothills — *Basin modeling aims at reconstructing the time evolution of a sedimentary basin in order to make quantitative predictions of geological phenomena leading to hydrocarbons accumulations. It accounts for porous medium compaction, heat transfer, hydrocarbon generation and fluid flow. Nevertheless, classical basin models handle simple geometry and are not usable in complex geometry area such as foothills. This is why, a numerical prototype Ceres, has been developed. This prototype is able to model three-phase flow in a 2D section of a basin, whose geometry changes through time accounting for deposition, compaction, erosion of the sediments, salt or mud creeping, and block displacement along faults.*

The classical flow chart to perform a case study is composed of three main steps. The first step is the building of the present day section. This is generally done with data coming from the seismic interpretation, wells, outcrops, and core data. At this stage, the section is generally balanced using software such as Locace.

The second step is the restoration of the section. The section at present day is restored back in the past for each of the defined layer until the substratum is reached. In order to balance the section during time and to constrain the eroded parts of the section, we have to use a forward kinematic modeling tool such as Thrustpack. Thus, from a section restored with Locace before thrusting, the use of Thrustpack allows to construct intermediate sections during time, which are consistent from a kinematics point of view.

The last step is the forward simulation. And, in order to solve the coupled equations that are generally used in basin models, we had to develop original numerical methods based on domain decomposition techniques.

The previous methodology is the fruit of tests performed for an Alberta (Canada) transect and an eastern Venezuelan transect. When considering the origin of the fluids in the Oligocene sandstones of the El Furrial structure or in the Mississipian dolomite of the Canadian foothills, a general scenario of squeegee fluids is recognized. Before the flexuring and thrusting period, fluids are in equilibrium with the sediments. They are being continuously expelled toward the surface during the compaction related dewatering. Then, because of the tilting and thrusting with the deposition of synflexural sediments, an episode of squeegee fluid is created with flow along the stratification that can reach velocity of tens of kilometers per million of years. This squeegee episode stops when the thrusts of the considered area become active.

INTRODUCTION

Basin modeling aims at reconstructing the time evolution of a sedimentary basin in order to make quantitative predictions of geological phenomena leading to pressure generation and hydrocarbons accumulations. It accounts for porous medium deformation, heat transfer, hydrocarbon formation and migration (*e.g.* Schneider *et al.*, 2000b).

Since the pioneer works on the modeling of hydrocarbon formation (Tissot, 1969), several generation of basin model were built. First 1D model such as Genex were used to simulate the temperature evolution and the maturation of the organic matter. In sedimentary basin context, the 1D approach is satisfactory for these purposes because the thermal transfers are mostly vertical. Indeed, thermal convective transfers are most frequently negligible. Then 2D model such as Temispack (Doligez *et al.*, 1986; Ungerer *et al.*, 1990) were built. These 2D models provide the possibilities of performing an evaluation of the pressure history and appreciating the hydrocarbon migration and reservoir filling. However, these evaluations can only be qualitative because fluid (water, oil, and gas) migration is mainly convective and therefore sensitive to 3D geometry and anisotropy. For these reason, a generation of 3D basin model was designed such as Temis3D (Schneider *et al.*, 2000b).

The previous sedimentary basin models are able to handle relatively simple geometries resulting from deposition, erosion and vertical compaction. But, as the exploration is now focussing in area such as foothills, a new generation of models able to handle faults was needed.

This is why the Ceres project was launched. The aim of this project was to build a prototype that is able to model three-phase flow in a 2D section of a basin, whose geometry changes through time accounting for deposition, compaction, erosion of the sediments, salt or mud creeping, and block displacement along faults.

In the first part of the paper, Ceres is presented according to the classical workflow used to perform case studies. The second part of the paper is dedicated to a quick review of some cases performed with Ceres.

1 CERES PROTOTYPE

The software is composed of several modules. The main modules are a section editor, a restoration module and a forward simulation simulator. The other modules are a chronostratigraphy editor, a lithology editor, a kerogen editor, fluid editors, a mesh editor, a run editor, and visualization modules. A study browser manages all these modules.

The main steps to carry out a case study are:

- the edition of the initial section;
- the restoration of the section;
- the forward simulations.

1.1 Edition of the Initial Section

The initial section is composed of a chronostratigraphic column and a 2D cross section. In the study browser, the root of a tree represents this first composite object. The number of horizons, the corresponding number layers, the ages of the horizons, and the values of the eustatism level are edited with the chronostratigraphic editor.

The section can be edited directly from scratch or its geometry can be imported from other softwares. It is recommended, at this stage to use structural software such as Locace (Moretti *et al.*, 1988; Moretti and Larrere, 1989) that is able to balance the section. The edition of the section is split into three steps (Schneider *et al.*, 2002b). First, the geometry of the section is defined. Second, the geological attributes are affected. These attributes may be horizon, fault, section boundary, facies change limit. At this stage a decollement level is defined by superimposing a horizon with a fault. At the end of this second step, the section is filled with the lithologies. The third step of the edition of the section, is the numerical step. At this stage, the blocks, which represent the smallest kinematics units, are defined. Then, each of the blocks is meshed. One specificity of the gridding is that each block holds its own grid with no constraint coming from the other blocks. The faults are not gridded in the initial section because their grids are created dynamically during the forward simulation.

The initial section *(Fig. 1)*, once edited, holds the upper mantle, the ductile lower crust, the brittle upper crust and the considered sedimentary part. The faults can exist in the sediments and in the brittle upper crust; they may be rooted at the interface of the brittle upper crust with the ductile lower crust or at predefined decollement levels.

1.2 Restoration of the section

Once the edition of the initial section is terminated, a backward process, which includes a kinematics restoration, backstripping, and thickness modification, allows determining the input data for the forward calculator. During the backward process all the periods defined in the chronostratigraphic column should be restored. For this step, it is recommended, especially for complex area, to use a structural forward model such as Thrustpack (Deville and Sassi, 1996; Sassi *et al.*, 1998) in order to constrain the geometry through time and to provide a reliable estimation of the eroded parts. At the study browser level, a branch of the tree represents a restoration scenario from the initial section to the final section, only composed of the basement. Each leaf of the branch is a section at the corresponding age. For each restoration, the backward process is composed of different steps (Schneider *et al.*, 2002b).

The first step is the edition of the eroded parts if erosion has occurred during the considered period. Importing templates may help this edition. The second step is an automatic popping of the section. It allows taking off what has been sedimented during the considered period. Once the erosion and the sedimentation accounted for, the resulting section is decompacted using porosity depth relationships.

The backstripped section is then restored from a kinematics point of view. At this stage, the displacements along faults are accounted using translations and vertical shear. As for the edition of the eroded parts, this operation may be constrained with the use of templates.

Once the section is restored, the last step of the backward process is the thickness modifications. This step allows accounting for salt or mud tectonics. Correction of the edition of the erosion may be done at this stage. Nevertheless, the main use of this editor is to account for thickness modifications due to salt tectonics or mud creeping.

These steps should be performed for each layer initially defined in the section at present day. The result is a scenario of restoration *(Fig. 2)* that should be validated against the previous structural study.

1.3 Forward Simulations

In these complex geometries, faults cut the basin into blocks that naturally define computational subdomains. In the blocks, the model accounts for the porous medium compaction, erosion, heat transfer, hydrocarbon formation and migration (Schneider *et al.*, 2002b). The equations are mass conservation of solid and fluids (water, oil and gas) coupled with the Darcy's law and a compaction law. The faults have a constant thickness and their permeability may evolve with time. The prototype allows using three permeability models for the faults: faults may be impervious, pervious or may have a permeability that is a function of the neighboring lithologies and of the strain rate.

In order to solve the three-phase flow in a basin cut by faults along which block displacements can occur, Domain Decomposition Methods (DDM) are used (Willien *et al.*, 1998a, 1998b). The classical techniques used in the first version have been improved by using optimized interface conditions (Faille *et al.*, 2002). In these methods, the faults are considered as subdomains with their own geological properties. However, because of the very small width of the

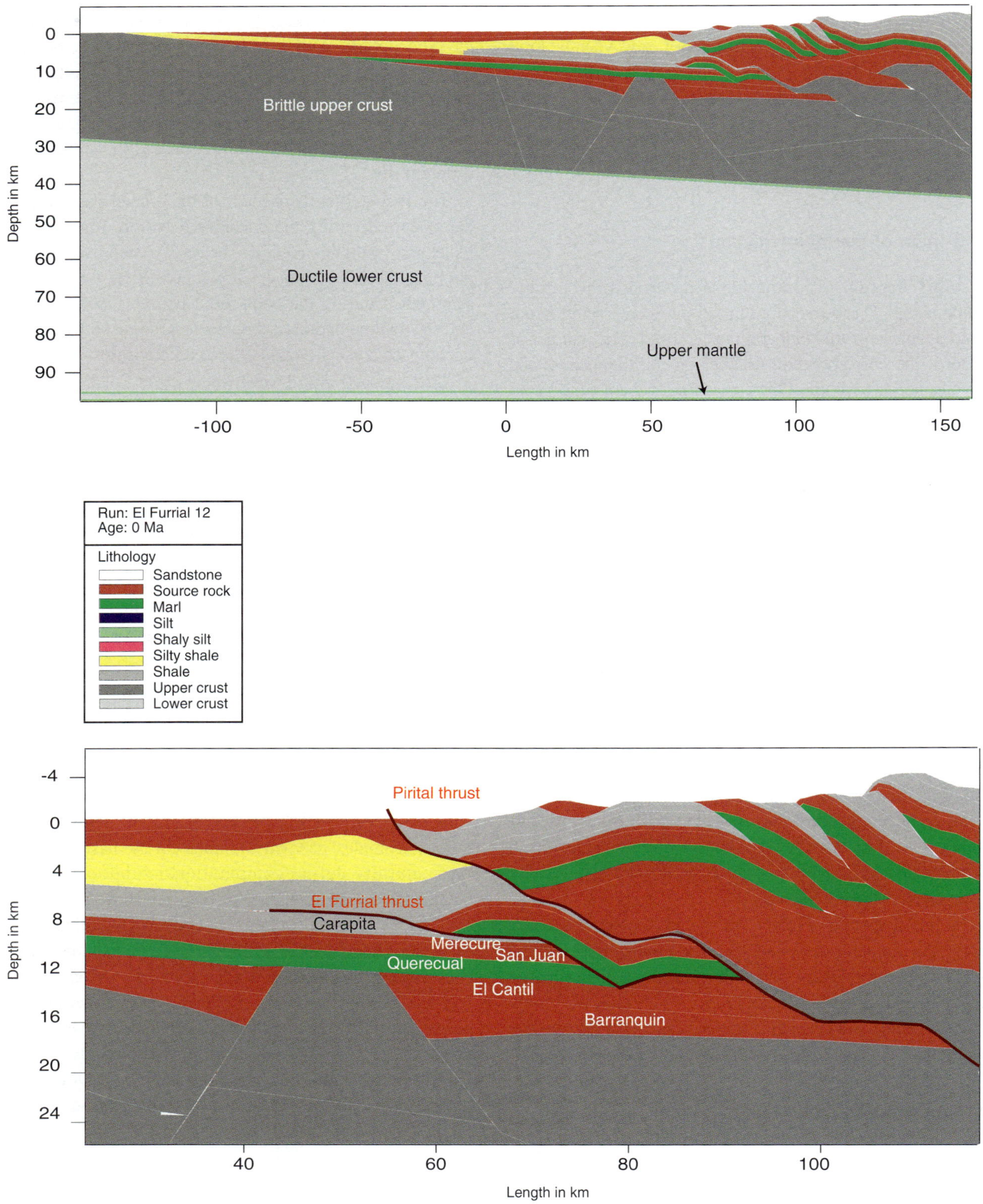

Figure 1

Geometry and lithology distribution of the Venezuelan section at present day. This section shows the ductile lower crust, the brittle upper crust with some faults rooted at the interface. In the upper part of the section are the "sediments". The upper mantle, used for the thermal simulation is not represented. The white dot indicates the El Furrial structure.

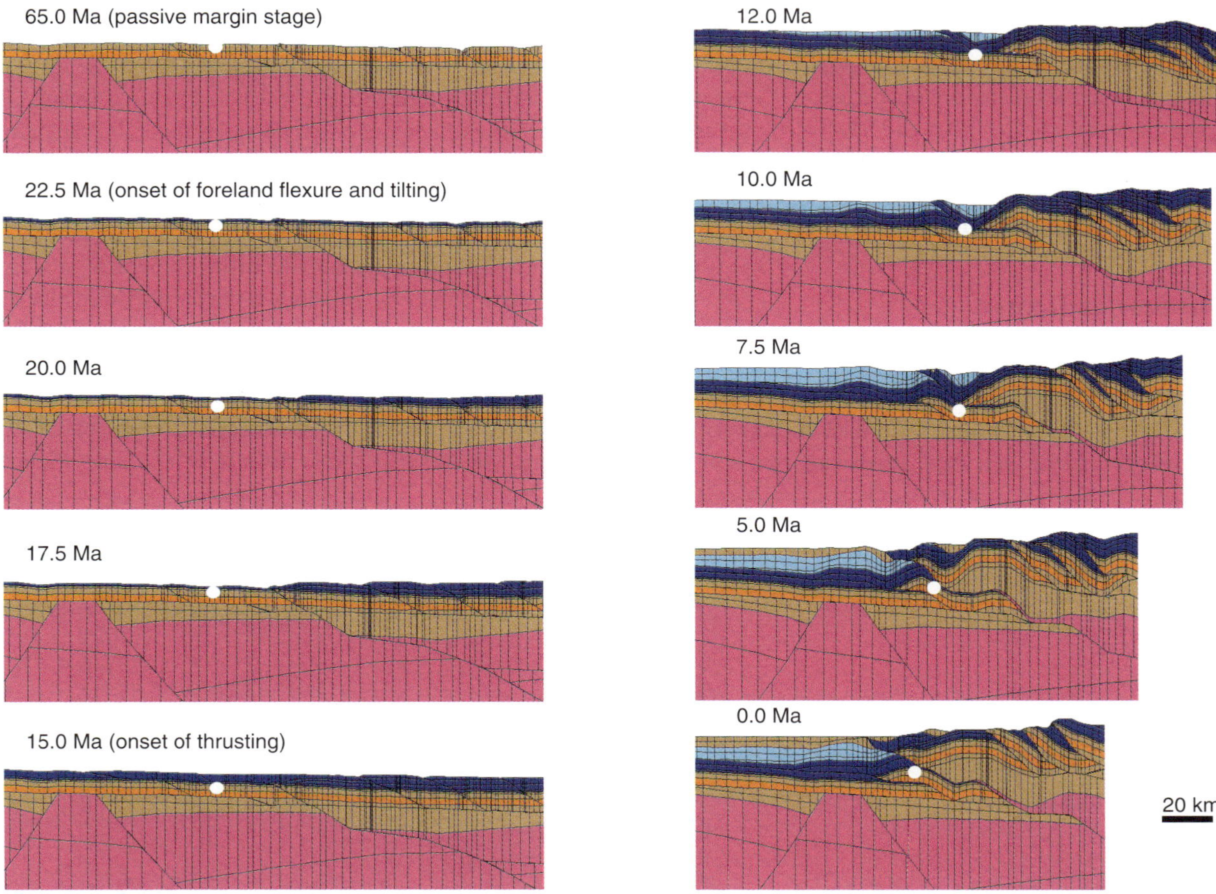

Figure 2

Geometrical evolution of the section from –65 Ma to present day. The flexure and the tilting of the foreland begins at 22.5 Ma. The thrusting begins at 15 Ma. The white dot indicates the El Furrial structure.

faults in comparison with the size of the basin, we are studying another approach where the faults are characterized as interfaces between blocks (Faille *et al.*, accepted). In all the cases, the equations are discretized using a cell-centered finite volume scheme in space. Finite volume methods are known to be robust and cheap methods for the discretization of conservation laws in heterogeneous media and have the important property to be conservative.

Different time discretizations coupled with several DDM algorithms have been tested. In a first stage, a classical IMPES scheme (Implicitly advances the Pressure and Explicitly updates the Saturation in time) was used. But, it is well known that this scheme needs time step limitations for stability reasons. In order to use larger time steps, we have tested (Flauraud *et al.*, 2000) a sequential scheme, called IMPIMS (Implicitly advances the Pressure and Implicitly advances the Saturation in time). It consists in solving two nonlinear systems, one for the pressure (as in the IMPES scheme) and one for the saturation.

2 CASE STUDIES

This prototype has now been used to study petroleum systems all around the world. In the Bolivian foothills were faults are supposed to be the most important carrier beds for hydrocarbon migration. This prototype was used to perform a sensitivity study on fault permeability (Schneider *et al.*, 1999). In the Congo offshore, it was used to study the impact of faulting on maturity of the organic matter and hydrocarbon migration (Schneider *et al.*, 1999). In the Gulf of Mexico, it was tested for studying the impact of the timing of salt withdrawal on hydrocarbon migration (Schneider *et al.*, 2000a).

More recently, Ceres was tested within the frame of the *IFP* subtrap consortium in thrust area such as the Canadian foothills and the eastern Venezuelan foothills (Schneider *et al.*, 2001, 2002a). The prototype has been used to quantify the fluid flow and to reconstruct the pore fluid history of the subthrust reservoirs. The use of Ceres implies a backward restoration. Although the backward restoration is fully appropriate for modeling an extensional basin well balanced, it may turn difficult to constrain successive intermediate evolutionary stages, as far as the topography (prediction of erosion) and thrust geometry are concerned. In consequence, we have decided to use also the Thrustpack forward kinematic software to elaborate a better structural scenario. These successive intermediate geometries were then used as templates for the Ceres modeling.

2.1 Venezuelan Transect

As defined previously, the first task to perform at the beginning of a Ceres study is the definition of the initial section *(Fig. 1)*. This was done with data coming from the seismic interpretation, well data, field data, and core data. Then, the available data and the Thrustpack scenario have been used to build the Ceres dataset.

From the available data, it is clear that the El Furrial structure (*e.g.* Parnaud *et al.*, 1995) is closed today as it is overpressured while the equivalent layers of the southern part have nearly hydrostatic pressures.

The El Furrial trend from the eastern Venezuelan transect displays two distinct episodes of quartz cementation. The first generation of quartz overgrowths, which account for more than 90% of the cements, is developed during the active dewatering processes in the Oligocene strata of the underthrust foreland. This is evidenced by the $\delta^{18}O$ signature and by the Ceres fluid flow simulations (Schneider *et al.*, 2001, submitted). In contrast, the second generation of quartz overgrowths displays quite distinct $\delta^{18}O$ values (Schneider

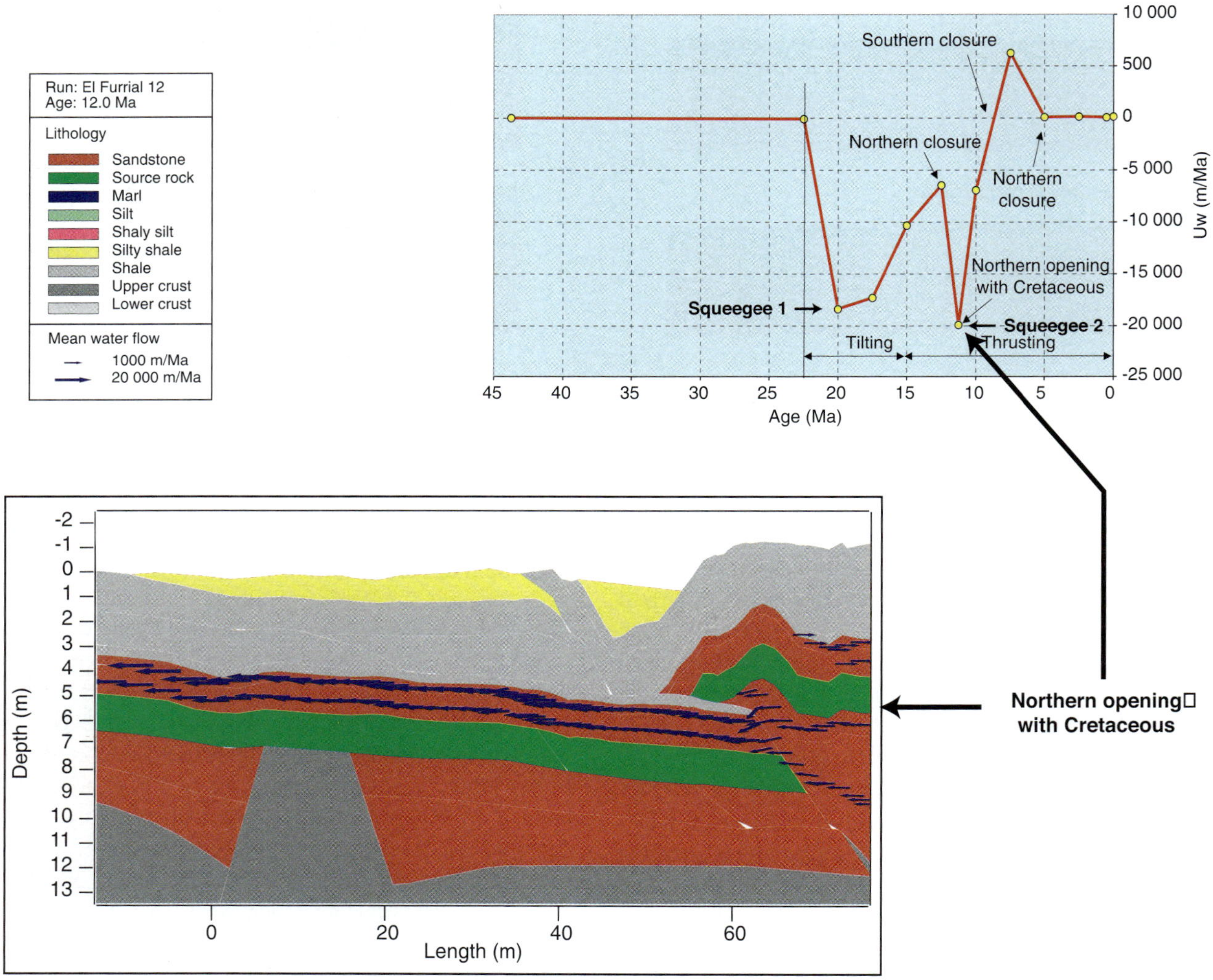

Figure 3

Computed fluid velocity evolution along the stratification in the Oligocene sandstones from the El Furrial structure. The evolution is characterized by two squeegee episodes. The first one is related to the beginning of the tilting. The second one is due to thrusting that allows the fluids to communicate between the Cretaceous sandstones and the Oligocene sandstones.

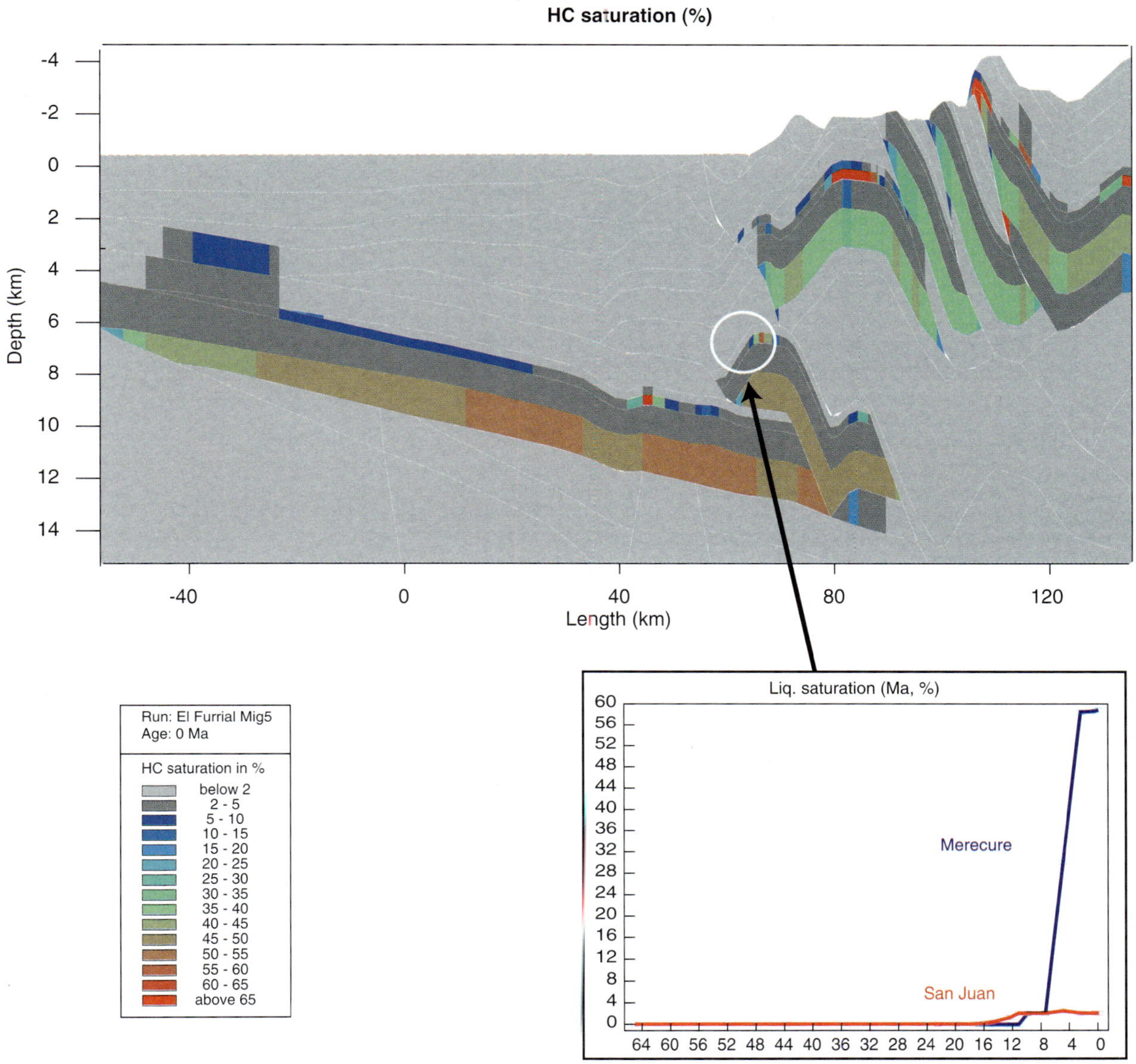

Figure 4

Computed hydrocarbon saturation at present day for the Venezuelan section. The computed evolution of the oil saturation shows that the hydrocarbons begins to flow 12 Ma ago. The filling of the reservoir is correlated with the creation of the El Furrial structure and more precisely with the southern closure of the structure.

et al., submitted). These cements are originated when the deeper aquifer of the Lower Cretaceous Barranquin formation from the Pirital hanginwall started to expel its fluids towards the Oligocene reservoir of the adjacent El Furrial footwall unit. This is well evidenced by the Ceres fluid flow simulations *(Fig. 3)*.

When considering the origin of the fluids in the late Cretaceous-Oligocene sandstone reservoirs of the El Furrial structure, four steps should be considered *(Fig. 3)*.

- From -65 to -20 Ma: fluids were at thermal equilibrium with the sediments. They were continuously expelled vertically toward the surface during compaction-driven dewatering processes.
- From –20 to –12 Ma: as a result of the regional flexuring and tilting and the deposition of the synflexural Naricual Formation, the Oligocene sandstones of the El Furrial structure became efficient conducts for fluids circulating from the north. These fluids (squeegee 1 in *Fig. 3*) were at

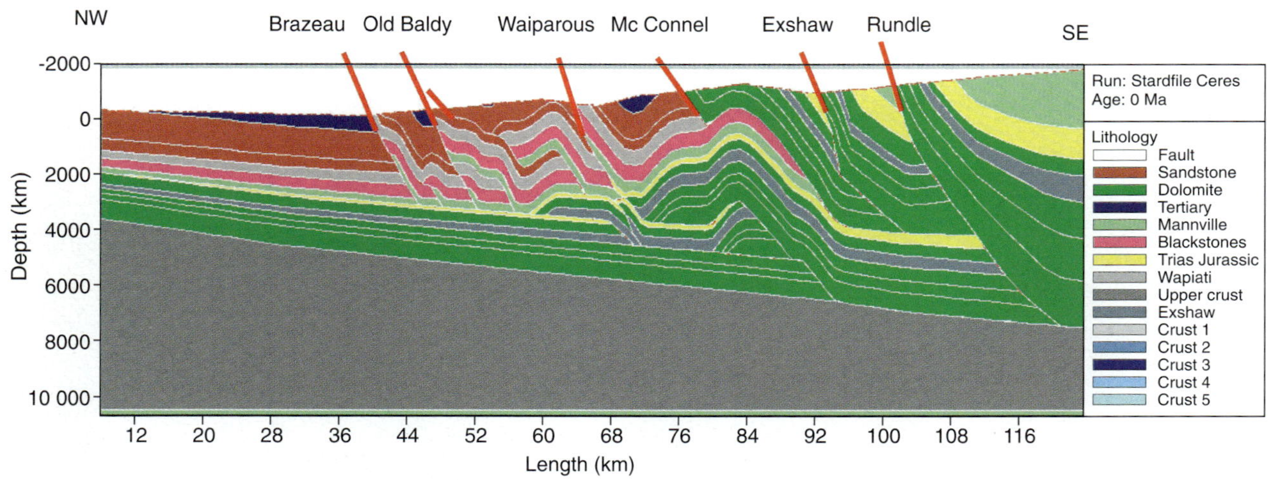

Figure 5

Geometry and lithology distribution of the Canadian section at present day. This section shows the main thrust. The main source rock is the Exshaw formation. The mains reservoirs are located in the Mississipian dolomites.

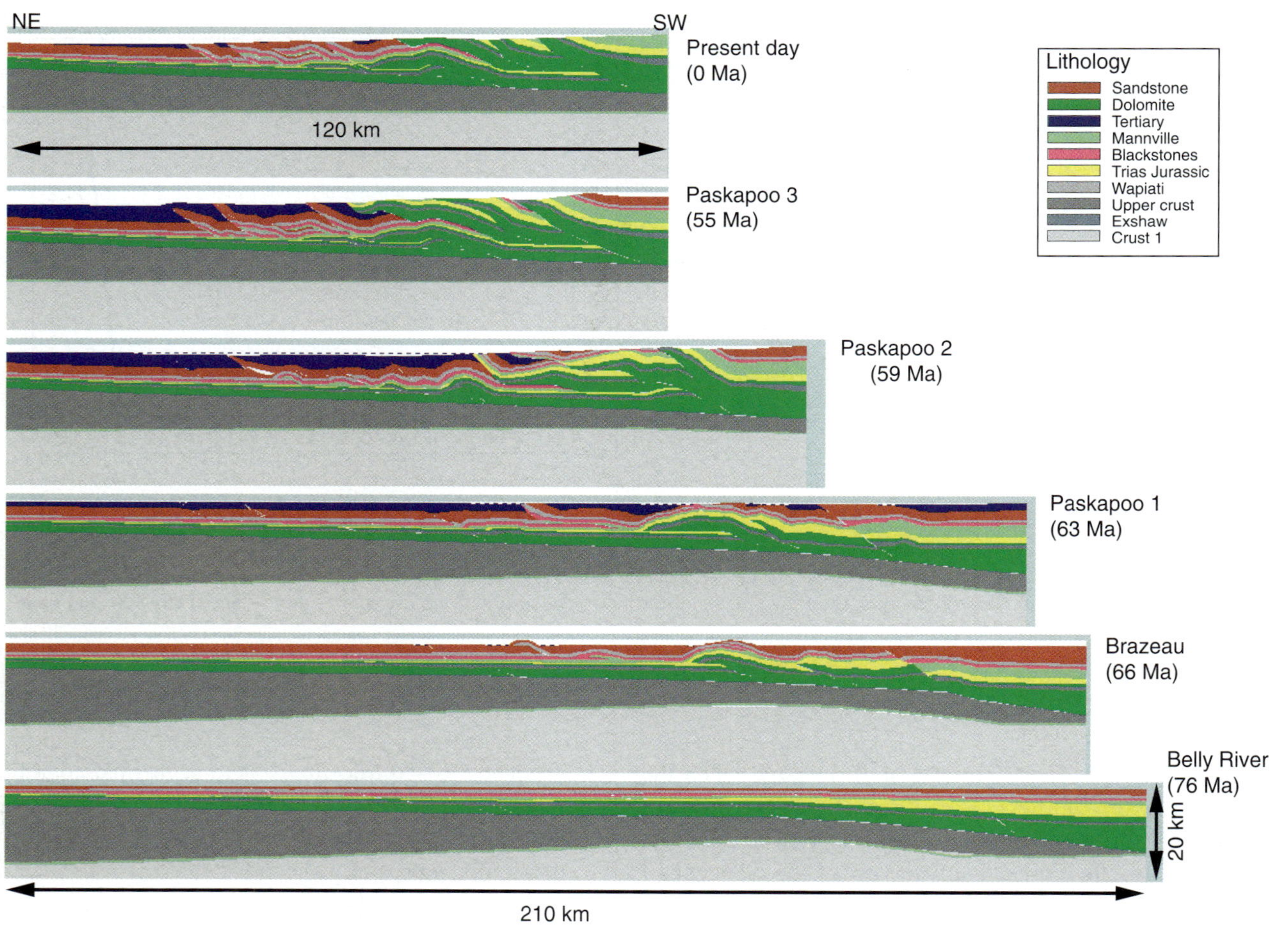

Figure 6

Geometrical evolution of the Canadian section from –76 Ma to present day. In this scenario, the flexure and the tilting begins at 76 Ma. The thrusting begins at 66 Ma.

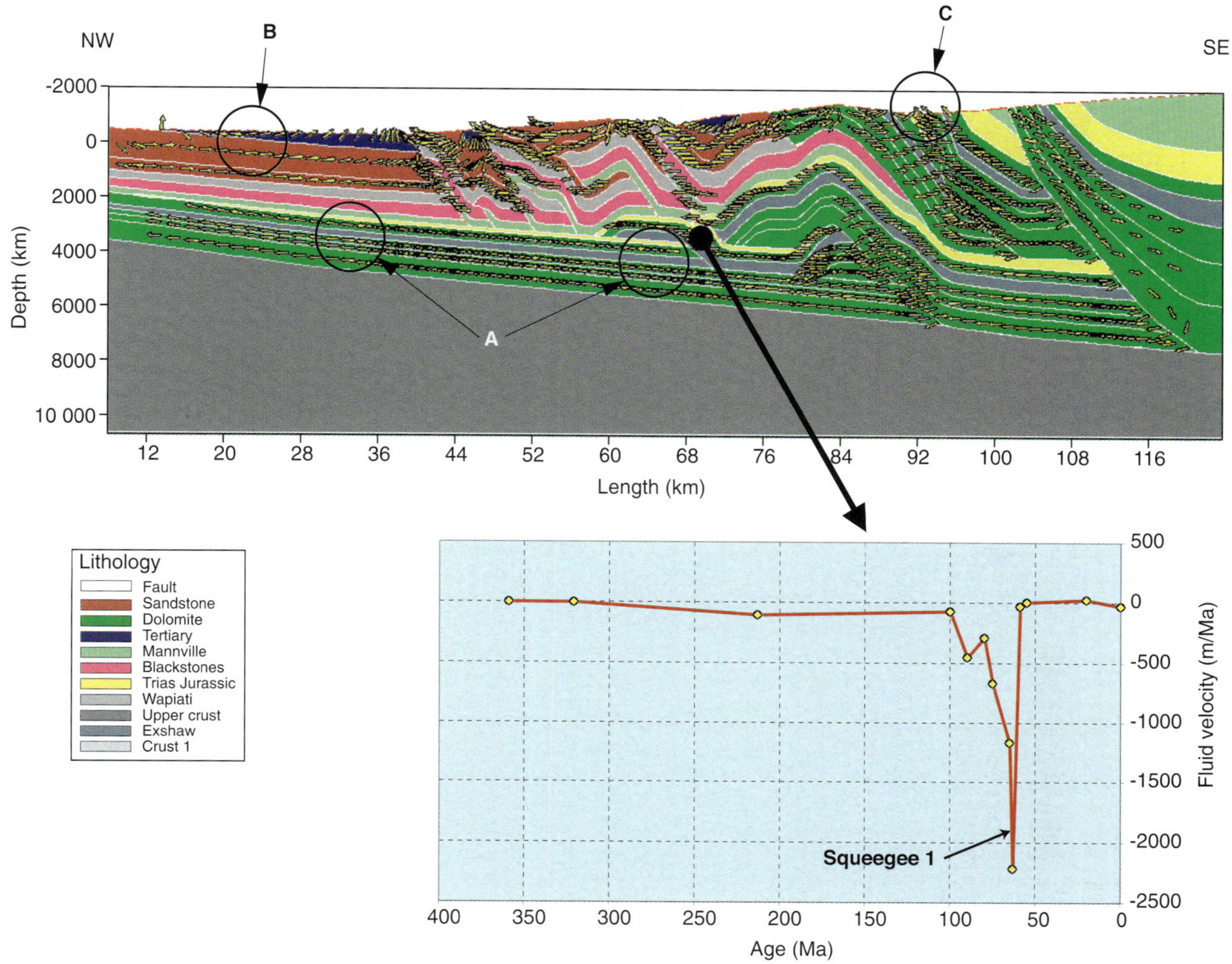

Figure 7

Canadian section at present day showing the computed fluid flow. A: water flow toward the NW; B: water flow toward the SE; C: simulated "hot springs". The diagram below shows the fluid velocity evolution for the Mississipian reservoir. One of the characteristics of this evolution is the squeegee episode related to the flexure and the tilting period.

thermal equilibrium with the Oligocene sandstones in the north, but recorded higher temperatures than adjacent sediments in the El Furrial area. The average Darcy velocity of these fluids decreased from 20 to 5 km/Ma.

This episode is characterised by an increase of the overpressure in the Oligocene and Upper Cretaceous sandstones correlated with a hydraulic fracturing of the sealing Carapita black shales.

- From –12 to –8 Ma: from the Ceres model, due to the tectonic accretion, deeper sedimentary units (*e.g.* early Cretaceous series) were in contact with shallower units (*e.g.*, Oligocene and late Cretaceous series), giving rise to an exchange and mixture of fluids (Squeegee 2 in *Fig.* 3) with different temperatures and compositions, that probably resulted in additional, but minor, episodes of quartz precipitation. This hypothesis is consistent with the information obtained from oxygen isotope analyses, which suggest that subsequent generations of quartz cement probably formed from evolved basinal fluids.
- A reduction of the intensity of the flow and then an inversion of this flow mark the sealing of the southern structural closure of the structure at around –8 Ma. Then the closure of the northern flank occurs at around –5 Ma as indicated by a present velocity of the fluids close to zero in the El Furrial reservoirs.

The hydrocarbons filling history shown in Figure 4 is consistent with the fluid inclusion studies, which show that hydrocarbons were present during the quartz cementation of the San Juan formation but they were not present during the quartz cementation of the Merecure formation.

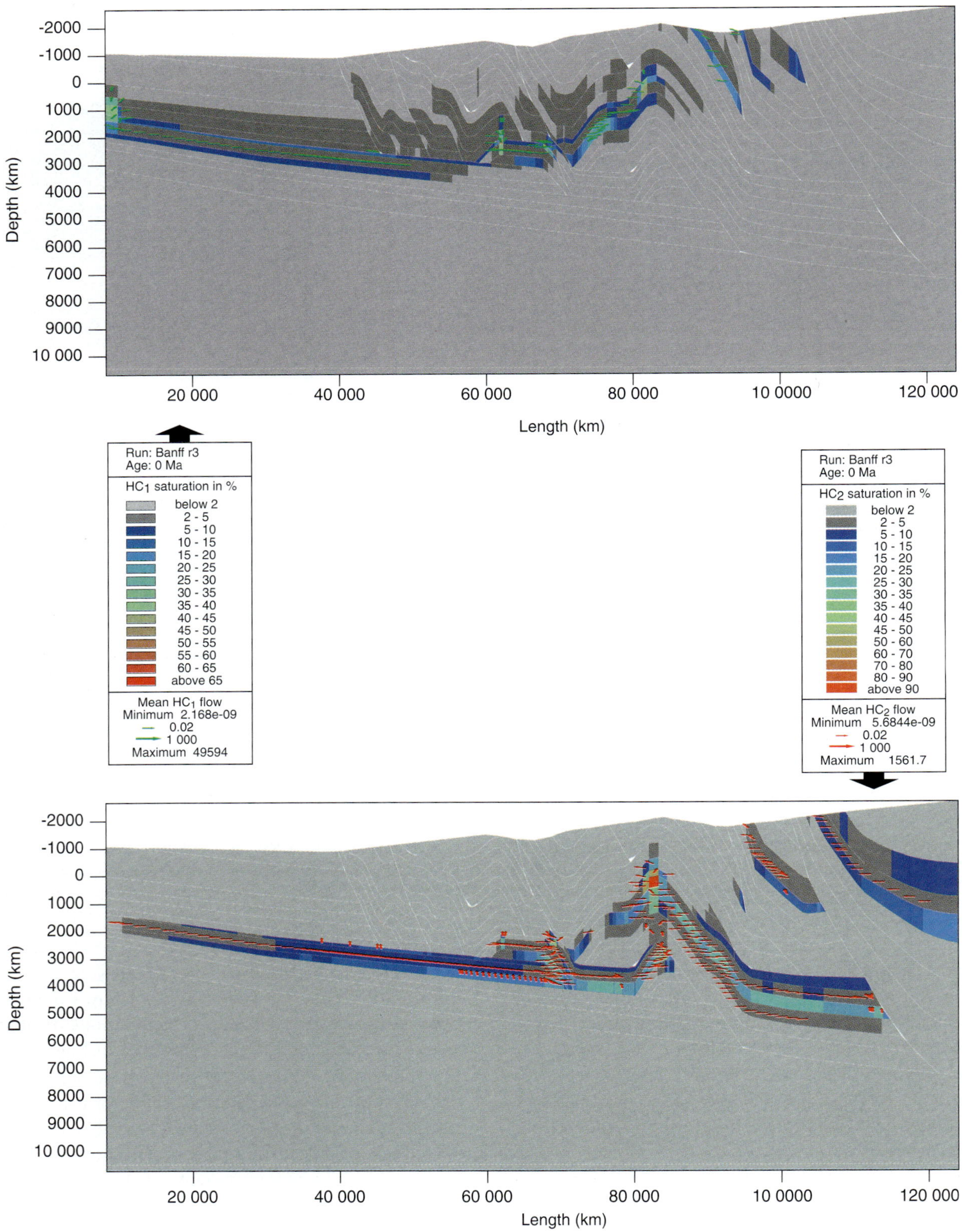

Figure 8

Computed hydrocarbon saturation at present day for the Canadian section. HC_1 means oil and HC_2 means gas.

2.2 Canadian Transect

As for the Venezuelan transect, the first work to perform at the beginning of a Ceres study is the definition of the initial section *(Fig. 5)*. Then, one scenario for the geometry evolution has been defined *(Fig. 6)*. It is characterized by five major episodes in the geodynamic evolution:

- pre-flexural deposition during the Paleozoic and Lower Mesozoic times on a Precambrian substratum;
- deposition of synflexural formation from –100 to –76 Ma;
- a main thrusting phase between –76 and –55 Ma;
- a strong erosion from –55 to –20 Ma;
- since –20 Ma the area is uplifted.

Unlike in Venezuela, where modern fluid flow is still controlled by the compressive evolution of the foothills, the Ceres fluid flow simulations shown that a dramatic change occurred since the end of the Laramian Orogeny. Indeed, more than 4 km of Mesozoic and Cenozoic series have been eroded and the hinterland and the foreland were progressively uplifted. Thus, the present day topography and the geometry of the aquifer *(Fig. 7)* mainly control fluid flow (Tóth, 1988; Garven, 1989; Bachu, 1999).

When considering the origin of the fluids in the Mississipian dolomite of the area, three steps should be considered *(Fig. 7)*:

- From deposition to –100 Ma: fluids are in equilibrium with the sediments. They are continuously expelled toward the surface during the compaction dewatering.
- From –100 to –63 Ma: because of the tilting and thrusting with the deposition of synflexural sediments, the Mississipian sediments of a considered area will receive fluids from the West with velocity reaching 2 km/Ma. These fluids are in equilibrium with the Mississipian dolomite at the West and should have higher temperatures than the considered sediments.
- From –59 Ma to present: due to the thrusting, the structure became closed and there is no flow any more.

Figure 8 shows preliminary results concerning hydrocarbon migration in this area. HC_1 and HC_2 represent respectively oil and gas.

CONCLUSION

Ceres is a new tool that enables to perform basin modeling in complex area where faults or salt tectonics may constrain the petroleum system. This prototype was tested successfully worldwide. It may now be used to understand petroleum systems in much more complex tectonic environment as the ones generally addressed with classical basin models.

The Ceres software, when coupled with a forward kinematics tool, such as Thrustpack, can be used for the simulation of the very complex structures of the foothills. It can provide reliable values for the velocity of the paleo-fluid circulations and for the paleo-fluid pressure. In the two last case studies performed, it was used to evidence a squeegee phase (call squeegee 1) during the tilting period. For the Venezuelan case study a second phase of squeegee (call squeegee 2) was simulated during the thrusting period.

It should be noticed that for these studies, it was of real importance to work closely with the structural geologists in order to define the initial section and the scenario for the kinematics restoration.

REFERENCES

Bachu, S. (1999) Flow Systems in the Alberta Basin; Patterns, Types and Driving Mechanisms. *Bulletin of Canadian Petroleum Geology*, **47**, 4, 455-474.

Deville, E. and Sassi, W. (1996) *Thrustpack; un logiciel de modélisation intégrée (cinématique, thermique et géochimique) ; application aux Alpes nord-occidentales*. In *L'exploration pétroliere dans les fronts compressifs. Pétrole et Techniques*, 401, p.77.

Doligez, B., Bessis, F., Burrus, J., Ungerer, Ph., and Chénet, P.Y. (1986) Integrated Numerical Simulation of Sedimentation, Heat Transfer, Hydrocarbon Formation and Fluid Migration in a Sedimentary Basin: The Temis Model. In *Thermal Modeling in Sedimentary Basins, Proceeding of the 1st IFP Research Conference on Exploration*, Carcans, June 3-7, 1985, 173-195.

Faille, I., Flauraud, E., Nataf, F., Schneider, F. and Willien, F. (2002) Optimized Interface Conditions for Sedimentary Basin Modeling. In *Proceedings of the Thirteenth International Conference on Domain Decomposition Methods*, N. Debit, M. Garbey, R. Hoppe, J. Periaux, D. Keyes and Y. Kuznetsov editors, Lyon, France, 2000, 453-460.

Faille, I., Flauraud, E., Nataf, F., Pégaz-Fiornet, S., Schneider, F. and Willien, F. (2002) A New Fault Model in Geological Basin Modelling. Application of a Finite Volume Scheme and Domain Decomposition Methods. *Third International Symposium on Finite Volumes for Complex Applications - Problems and Perspectives*, June 24-28, 2002, Porquerolles, France.

Flauraud, E., Faille, I., Willien, F., Nataf, F. and Schneider F. (2000) Implicit Scheme and Domain Decomposition Method for Multiphase Flow in Porous Medium. *Proceedings of the 7th European Conference on the Mathematics of Oil Recovery (ECMOR VII)*, Dino Hotel, Italy, 5-8 September.

Garven, G. (1989) A Hydrogeologic Model for the Formation of the Giant Oil Sands Deposits of the Western Canada Sedimentary Basin. *American Journal of Science*, **289**, 2, 105-166.

Moretti, I., Bessis, F., Colletta, B., Cuvillier, T., Larrere, M., Triboulet, S., Letouzey, J., Maisondieu, V., and Werner, P. (1988) *Locace Program; A New Tool for Balancing Cross Section*. In *Mediterranean Basins Conference and Exhibition, AAPG Bulletin*, **72**, 8, 1015.

Moretti, I. and Larrere, M. (1989) Locace; Computer-Aided Construction of Balanced Geological Cross Sections. *Geobyte*, **4**, 5, 16-24.

Parnaud, F., Gou, Y., Pascual, J.C., Truskowski, I., Gallango, O. and Passalacqua, H. (1995) Petroleum Geology of the Central Part of the Eastern Venezuelan Basin. In *Petroleum Basins of South America*, Tankard A.J., Soruco R.S. and Welsink H.J. editors, *AAPG Memoir*, 62, 741-756.

Sassi, W., Rudkiewicz, J.L., and Divies, R. (1998) New Methods for Integrated Modeling of Deformation and Petroleum Generation in Fold and Thrust Belts. In *American Association of Petroleum Geologist 1998 Annual Meeting* (abstract).

Schneider, F., Devoitine, H., Faille, I., Flauraud, E. and Willien, F. (1999) A New 2D Basin Modeling Tool for HC Potential Evaluation in Faulted Area. Application to the Congo Offshore and to the Bolivian Sub Andean Zone. *AAPG Hedberg Research Conference*, Colorado Spring, May 9-13.

Schneider, F., Faille, I., Flauraud E. and Willien, F. (2000a) A 2D Basin Modeling Tool for HC Potential Evaluation in Complex Area. *62nd EAGE*, Glasgow, May 29-June 2.

Schneider, F., Wolf, S., Faille, I. and Pot, D. (2000b) A 3D-Basin Model for Hydrocarbon Potential Evaluation: Application to Congo Offshore. *Oil & Gas Science and Technology - Revue de l'Institut Français du Pétrole*, **55**, 1, 3-13.

Schneider, F., Hernández, H. and Roure, F. (2001) Ceres Fluid-Flow and Pressure Modeling of an Eastern Venezuelan Transect. *AAPG Hedberg Conference "New Technologies and New Play Concepts in Latin America"*, November 4-9, 2001, Mendoza, Argentina.

Schneider, F., Faure, J.L. and Roure, F. (2002a) Methodology for Basin Modeling in Complex Area: Examples from Eastern Venezuelan and Canadian Foothills. *AAPG Hedberg Conference "Deformation History, Fluid Flow Reconstruction and Reservoir Appraisal in Foreland Fold and Thrust Belts"*, May 14-18, 2002, Palermo-Mondello, Sicily, Italy.

Schneider, F., Devoitine, H., Faille, I., Flauraud, E. and Willien, F. (2002b) Ceres2D: A Numerical Prototype for HC Potential Evaluation in Complex Area. *Oil & Gas Science and Technology - Revue de l'Institut Français du Pétrole*, **57**, 6, 607-619.

Schneider, F., Pagel, M. and Hernandez, E. (submitted) Fluid Flow History and Quartz Cementation in the Eastern Venezuelan Foothill. *AAPG Memoir*.

Tissot, B. (1969) Premières données sur les mécanismes et la cinétique de la formation du pétrole dans les sédiments. Simulation d'un schéma réactionnel sur ordinateur. *Revue de l'Institut Français du Pétrole*, **24**, 6, 470-501.

Ungerer, Ph., Burrus, J., Doligez, B. and Bessis, F. (1990) Basin Evaluation by Integrated Two-Dimensional Modeling of Heat Transfer, Fluid Flow, Hydrocarbon Generation and Migration. *AAPG Bull.*, **74**, 3, 309-335.

Tóth, J. (1988) Ground Water and Hydrocarbon Migration. In *The Geology of North America*, Back W., Rosenshein J.S. and Seaber P. R. editors, O-2, 485-502.

Willien, F., Faille, I. and Schneider, F. (1998a) Domain Decomposition Methods Applied To Sedimentary Basin Modeling. In *Ninth International Conference on Domain Decomposition Methods*, P.E. Bjorstad, M.S. Espedal and D.E. Keyes editors, 736-744.

Willien, F., Faille, I., Nataf, F. and Schneider, F. (1998b) Domain Decomposition Methods for Fluid Flow in Porous Medium. *Proceedings of the 6th European Conference on the Mathematics of Oil Recovery (ECMOR VI)*, Peebles, GB, 7-11 September.

Final manuscript received in March 2003

Editions TECHNIP • IFP Publications
Nouveautés
PUBLICATIONS DE L'INSTITUT FRANÇAIS DU PÉTROLE
LES HUILES USAGÉES
RERAFFINAGE ET VALORISATION ÉNERGÉTIQUE
FRANÇOIS AUDIBERT
Editions TECHNIP
Catalyse acido-basique
Application au raffinage et à la pétrochimie | Volume 1
Christian Marcilly
Editions TECHNIP
Catalyse acido-basique
Application au raffinage et à la pétrochimie | Volume 2
Christian Marcilly
Editions TECHNIP
INSTITUT FRANÇAIS DU PÉTROLE PUBLICATIONS
JEAN VIDAL
THERMODYNAMICS
APPLICATIONS IN CHEMICAL ENGINEERING AND THE PETROLEUM INDUSTRY
Editions TECHNIP
New Titles
Les Rencontres scientifiques de l'IFP • December 5-7, 2001 • Rueil-Malmaison, France
GEOMECHANICS IN RESERVOIR SIMULATION
Edited by Pascal Longuemare
Editions TECHNIP
www.editionstechnip.com
Institut Français du Pétrole Publications

ifp
EXCELLENCE IN PETROLEUM SCIENCE
INSTITUT FRANÇAIS DU PÉTROLE PUBLICATIONS
H. CHOLET, EDITOR
WELL PRODUCTION PRACTICAL HANDBOOK
Editions TECHNIP
INSTITUT FRANÇAIS DU PÉTROLE PUBLICATIONS
J. DENIS • J. BRIANT • J.C. HIPEAUX
LUBRICANT PROPERTIES ANALYSIS & TESTING
Editions TECHNIP
INSTITUT FRANÇAIS DU PÉTROLE PUBLICATIONS
B. BIJU-DUVAL
NEW
SEDIMENTARY GEOLOGY
Sedimentary Basins, Depositional Environments, Petroleum Formation
Editions TECHNIP
Les Entretiens IFP • November 9, 1999 • Rueil-Malmaison, France
INTERACTIVE DRILLING FOR FAST TRACK OILFIELD DEVELOPMENT
Edited by Jacqueline Lecourtier
Editions TECHNIP
INSTITUT FRANÇAIS DU PÉTROLE PUBLICATIONS
Luca Cosentino
Integrated Reservoir Studies
Editions TECHNIP
IFP International Congress • November 26-27, 2001 • Rueil-Malmaison, France
A NEW GENERATION OF ENGINE COMBUSTION PROCESSES FOR THE FUTURE?
NEW
Edited by Pierre Duret
Editions TECHNIP
INSTITUT FRANÇAIS DU PÉTROLE PUBLICATIONS
PETROLEUM REFINING
5
REFINERY OPERATIONS AND MANAGEMENT
J.-P. Favennec
Editions TECHNIP
P. Arpentinier
F. Cavani • F. Trifirò
The Technology of Catalytic Oxidations
2
Safety aspects
Editions TECHNIP
P. Arpentinier
F. Cavani • F. Trifirò
The Technology of Catalytic Oxidations
1
Chemical, catalytic & engineering aspects
Editions TECHNIP
www.editionstechnip.com
Welcome on our Website
Institut Français du Pétrole Publications

Editions TECHNIP

27, rue Ginoux, 75737 PARIS Cedex 15 • FRANCE • Tél. 33 (0)1 45 78 33 80 • Fax 33 (0)1 45 75 37 11
E-mail : infos@editionstechnip.com

Publications de l'Institut Français du Pétrole

Rédacteur en chef : J. LYNCH
Réalisation : GraphiDoc
92852 Rueil-Malmaison Cedex - France
Imprimerie JOUVE 75001 PARIS
N° d'impression : 327 036 B – Juin 2003
Éditeur : n° 244 CPPAP : n° 59 260
IMPRIMÉ EN FRANCE